F. and R. Nevanlinna

Absolute Analysis

Translated from the German
by Phillip Emig

With 5 Figures

Springer-Verlag
New York Heidelberg Berlin 1973

F. and R. Nevanlinna
Department of Mathematics, University,
Helsinki/Finland

Translator P. Emig
Granada Hills, CA 91344/USA

Geschäftsführende Herausgeber
B. Eckmann
Eidgenössische Technische Hochschule Zürich

B. L. van der Waerden
Mathematisches Institut der Universität Zürich

AMS Subject Classifications (1970): 26 A 60

ISBN 0-387-05917-2 Springer-Verlag New York Heidelberg Berlin
ISBN 3-540-05917-2 Springer-Verlag Berlin Heidelberg New York

---

To the Memory of our father
Otto Nevanlinna

Foreword

The first edition of this book, published in German, came into being as the result of lectures which the authors held over a period of several years since 1953 at the Universities of Helsinki and Zürich. The Introduction, which follows, provides information on what motivated our presentation of an absolute, coordinate- and dimension-free infinitesimal calculus.

Little previous knowledge is presumed of the reader. It can be recommended to students familiar with the usual structure, based on coordinates, of the elements of analytic geometry, differential and integral calculus and of the theory of differential equations.

We are indebted to H. Keller, T. Klemola, T. Nieminen, Ph. Tondeur and K. I. Virtanen, who read our presentation in our first manuscript, for important critical remarks.

The present new English edition deviates at several points from the first edition (cf. Introduction). Professor I. S. Louhivaara has from the beginning to the end taken part in the production of the new edition and has advanced our work by suggestions on both content and form. For his important support we wish to express our hearty thanks. We are indebted also to W. Greub and to H. Haahti for various valuable remarks.

Our manuscript for this new edition has been translated into English by Doctor P. Emig. We express to him our gratitude for his careful interest and skillful attention during this work.

Our thanks are also due to Professor F. K. Schmidt and to Springer-Verlag for including this new edition in the ,,Grundlehren der mathematischen Wissenschaften" series.

Helsinki, May 1973 THE AUTHORS

Table of Contents

Introduction

In the modern development of functional analysis, inspired by the pioneering work of David Hilbert, Erhard Schmidt and Friedrich Riesz, the appearance of J. von Neumann (1928) signaled a decisive change. Before him the theory of linear operators and of quadratic and Hermitian forms was tied in an essential way to the coordinate representation of the vector spaces considered and to the matrix calculus. Von Neumann's investigations brought about an essentially new situation. Linear and quadratic analysis were freed from these restrictions and shaped into an "absolute" theory, independent of coordinate representations and also largely of the dimension of the vector spaces. It was only on the basis of the general axiomatic foundation created by von Neumann that the geometric points of view, crucial to Hilbert's conception of functional analysis, were able to prevail. It is not necessary here to recall in more detail the enormous development to which von Neumann's ideas opened the way.

In this work an attempt is made to present a systematic basis for a general, absolute, coordinate and dimension free infinitesimal vector calculus. The beginnings for such a calculus appear in the literature quite early. Above all, we should mention the works of M. Fréchet, in which the notion of a differential was introduced in a function space. This same trend, the translation of differential calculus to functional analysis, is pursued in a number of later investigations (Gateaux, Hildebrandt, Fischer, Graves, Keller, Kerner, Michal, Elconin, Taylor, Rothe, Sebastião e Silva, Laugwitz, Bartle, Whitney, Fischer and others)[1]. In all of these less attention was paid to classical analysis, the theory of finite dimensional spaces. And yet it seems that already here the absolute point of view offers essential advantages. The elimination of coordinates signifies a gain not only in a formal sense. It leads to a greater unity and simplicity in the theory of functions of arbitrarily many variables, the algebraic structure of analysis is clarified, and at the same time the geometric aspects of linear algebra become more prominent, which simplifies one's ability to comprehend the overall structures and promotes the formation of new ideas and methods.

---

[1] Cf., e. g., the book of E. Hille and R. S. Phillips [1] as well as the bibliography in this book.

Since this way of viewing things is just as valid for classical analysis as for general functional analysis, our presentation is restricted for the most part to the finite dimensional case, that is, to the theory of finitely many variables. But it lies in the nature of the methods that they can be applied, either directly or with certain modifications, which in general are readily apparent, to the case of infinitely many dimensions (for Hilbert or Banach spaces).

Our presentation starts with a chapter on linear algebra and the analytic geometry of finite dimensional spaces. As is well known, a great number of works are available on this basic topic, among them some in which the general points of view discussed above are fully considered. In this regard the fundamental presentation of N. Bourbaki [1] deserves special attention. Nevertheless, we have deemed it necessary to enter into a thorough discussion of linear algebra, in order to introduce the basic concepts and the notations that have fundamental significance for infinitesimal analysis.

Our presentation of the infinitesimal calculus begins in the second chapter, where the central problems of differential calculus are treated briefly. Among those problems for which the advantages of absolute analysis are particularly apparent, the theorems on the commutativity of differentiation and the theory of implicit functions merit special attention. The coordinate free method together with the application of an extension of the classical mean value theorem leads in a natural way to a uniqueness result for the second problem that is complete as regards the precision of the domain of validity for the inverse of a function or the solution of an equation.

The following Chapter III is devoted to integral calculus. The integration of a multilinear alternating differential form constitutes the central problem. The "affine integral" introduced here is essentially the same as the Grassmann-Cartan integral of an "exterior differential". As an application, the problem of integrating a skew symmetric tensor field is solved.

In the theory of differential equations (Chapter IV), the "absolute" mode of viewing things also produces order and unity. After a preparatory treatment of normal systems, there follows the solution of partial differential equations by two different methods, the second of which leads to a sharpening of the conditions under which the problem, is usually solved.

In Chapter V the basic features of the theory of curves and of the Gaussian theory of surfaces are presented. Although this chapter offers nothing new as far as content is concerned, the usefulness of the absolute coordinate free point of view is also shown in this case. In the theory of surfaces we restrict oursdves to the case of a m-dimensional surface embedded in an $(m + 1)$-dimensional euclidean space. In accordance with the basic theme of our work, the "inner geometry" is heavily emphasized, and the theory is so constructed that it also includes an independent presentation of Riemannian geometry and of affine differential geometry. To this end it was necessary for tensor calculus to receive special attention. This latter theory is also developed in a coordinate free way. The elimination of coordinates and indices, which in the usual treatment of tensors is already typographically burdensome, simplifies the notation. On the other hand, very extensive abstractions are avoided, as occur in Bourbaki, for example. It was our goal to so shape the tensor calculus that its connection with the classical version is not broken and that it retains the character of an automatic calculus. It seems to us that the thus modified calculus, as indeed the absolute infinitesimal calculus, can be used to advantage not only in mathematics, but also in theoretical physics.

From many quarters the wish has been expressed that the authors append a synopsis of the elements of Riemannian geometry. Such a survey is contained in the last chapter (Chapter VI) of this new edition.

Otherwise this edition also deviates at several points from the first. We refer in particular to the theory of implicit function, which is presented in Chapter II following two different methods, and to Chapter IV, on differential equations, which has been substantially reworked and extended.

I. Linear Algebra

§ 1. The Linear Space with Real Multiplier Domain

1.1. The basic relations of linearity. Let R be a set whose elements we denote by $a, b, c \ldots, x, y, z \ldots$

We assume that addition and multiplication by a real scalar λ are defined so as to satisfy the following *rules of linearity*:

Sum. Corresponding to each pair of elements a, b there is a unique element $a + b$ in R, the sum of a and b, with the following properties:

I.1. The sum is associative,

$$a + (b + c) = (a + b) + c .$$

I.2. There exists in R a unique element 0, the zero, so that for every a

$$a + 0 = 0 + a = a .$$

I.3. Each element a has a unique inverse, $-a$, in R with the property

$$a + (-a) = (-a) + a = 0 .$$

I.4. The sum is commutative.

$$a + b = b + a .$$

The first three axioms state that with respect to addition R is a group, and as a consequence of I.4, a commutative or abelian group. They imply the existence of a unique difference $a - b$ with the property

$$b + (a - b) = (a - b) + b = a;$$

namely

$$a - b = a + (-b) = (-b) + a .$$

Product. Uniquely corresponding to each real λ and each element $a \in R$ there is in R an element λa, the product of λ and a, with the following properties:

II.1. $1 a = a$.

II.2. The product is associative,

$$\lambda(\mu a) = (\lambda \mu) a .$$

II.3. The product is distributive,

$$\lambda (a + b) = \lambda a + \lambda b , \qquad (\lambda + \mu) a = \lambda a + \mu a .$$

It follows from the distributive laws for $b = 0$ and $\mu = 0$ that

$$\lambda\, 0 = 0\, a = 0\,.$$[1]

Conversely, from $\lambda a = 0$ one concludes, provided $\lambda \neq 0$, that

$$a = 1a = \left(\frac{1}{\lambda}\lambda\right) a = \frac{1}{\lambda}(\lambda a) = 0\,.$$

The product therefore vanishes if and only if one of the factors is zero. Since for $a \neq 0$ the equation $\lambda a = \mu a$ holds only for $\lambda = \mu$, the additive abelian group R either reduces to the element zero or is of infinite order; for with $a \neq 0$ R includes all of the elements λa, which for different scalars λ are different.

Further, note that as a consequence of II.3 and II.1 we have for an integer $\lambda = m$,

$$m\, a = \underbrace{(1 + \cdots + 1)}_{m}\, a = \underbrace{1\, a + \cdots + 1\, a}_{m} = \underbrace{a + \cdots + a}_{m};$$

consequently, $m\, a$ is equal to the multiple $m \cdot a$. From here it follows that

$$a = 1\, a = \left(m \cdot \frac{1}{m}\right) a = m \cdot \left(\frac{1}{m} a\right),$$

and hence $\frac{1}{m} a$ ist equal to the quotient a/m. Thus for a positive rational $\lambda = p/q$

$$\frac{p}{q} a = \left(p \cdot \frac{1}{q}\right) a = p \cdot \left(\frac{1}{q} a\right) = p \cdot \frac{a}{q}\,.$$

1.2. Linear dependence. Dimension. A set R whose elements satisfy the axioms of addition and of multiplication listed above is called a *linear space over the real multiplier domain.* If $a_1, \ldots, a_n$ are arbitrary elements of the space and $\lambda_1, \ldots, \lambda_n$ are arbitrary real numbers, the linear combination

$$\lambda_1 a_1 + \cdots + \lambda_n a_n$$

also has meaning and is contained in R.

The elements $a_1, \ldots, a_n$ are said to be *linearly independent* if the above linear combination is equal to the zero element only for $\lambda_1 = \cdots = \lambda_n = 0$; otherwise, they are *linearly dependent.*

It follows immediately from this definition that every subset of linearly independent elements is likewise linearly independent. Because the linear independence of one single element a is synonymous with $a \neq 0$, this implies in particular that linearly independent elements are different from zero.

[1] We use the symbol 0 for *every* zero element. In the above formula, 0 on the right and left stands for the zero element in the linear space, in the middle for the number zero.

With regard to the linear dependence of the elements of a linear space R there are two possibilities: Either there exists a positive integer m such that $m + 1$ elements are always linearly dependent, while on the other hand at least one system of m linearly independent elements can be found. Or there is no such maximal system: for every m, no matter how large, there are always $m + 1$ linearly independent elements.

In the first case m is called the *dimension* of the linear space R; in the second case the dimension is infinite. We shall be concerned primarily with the first far simpler case. However we wish to at first discuss a few concepts which are independent of the dimension.

1.3. Subspaces. Congruences. A subset U of the linear space R that with respect to the given fundamental relations is itself a linear space is called a *subspace* of R. For this it is necessary and sufficient that U contains every linear combination of its elements.

Every subspace contains the zero element of the given space R, and this element by itself is a subspace.

Let U be a subspace, a and b two elements in R.

The elements a and b are said to be *congruent modulo* U,

$$a \equiv b \pmod{U},$$

if $b - a$ is contained in U.

It follows from the notion of a subspace, first of all, that congruence is an *equivalence*:

1. $a \equiv a$;
2. $a \equiv b$ implies $b \equiv a$;
3. $a \equiv b$ and $b \equiv c$ imply $a \equiv c$;

all modulo U. Further, if $a \equiv b$ and $c \equiv d$ the congruence

$$a \pm c \equiv b \pm d$$

and for every λ the congruence

$$\lambda a \equiv \lambda b$$

hold modulo U.

Conversely, an equivalence $a \sim b$ in R having the last two operational properties is a congruence, modulo a definite subspace U which contains precisely those elements of the space R that are equivalent to zero (cf. 1.6, exercise 1).

Therefore, a congruence in the space R can be defined as an equivalence that has the above two operational properties.

1.4. Hyperplanes. Factor spaces. Any equivalence in R decomposes all of the R-elements into disjoint classes so that two elements belong to the same equivalence class precisely when they are equivalent. If the equivalence is in particular a congruence modulo a subspace U, then two elements belong to the same congruence class provided their difference is an element of U.

In conformity with the geometric terminology introduced in the next section, we call these congruence classes modulo U *hyperplanes* (or more briefly *planes*) that are "parallel" to U. If a is an arbitrary element of such a hyperplane parallel to U it contains exactly all of the elements $a + U$. Of these parallel hyperplanes only U, which contains the zero element, is a subspace.

Retaining the basic relations of linearity, replace the original identity of elements with congruence modulo U. By means of this identification modulo U there results a new linear space the *factor space* of R modulo U:

$$R_U = R/U .$$

As elements of the latter factor space, the elements of the hyperplane $a + U$ are all equal to a.

1.5. The affine vector space. Parallel displacement. Up until now we have spoken of "elements" of the "space" R without worrying about concrete interpretations of these concepts. It is however useful to think of the abstractly defined linear space as a generalization of the concrete 1-, 2- and 3-dimensional spaces and to introduce a corresponding geometric terminology.

An ordered pair of elements $[a, b]$ of the linear space R is called a *vector*; a is the "tail", b the "head" of the vector. Using the basic linear relations defined in R, we define

$$[a, b] + [c, d] = [a + c,\ b + d]$$

and for every real λ

$$\lambda\, [a, b] = [\lambda\, a,\ \lambda\, b] .$$

One verifies at once that the vectors together with these definitions form a linear space, the *affine vector space* associated with R. The zero element of this vector space is the vector $[0, 0]$, and the vector inverse of $[a, b]$ is

$$-\, [a, b] = [-a,\ -b] .$$

If one sets

$$[a, b] \equiv [c, d]$$

whenever $b - a = d - c$, an equivalence (indeed a *congruence*) is defined in this vector space. For it follows from $[a, b] \equiv [c, d]$ and $[a', b'] \equiv [c', d']$, by virtue of the above definitions, that also

$$[a, b] + [a', b'] \equiv [c, d] + [c', d']$$

and

$$\lambda[a, b] \equiv \lambda[c, d] .$$

The modulus of this congruence is the subspace of the affine vector space consisting of all vectors $[a, a]$, and the factor space that corresponds to this subspace consists of all vectors with an arbitrarily fixed tail, for example, of all vectors $[0, x]$ with the fixed tail 0. This factor space is hence isomorphic to the original linear space R: in the one-to-one correspondence

$$[0, x] \leftrightarrow x ,$$

$x + y$ and $[0, x] + [0, y] = [0, x + y]$ and likewise λx and $\lambda[0, x] = [0, \lambda x]$ are image elements. These elements can therefore be thought of as either a vector congruent to $[0, x]$ or as a point, namely as the point at the head of the vector $[0, x]$.

From the congruence $[a, b] \equiv [c, d]$ it follows that $c - a = d - b$ which implies the congruence

$$[a, c] \equiv [b, d] ,$$

and conversely. The congruent vectors $[a, b]$, $[c, d]$ and $[a, c]$, $[b, d]$ are the "parallel" sides of the parallelogram $a\,b\,c\,d$. One says that $[c, d]$ is obtained from $[a, b]$ through *parallel displacement* by the vector $[a, c] \equiv [b, d]$, and $[b, d]$ from $[a, c]$ through parallel displacement by the vector $[a, b] \equiv [c, d]$.

According to the definitions we have established,

$$[a, b] + [b, c] = [a + b,\ b + c] \equiv [a, c];$$

this is the elementary geometric rule for the "combining" of two vectors, which more generally implies that

$$[a_1, b_1] + [b_1, b_2] + \cdots + [b_{n-2}, b_{n-1}] + [b_{n-1}, c] \equiv [a, c] .$$

1.6. Exercises. 1. Let $\sim$ be a given equivalence in the linear space R with the following properties: from $a \sim b$ and $c \sim d$, it follows that

$$a + c \sim b + d ,$$

and for every real λ

$$\lambda a \sim \lambda b .$$

Prove that the set of all elements $u \sim 0$ is a subspace U of R and further that the equivalence $a \sim b$ is the same as the congruence

$$a \equiv b \pmod{U} .$$

2. Let $M_1, \ldots, M_k$ be arbitrary point sets of the linear space R and

$$(M_1, \ldots, M_k)$$

the set of all finite linear combinations of elements in the sets.

Prove that this set is a subspace. This subspace of R is *generated* or *spanned* by the sets $M_1, \ldots, M_k$.

3. Suppose that the sets in the previous exercise are in particular subspaces $U_1, \ldots, U_k$ of R and that

$$U = (U_1, \ldots, U_k)$$

is the subspace of R spanned by these subspaces.

Show that every element in U can be represented uniquely as a sum of elements from the subspaces U_i precisely when the equation

$$u_1 + \cdots + u_k = 0$$

with $u_i \in U_i$ holds only for $u_1 = \cdots = u_k = 0$.

In this case the subspaces U_i are called *linearly independent* and U is written as the *direct sum*

$$U = U_1 + \cdots + U_k .$$

4. Show, keeping the above notation, that the *intersection*

$$[U_1, \ldots, U_k] ,$$

i.e., the set of elements common to all of the subspaces U_i, is a subspace.

5. Prove: In order that the subspaces $U_1, \ldots, U_k$ of the space R be linearly independent it is necessary and sufficient that the k intersections

$$[U_i, (U_1, \ldots, U_{i-1}, U_{i+1}, \ldots, U_k)] = 0 \qquad (i = 1, \ldots, k) .$$

Thus in particular two subspaces U_1 and U_2 are linearly independent precisely when they are disjoint except for the zero point.

6. Let R_x and R_y be two linear spaces over the real multiplier domain. We consider the set of all ordered pairs of points $[x, y]$ and define:

$$[x_1, y_1] = [x_2, y_2]$$

if and only if $x_1 = x_2$ in R_x and $y_1 = y_2$ in R_y. Further, let

$$[x_1, y_1] + [x_2, y_2] = [x_1 + x_2, y_1 + y_2]$$

and for each λ

$$\lambda[x, y] = [\lambda x, \lambda y] .$$

Prove that with these definitions the elements $[x, y]$ form a linear space, the *product space*

$$R_x \times R_y .$$

Construct more generally the product space of k linear spaces $R_1, \ldots, R_k$.

7. In the above construction, identify $[x, 0]$ with x and $[0, y]$ with y, and show that R_x and R_y are then linearly independent subspaces of the product space and that

$$R_x \times R_y = R_x + R_y .$$

§ 2. Finite Dimensional Linear Spaces

2.1. Linear coordinate systems. Having introduced the basic concepts of general linear spaces we now consider those of finite dimension.

Let R^m be a linear space of dimension m: Every system of $m + 1$ vectors is linearly dependent, while there exists at least one system of m linearly independent vectors $a_1, \ldots, a_m$.

It follows from this that any vector x in the space can be uniquely represented as a linear combination of these vectors

$$x = \sum_{i=1}^{m} \xi^i a_i .$$

The linearly independent vectors a_i consequently generate the entire space $R^m = (a_1, \ldots, a_m)$. They form a *basis* or a linear *coordinate system* for the space. The unique real numbers ξ^i are the m linear coordinates of the point x in the former coordinate system. We claim:

Any coordinate system of an m-dimensional space contains precisely m linearly independent vectors.

Let $b_1, \ldots, b_n$ be a second basis, in addition to $a_1, \ldots, a_m$. Since these vectors are linearly independent, it is necessarily true that $n \leqq m$; we maintain that $n = m$.

Since the vectors b_j by hypothesis generate the entire space R^m, the vectors a_i are, in particular, unique linear combinations of the vectors b_j. Since $a_1 \neq 0$, the coefficients in the representation of a_1 cannot all vanish. This equation can therefore be solved for some b_j, say for b_1, and this expression for b_1 substituted into those for $a_2, \ldots, a_m$, which then become linear combinations of a_1 and $b_2, \ldots, b_n$. Because of the linear independence of a_1 and a_2, not all of the coefficients of $b_2, \ldots, b_n$ vanish in the expression thus obtained for a_2, and the equation therefore can be solved for b_2, say, and b_2 can be eliminated from the expressions for $a_3, \ldots, a_m$.

Now if it were true that $n < m$, then the continuation of this elimination process would after n steps lead to an equation of the form

$$a_{n+1} = \sum_{i=1}^{n} \lambda^i a_i ,$$

which contradicts the linear independence of the vectors a_i. Consequently $n = m$, and the assertion is proved.

2.2. The monomorphism of the space R^m. It follows from the above theorem that with respect to the relations of linearity all m-dimensional linear spaces are isomorphic and consequently have the same linear structure. In fact:

Let R^m and $\bar{R}^m$ be two m-dimensional linear spaces with coordinate systems a_i and $\bar{a}_i$. Two arbitrary points x and $\bar{x}$ in these spaces then have unique coordinates,

$$x = \sum_{i=1}^{m} \xi^i a_i , \qquad \bar{x} = \sum_{i=1}^{m} \bar{\xi}^i \bar{a}_i .$$

If one now lets x and $\bar{x}$ correspond if and only if $\xi^i = \bar{\xi}^i$ $(i = 1, \ldots, m)$, a one-to-one mapping

$$x \leftrightarrow \bar{x}$$

is obtained that has the following properties:

1. From $x + y = z$ in R^m it follows that $\bar{x} + \bar{y} = \bar{z}$ in $\bar{R}^m$.
2. From $y = \lambda x$ in R^m it follows that $\bar{y} = \lambda \bar{x}$ in $\bar{R}^m$.

The one-to-one mapping just defined consequently preserves the basic relations of linearity to which any linear statement can be reduced. Every linear statement which is correct in R^m remains correct when the points which appear are replaced by their image points in $\bar{R}^m$, and conversely. This means that spaces of equal dimension have the same linear structure, they are linearly isomorphic.

Conversely, it obviously follows from a given isomorphic mapping

$$R^m \leftrightarrow \bar{R}^m$$

that there is corresponding to any coordinate system a_i in R^m a uniquely determined coordinate system $\bar{a}_i \leftrightarrow a_i$ in $\bar{R}^m$. We shall come back to the determination of this coordinate system and the corresponding isomorphisms later in connection with linear mappings.

2.3. Subspaces and factor spaces in R^m. Let U^d be a subspace of R^m of dimension d; thus $0 \leqq d \leqq m$. For $d = 0$ U^d reduces to the zero point, for $d = m$ it spans R^m. Let now U^d be a subspace with $0 < d < m$.

If $a_1, \ldots, a_d$ is then a basis for U^d it can be completed to a coordinate system for the entire space R^m. For this choose new basis vectors $a_{d+1}, \ldots, a_m$ so that a_{d+i+1} is not contained in the subspace $(a_1, \ldots, a_{d+i})$ generated by the preceding vectors.

By virtue of this completion one can write for any x in R^m

$$x = \sum_{i=1}^{d} \xi^i a_i + \sum_{i=d+1}^{m} \xi^i a_i ,$$

or

$$x \equiv \sum_{i=d+1}^{m} \xi^i a_i \pmod{U^d} .$$

From the linear independence of the vectors $a_1, \ldots, a_m$ it follows that the $m - d$ basis vectors on the right are even independent modulo U^d, and consequently they form a basis for the factor space R^m/U^d. This factor space thus has dimension $m - d$.

As vectors in the original space R^m the vectors $a_{d+1}, \ldots, a_m$ generate a subspace V^{m-d} of dimension $m - d$. The subspaces U^d and V^{m-d} are obviously linearly independent, and

$$R^m = U^d + V^{m-d} .$$

These subspaces are linearly independent *complements* of one another in R^m.

It can be seen from the above that a linearly independent complement V^{m-d} for a given space U^d can be generated in many ways. They are all of dimension $m - d$ and are linearly isomorphic to each other and to the factor space R^m/U^d.

In general, the following holds: If the linearly independent subspaces $U_i^{d_i}$ $(i = 1, \ldots, k)$ are, respectively, of dimension d_i and generate R^m, then

$$\sum_{i=1}^{k} d_i = m .$$

Then if $a_1^i, \ldots, a_{d_i}^i$ is an arbitrary basis for U^{d_i}, the set of these m basis vectors provides a coordinate system in R^m.

2.4. Hyperplanes in R^m. We have defined hyperplanes in 1.4 as congruence classes modulo the subspaces. A subspace U^d of dimension $d \leqq m$ and a point x_0 of R^m determine a d-dimensional hyperplane E^d through x_0 parallel to U^d and containing the set of points x with the property

$$x \equiv x_0 \pmod{U^d} .$$

For $d = 0$ E^d reduces to the point x_0; for $d = 1$ E^d is a line through x_0, etc. Two points x_0, x_1 uniquely determine a line containing these points and three points not lying on the same line, a plane through these points. We prove:

If $x_0, x_1, \ldots, x_d$ $(d \leqq m)$ are points in the space R^m such that the d differences

$$x_1 - x_0, \ldots, x_d - x_0$$

are linearly independent, then there is one and only one d-dimensional plane E^d through these points; E^d then contains precisely the set of points

$$x = \sum_{i=0}^{d} \mu^i x_i \qquad \text{with} \qquad \sum_{i=0}^{d} \mu^i = 1 ,$$

where the numbers μ^i are uniquely determined for any x in the plane.

For the proof, observe that it makes no difference which of the $d + 1$ vectors x_j appears in the above differences as the subtrahend. For if they are linearly independent, the same is obviously true for the differences

$$x_0 - x_j, \ldots, x_{j-1} - x_j, x_{j+1} - x_j, \ldots, x_d - x_j$$

for every $j = 1, \ldots, d$. They determine the same d-dimensional subspace U^d.

The desired plane E^d through the given points is parallel to the subspace U^d and consequently contains, since it goes through x_0, all points

$$x = x_0 + \sum_{i=1}^{d} \xi^i (x_i - x_0) .$$

If we set

$$\mu^0 = 1 - \sum_{i=1}^{d} \xi^i , \qquad \mu^i = \xi^i \quad (i = 1, \ldots, d) ,$$

we obtain the equation for the plane E^d in the above form.

Conversely, this point set obviously defines a hyperplane E^d through the given points parallel to the subspace U^d, where the numbers μ^i are uniquely determined for a given x in E^d. This proves the assertion.

The plane E^d discussed above contains the origin and coincides with the subspace U^d if there exists a system of numbers μ^i with $\sum_{i=0}^{d} \mu^i = 1$ such that

$$\sum_{i=0}^{d} \mu^i x_i = 0 ,$$

that is, if the $d + 1$ vectors x_i are linearly dependent. This is for $d = m$, naturally, always the case, and then $E^m = U^m = R^m$.

2.5. Simplexes. Barycentric coordinates. A configuration of the kind considered above, which consists of $d + 1$ arbitrary points $x_0, x_1, \ldots, x_d$ $(d \leq m)$ with linearly independent differences $x_i - x_0$, is called a d-dimensional *simplex*.

As was shown above, the simplex determines a plane E^d in which the simplex lies, and the points of this plane are uniquely given by

$$x = \sum_{i=0}^{d} \mu^i x_i \qquad \text{with} \qquad \sum_{i=0}^{d} \mu^i = 1 .$$

This representation can be interpreted in the following way:

If the total mass 1 is to be distributed among the vertices x_i in such a way that the center of mass of the system lies at x, then the point x_i must be given precisely the mass μ^i. For this reason the numbers μ^i are called *barycentric* coordinates for the E^d-point x with respect to the simplex that determines E^d.

If the barycentric coordinates of the point x are all positive, x is called an *interior* point of the simplex. If one or more of the coordinates is zero, the remainder positive, then x is a *boundary point*. Finally, if at least one of these coordinates is negative, x is called an *exterior* point of the simplex.

The vertices x_j of the d-dimensional simplex are obtained for $\mu^j = 1$ and $\mu^i = 0$ $(i \neq j)$, and hence, using the Kronecker delta δ^i_j, for $\mu^i = \delta^i_j$.

These are the side simplexes of dimension zero. In general, a d-dimensional simplex has

$$\binom{d+1}{p+1}$$

side simplexes of dimension p whose interior points are obtained whenever $d-p$ of the barycentric coordinates are zero and the remaining $p+1$ positive.

2.6. Exercises. 1. Let U^p and V^q be subspaces of the linear space R^m of dimensions p and q, respectively. Suppose the space (U^p, V^q) generated by these subspaces has dimension s and the intersection $[U^p, V^q]$ dimension r. Show that

$$s + r = p + q ,$$

and determine r in the case where the subspaces $U^p \neq V^q$ are of dimension $p = q = m - 1$.

2. Let

$$s^m = s^m(x_0, \ldots, x_m) , \qquad \bar{s}^m = \bar{s}^m(\bar{x}_0, \ldots, \bar{x}_m)$$

be two m-dimensional simplexes in the same m-dimensional plane E^m and

$$\bar{x}_j = \sum_{i=0}^{m} \mu_j^i \, x_i \qquad (j = 0, \ldots, m)$$

the barycentric representation of the points $\bar{x}_j$ with respect to s^m. Further let x be a point in the plane E^m and

$$x = \sum_{j=0}^{m} \lambda^j \bar{x}_j$$

the barycentric representation of x with respect to $\bar{s}_m$.

Determine the barycentric representation of this point with respect to s^m.

3. In the preceding exercise suppose that $\bar{x}_j = x_j$ for $j = 1, \ldots, m$ and

$$\bar{x}_0 = \mu \, x_0 + \sum_{i=1}^{m} \mu^i \, x_i$$

is the barycentric representation of the vertex $\bar{x}_0$ with respect to s^m and, conversely, as a consequence,

$$x_0 = \frac{1}{\mu} \bar{x}_0 - \sum_{i=1}^{m} \frac{\mu^i}{\mu} \bar{x}_i$$

the barycentric representation of x_0 with respect to $\bar{s}^m$. Prove:

The simplexes

$$s^m(x_0, x_1, \ldots, x_m) \quad \text{and} \quad \bar{s}^m(\bar{x}_0, \bar{x}_1 \ldots, \bar{x}_m)$$

in the plane E^m with the common $(m-1)$-dimensional side simplex

$$s_0^{m-1}(x_1, \ldots, x_m),$$

then have no common interior points if and only if $\mu < 0$.

§ 3. Linear Mappings

3.1. Definition. Let R_x^m and R_y^n be two linear spaces of dimensions m and n, respectively, and G_x a point set in R_x^m which for the time being is arbitrary. Then if corresponding to each point x in G_x there is a unique point

$$x \to y = y(x)$$

in the space R_y^n, a mapping of the set G_x in R_x^m into R_y^n is defined. G_x is the *domain* and the set $y(G_x)$ of image points in R_y^n the *range* of the mapping *vector function* $y(x)$.

We shall thoroughly investigate such general vector functions later. Here only the simplest, namely the *linear mappings*, are to be discussed.

The mapping $x \to y = y(x)$ is said to be linear if $y(x_1 + x_2) = y(x_1) + y(x_2)$ and $y(\lambda x) = \lambda y(x)$ for every real λ, and consequently

$$y\left(\sum_{i=1}^{k} \lambda_i x_i\right) = \sum_{i=1}^{k} \lambda_i y(x_i).$$

We shall in what follows denote the vector function $y(x)$ by $A(x)$ and in general omit the parentheses for linear mappings. We write instead of $y = A(x)$, $y = Ax$ for short.

3.2. Domain of definition and range of linear mappings. In order for the above definition of a linear mapping to make sense, the domain, the point set G_x, must contain all linear combinations of its vectors and therefore must be a subspace of R_x^m. In the following we restrict ourselves to this subspace and denote it from the start by R_x^m.

Let

$$y = Ax$$

be a linear mapping from R_x^m into R_y^n. We prove the following theorem:

In R_y^n the image AU_x of a subspace U_x of R_x^m, and thus in particular the entire range AR_x^m, is a subspace of R_y^n whose dimension is at most equal to the dimension of the preimage U_x.

Let $x = \sum_{i=1}^{k} \lambda_i x_i$ be a finite linear combination of vectors x_i in U_x and $Ax_i = y_i$. Then

$$y = Ax = A\left(\sum_{i=1}^{k} \lambda_i x_i\right) = \sum_{i=1}^{k} \lambda_i Ax_i = \sum_{i=1}^{k} \lambda_i y_i.$$

Conversely, if $y_1, \ldots, y_k$ are arbitrary points in the image $A U_x$, then there are in U_x points x_i such that the above equations hold. Since U_x is a subspace, x also lies in U_x and therefore y in $A U_x$. Accordingly, the latter set contains every finite linear combination of its vectors and it is therefore a subspace of R_y^n.

The image of the most strict subspace, which contains only the zero point of R_x^m, is a subspace; since every subspace contains the zero point, it thus follows that $A0 = 0$. This equality also follows, by the way, directly from $A(\lambda x) = \lambda A x$ for $\lambda = 0$.

It follows from this remark and from the above equations that the image vectors $y_i = Ax_i$ of linearly dependent vectors x_i are likewise linearly dependent. The dimension of the image space $A U_x$ can therefore not exceed the dimension of the preimage.

3.3. Regular and nonregular linear mappings. The set of those vectors in the original space R_x^m which in the linear mapping A are mapped onto the zero vector of the image space R_y^n obviously contains all finite linear combinations of its vectors and is therefore a subspace $K^p = K^p(A)$ of R_x^m. This subspace, whose dimension we assume is p ($\leqq m$), is called the *kernel* of A.

If $p = 0$, then $Ax = 0$ only for $x = 0$, and hence $Ax_1 = Ax_2$ only for $x_1 = x_2$. The range AR_x^m in R_y^n is then "simple"; the mapping is said to be *regular* in this case.

Since for a regular mapping the linear dependence of the vector x_i follows from the linear dependence of the image vectors $y_i = Ax_i$, the dimension of the preimage U_x is no larger than the dimension of the image space $A U_x$. These dimensions are thus equal for a regular linear mapping. In particular, the dimension of the range AR_x^m is also equal to m and therefore necessarily $m \leqq n$.

If the dimension of the kernel $K^p(A)$ is not $= 0$, then the mapping is said to be *nonregular*.

To determine the dimension of the image $A U^d$ of a d-dimensional subspace U^d of R_x^m in this case, observe that A considered as a mapping of the space U^d has as kernel the intersection

$$[U^d, K^p] = K_0 ,$$

whose dimension we suppose is p_0 ($0 \leqq p_0 \leqq d, p$). If we now go over to the factor space

$$U_{K_0} = U^d/K_0 ,$$

replacing the original identity relation in U^d by congruence modulo K_0, A obviously becomes regular in U_{K_0}. Since nothing was changed in the image space R_y^n, $A U_{K_0} = A U^d$, and the image space has by the above the same dimension as U_{K_0}, i.e., $d - p_0$.

If U^d and K^p are linearly independent, then $p_0 = 0$, and the dimension of $A\,U^d$ is equal to d. If K^p is a subspace of U^d, this dimension is $d - p$. Finally, if U^d is a subspace of the kernel K^p, then AU^d is also of dimension 0 and reduces to the zero point.

From this simple argument it follows in particular that the dimension of the range AR_x^m is equal to

$$r = m - p\,.$$

For a linear mapping of R_x^m into R_y^n, therefore, we must always have $m - p \leqq n$, and consequently the dimension of the kernel has to be

$$p \geqq m - n\,.$$

3.4. Matrices. We fix two coordinate systems $a_1, \ldots, a_m$ and $b_1, \ldots, b_n$ in the spaces R_x^m and R_y^n. Then

$$x = \sum_{i=1}^{m} \xi^i a_i \qquad \text{and} \qquad y = Ax = \sum_{j=1}^{n} \eta^j b_j\,.$$

In particular, let

$$Aa_i = \sum_{j=1}^{n} \alpha_i^j b_j;$$

as a consequence

$$Ax = A\left(\sum_{i=1}^{m} \xi^i a_i\right) = \sum_{i=1}^{m} \xi^i Aa_i = \sum_{j=1}^{n}\left(\sum_{i=1}^{m} \alpha_i^j \xi^i\right) b_j\,.$$

Hence

$$\eta_j = \sum_{i=1}^{m} \alpha_i^j \xi^i \qquad (j = 1, \ldots, n)\,.$$

We see:

Corresponding to a linear mapping or, as we also wish to say, a linear *operator* A from R_x^m into R_y^n there is, with respect to two fixed coordinate systems for these spaces, a system of equations that gives the n coordinates of $y = Ax$ as linear homogeneous expressions in the m coordinates of x. The coefficients in this system of equations form a *matrix*

$$\begin{pmatrix} \alpha_1^1 & \ldots & \alpha_m^1 \\ \vdots & & \vdots \\ \alpha_1^n & \ldots & \alpha_m^n \end{pmatrix} = (\alpha_i^j)$$

with n rows and m columns. Conversely, by means of the above system of equations such a matrix defines a linear mapping $y = Ax$ when we set

$$x = \sum_{i=1}^{m} \xi^i a_i \qquad \text{and} \qquad y = \sum_{j=1}^{n} \eta^j b_j\,.$$

If R_z^p is a third linear space and $z = By$ stands for a linear mapping of R_y^n into R_z^p with the matrix

$$\begin{pmatrix} \beta_1^1 \dots \beta_n^1 \\ \vdots \quad \vdots \\ \beta_1^p \dots \beta_n^p \end{pmatrix} = (\beta_j^k)$$

with respect to the above coordinate system $b_1, \dots, b_n$ in R_y^n and a coordinate system $c_1, \dots, c_p$ in R_z^p, the equation

$$z = B\,y = B\,A\,x$$

obviously defines a linear mapping from R_x^m into R_z^p which is produced by composition of the linear mappings A and B. With respect to the coordinate systems $a_1, \dots, a_m$ in R_x^m and $c_1, \dots, c_p$ in R_z^p this mapping has as matrix the product of the matrices (β_j^k) and (α_i^j)

$$(\gamma_i^k) = (\beta_j^k)\,(\alpha_i^j) = \left(\sum_{j=1}^{n} \beta_j^k \alpha_i^j\right)$$

with p rows and m columns.

3.5. The linear operator space. We now consider the set of all linear operators from R_x^m into R_y^n and are going to look at them as elements of a space, to which we give a linear structure by means of the following definitions.

If A and B are two linear operators from R_x^m to R_y^n we set $A = B$ if

$$A\,x = B\,x$$

identically in x. We define the sum $A + B$ by means of the identity

$$(A + B)\,x = A\,x + B\,x$$

and the product λA by

$$(\lambda A)\,x = \lambda\,Ax\,.$$

With these definitions the linear operators considered obviously form a linear space, the *linear operator space* from R_x^m into R_y^n, whose zero element is the identically vanishing linear mapping which maps all vectors in R_x^m onto the zero vector in R_y^n.

In order to determine the dimension of this linear operator space we consider an arbitrary operator A in the space and have, with the earlier notation, for every x in R_x^m

$$Ax = \sum_{j=1}^{n}\left(\sum_{i=1}^{m}\alpha_i^j\,\xi^i\right)b_j = \sum_{i=1}^{m}\sum_{j=1}^{n}\alpha_i^j\,\xi^i\,b_j = \sum_{i=1}^{m}\sum_{j=1}^{n}\alpha_i^j\,A_j^i\,x\,,$$

where

$$A_j^i\,x = \xi^i\,b_j \qquad (i = 1, \dots, m;\, j = 1, \dots, n)$$

are linear mappings of R_x^m into R_y^n. Since the above equations hold identically in x, we have in the operator space

$$A = \sum_{i=1}^{m} \sum_{j=1}^{n} \alpha_i^j A_j^i ,$$

from which it can be seen that the mn operators A_j^i span the entire operator space. Furthermore because

$$A a_i = \sum_{j=1}^{n} \alpha_i^j b_j$$

the coefficients α_i^j are uniquely determined by the operator A, and the generators A_j^i are consequently linearly independent and form a basis for the operator space. This space therefore has dimension mn.

3.6. The case $n = 1$. The dual space. The case $n = 1$, where R_y^n is a one-dimensional line

$$y = \eta\, b$$

which, if one wishes, can be identified with the real η-axis, merits special attention. The linear operator space from R_x^m into the real axis has dimension m according to the above and in this case is called the linear space *dual* to R_x^m.

This linear space is spanned by the linearly independent operators $A_j^i = A^i$:

$$A^i x = \xi^i b = \xi^i ,$$

where, as before, for any x in R_x^m

$$x = \sum_{i=1}^{m} \xi^i a_i .$$

If the *dual basis* A^i, dual to a_i, of the dual operator space is denoted by a^{*i} and the operators in general by x^*, one has for every operator x^* in the dual space

$$x^* = \sum_{i=1}^{m} \xi_i^* a^{*i} ,$$

with uniquely determined real coefficients $x^* a_i = \xi_i^*$. If this operator is applied to x, one obtains the linear mapping

$$y = x^* x = \sum_{i=1}^{m} \xi_i^* \xi^i$$

of the space R_x^m into the real y-axis.

According to this, to each vector x in R_x^m and to each operator or "vector" x^* in the dual space $R_{x^*}^m$ there corresponds a real number which depends linearly on x as well as on x^* and is therefore a real-valued bilinear function of these vectors. Corresponding to each fixed x_0^* of the dual operator space $R_{x^*}^m$ there is a real linear function

$x_0^* \, x$ of the vector x, and each fixed x_0 in R_x^m gives a real linear function $x^* \, x_0$ of the vector operator x^* of the dual space $R_{x^*}^m$, thus an element x_0^{**} of the likewise m-dimensional space $R_{x^{**}}^m$, dual to the former operator space. The correspondence

$$x_0 \leftrightarrow x_0^{**}$$

is one-to-one and linearly isomorphic. If these image elements are identified, $R_x^m = R_{x^{**}}^m$ becomes, conversely, the space dual to $R_{x^*}^m$; in this sense the duality is in fact symmetric, thus motivating the term "dual".

3.7. The case $n = m$. Linear transformations. For $n = m$ the spaces R_x^m and R_y^n are isomorphic: there exists a one-to-one mapping that preserves linear relationships.

Such an isomorphic mapping is obviously a regular linear mapping of R_x^m onto R_y^n. For if the well-defined image of x in R_y^m is denoted by $y = Ax$ it follows from the invariance of linear relations that A is linear; and since the mapping is one-to-one A is moreover regular. Conversely, any such linear mapping of R_x^m onto R_y^m provides an isomorphism between these spaces.

In the case at hand the *inverse* linear mapping $x = A^{-1} y$ from R_y^m onto R_x^m exists; it is likewise regular, and the identities

$$A^{-1} A \, x = x \, , \qquad A \, A^{-1} y = y$$

hold.

If the isomorphic spaces R_x^m and R_y^m are identified, the linear mappings become linear self-mappings of the m-dimensional linear space R^m. We are going to call such self-mappings of R^m *linear transformations* and denote them by $y = Tx$.

If such a transformation is regular, it maps R^m one-to-one onto itself, and the inverse linear transformation T^{-1} likewise exists and has the property

$$T^{-1} \, T = T \, T^{-1} = I \, ,$$

where I stands for the identity transformation $y = x$.

If on the other hand the transformation T is nonregular, with the p-dimensional kernel $K^p = K^p(T)$, then $T \, K^p = 0$, and the space R^m is mapped onto the $(m - p)$-dimensional subspace $T \, R^m$, which is isomorphic to the factor space R^m / K^p.

With regard to regular linear transformations of the linear space R^m, observe that they obviously form a group with respect to composition or multiplication. For if T_1 and T_2 are two regular linear transformations, then the composite transformation

$$T \, x = T_1 \, T_2 \, x$$

is also linear and regular. Furthermore, the identity transformation I is regular and

$$T\,I = I\,T = T\,.$$

Finally, as has already been remarked, every regular linear transformation T has a regular linear inverse transformation T^{-1}. This group of regular linear transformations is not commutative, if $m > 1$.

On the other hand, the set of all linear transformations of R^m does not form a group. Of course, the composite of two transformations is again a linear transformation, and the identity transformation does exist. But a nonregular transformation has no inverse.

3.8. Determination of all linear coordinate systems in R^m. A linear coordinate system $a_1, \ldots, a_m$ for R^m is mapped by a regular linear transformation T onto m vectors $b_i = T\,a_i$, which are likewise linearly independent and which therefore form a basis for R^m; and indeed one obtains all linear coordinate systems in this way. For if $b_1, \ldots, b_m$ are at first an arbitrary ordered system of vectors in R^m, there exists a unique linear transformation $y = T\,x$, namely

$$y = T\,x = \sum_{i=1}^{m} \xi^i\, b_i \qquad \text{for} \qquad x = \sum_{i=1}^{m} \xi^i\, a_i\,,$$

which transforms the a_i into the b_i. If the vectors b_i are in addition linearly independent, T is obviously regular: from $y = 0$ it follows that $x = 0$. We see:

The ordered coordinate systems for the space R^m, on the one hand, and the regular linear transformations of this space, on the other hand, are in one-to-one correspondence.

Let a_i and b_j be two coordinate systems for the space R^m. Then for an arbitrary x

$$x = \sum_{i=1}^{m} \xi^i\, a_i = \sum_{j=1}^{m} \eta^j\, b_j\,.$$

According to the above, a unique regular transformation exists such that

$$b_j = T\,a_j = \sum_{i=1}^{m} \alpha^i_j\, a_i\,, \qquad a_i = T^{-1}\,b_i = \sum_{j=1}^{m} \beta^j_i\, b\ .$$

where because $T^{-1}\,T = T\,T^{-1} = I$

$$\sum_{j=1}^{m} \beta^j_i\, \alpha^k_j = \sum_{j=1}^{m} \alpha^j_i\, \beta^k_j = \delta^k_i\,.$$

If the expressions for a_i and b_j are substituted into the above representations of x, the formulas for the linear transformation of the coordinates are obtained:

$$\eta^j = \sum_{i=1}^{m} \beta^j_i\, \xi^i\,, \qquad \xi^i = \sum_{j=1}^{m} \alpha^i_j\, \eta^j\,.$$

3.9. Affine transformations. Besides the linear transformations of the space R_x^m we consider somewhat more generally the *affine* transformations of this space. Such a transformation results from the composition of a linear transformation Tx with a *translation* of the space. A translation of the space R_x^m is a single-valued mapping $A(x)$ of this space onto itself such that every vector $[x_1, x_2]$, with initial point x_1 and end point x_2 is transformed into a congruent vector $[A(x_1), A(x_2)]$. Consequently, we have

$$A(x_1) - x_1 = A(x_2) - x_2 .$$

If one sets $x_1 = 0$, $x_2 = x$, it follows from this that the translation is necessarily of the form

$$A(x) = x_0 + x ,$$

where $x_0 = A(0)$.

This is the general form of a translation: for an arbitrary vector x_0 the expression $A(x) = x_0 + x$ possesses the property required by the definition.

The general affine transformation is thus

$$A(x) = x_0 + T x ,$$

where $T x$ is a linear transformation of the space R_x^m.

Those affine transformations for which the linear transformation T is regular form a group. In this group the set of translations is a commutative subgroup.

3.10. Exercises. 1. We start from the matrix introduced in 3.4

$$\begin{pmatrix} \alpha_1^1 \dots \alpha_m^1 \\ \vdots \quad\quad \vdots \\ \alpha_1^n \dots \alpha_m^n \end{pmatrix} = (\alpha_i^j)$$

and with the notation used there define the "column vectors"

$$y_i = \sum_{j=1}^{n} \alpha_i^j b_j \qquad (i = 1, \dots, m)$$

in R_y^n and the "row vectors"

$$x_j = \sum_{i=1}^{m} \alpha_j^i a_i \qquad (j = 1, \dots, n)$$

in R_x^m. The *transpose*

$$\begin{pmatrix} \alpha_1^1 \dots \alpha_1^n \\ \vdots \quad\quad \vdots \\ \alpha_m^1 \dots \alpha_m^n \end{pmatrix} = (\alpha_j^i)$$

has, on the other hand, the column vectors x_j and the row vectors y_i.

By means of the equations

$$A\,a_i = y_i \quad (i = 1, \ldots, m)\,, \qquad A^*\,b_j = x_j \quad (j = 1, \ldots, n)$$

we define in R^m_x and R^n_y the linear mappings A and A^* and denote their kernels of dimension $p \leqq m$ and $q \leqq n$, respectively, by $K^p_x = K^p_x(A)$ and $K^q_y = K^q_y(A^*)$; therefore, $A\,x = 0$ or $A^*\,y = 0$ precisely when x lies in K^p_x or y lies in K^q_y.

Prove that

$$\dim A\,R^m_x = m - p = n - q = \dim A^*\,R^n_y\,.$$

Proof. A is a regular linear mapping of the factor space R^m_x/K^p_x and consequently $\dim A\,R^m_x = \dim A(R^m_x/K^p_x) = m - p$; for the same reason $\dim A^*\,R^n_y = \dim A^*(R^n_y/K^q_y) = n - q$.

In order to prove that $m - p = n - q$, show that $A\,R^m_x$ and $K^q_y(A^*)$ are linearly independent subspaces in R^n_y. In fact: If y is a vector common to these subspaces, one has

$$y = A\,x = \sum_{i=1}^{m} \xi^i\,y_i = \sum_{j=1}^{n} \eta^j\,b_j \quad \text{with} \quad A^*\,y = \sum_{j=1}^{n} \eta^j\,x_j = 0\,,$$

and consequently

$$\eta^j = \sum_{i=1}^{m} \alpha^j_i\,\xi^i \quad \text{with} \quad \sum_{j=1}^{n} \alpha^j_i\,\eta^j = 0\,.$$

Thus

$$0 = \sum_{i=1}^{m} \xi^i \sum_{j=1}^{n} \alpha^j_i\,\eta^j = \sum_{j=1}^{n} \eta^j \sum_{=1}^{m} \alpha^j_i\,\xi^i = \sum_{j=1}^{n} (\eta^j)^2\,,$$

and therefore $y = 0$.

According to this it follows from $A^*\,Ax = 0$ that $Ax = 0$, and A is therefore a regular mapping of the factor space R^m_x/K^p_x into the factor space R^n_y/K^q_y, and thus $m - p \leqq n - q$.

The linear independence of the subspaces $A^*\,R^n_y$ and $K^p_x(A)$ in R^m_x is established in the same way, from which the converse $n - q \leqq m - p$ follows. Therefore, $m - p = n - q$, which was to be proved.

Remark 1. The matrix invariant

$$r = m - p = n - q\,,$$

which by the above is independent of the choice of the coordinate systems a_i and b_j, is called the *rank* of the matrix (α^j_i). It indicates the number of linearly independent column vectors and row vectors. If $m = n$, the matrix is said to be *square*. The square matrix (α^j_i) is *symmetric* if $\alpha^j_i = \alpha^i_j$, that is, if the matrix is identical with its transpose. Further, a square, not necessarily symmetric matrix is called *regular* if the rank of the matrix is $r = m$.

Remark 2. It follows from the above that $A\,R_x^m$ and $K_y^q(A^*)$ are linearly independent complements in R_y^n and that the same is true for $A^*\,R_y^n$ and $K_x^p(A)$ in R_x^m:

$$R_y^n = A\,R_x^m + K_y^q(A^*)\,, \qquad R_x^m = A^*\,R_y^n + K_x^p(A)\,.$$

2. The results of exercise 1 contain the complete theory of systems of real linear equations. Verify the following main theorems:

a. If the coefficient matrix in the linear homogeneous system of equations

$$\sum_{i=1}^{m} \alpha_i^j \xi^i = 0 \qquad (j = 1, \ldots, n)$$

is of rank r, the system has precisely $m - r$ linearly independent solution vectors $x = \sum_{i=8}^{m} \xi^i a_i$.

b. For the solvability of the corresponding nonhomogeneous system

$$\sum_{i=1}^{m} \alpha_i^j \xi^i = \beta^j \qquad (j = 1, \ldots, n)$$

it is necessary and sufficient that for every solution $y = \sum_{j=1}^{n} \eta^j b_j$ of the transposed homogeneous system

$$\sum_{j=1}^{n} \alpha_i^j \eta^j = 0 \qquad (i = 1, \ldots, m)$$

the equation

$$\sum_{j=1}^{n} \beta^j \eta^j = 0$$

holds. Then if $x_0 = \sum_{i=1}^{m} \xi_0^i a_i$ is a particular solution of the nonhomogeneous system, one obtains the general solution by adding the general solution of the corresponding homogeneous system of equations.

Proof. The condition named in theorem b is obviously necessary. That it also is sufficient follows from remark 2 of exercise 1, according to which $b = \sum_{j=1}^{n} \beta^j b_j$ can be uniquely decomposed into two components $b = c + d$ with $c = \sum_{j=1}^{n} \gamma^j b_j$ in $A\,R_x^m$ and $A^* d = 0$. Therefore, $d = \sum_{j=1}^{n} \delta^j b_j$ is a solution vector of the transposed homogeneous system of equations, and there exists an $x = \sum_{i=1}^{m} \xi^i a_i$ in R_x^m with the property

$c = A\,x$, i.e., $\gamma^j = \sum_{i=1}^{m} \alpha_i^j \xi^i$. Now since for every solution $y = \sum_{j=1}^{m} \eta^j b_j$ of the transposed homogeneous system, because

$$\sum_{j=1}^{n} \gamma^j \eta^j = \sum_{i=1}^{m} \xi^i \sum_{j=1}^{n} \alpha_i^j \eta^j = 0\,,$$

the equation

$$\sum_{j=1}^{n} \beta^j \eta^j = \sum_{j=1}^{n} \delta^j \eta^j$$

holds and the left hand side is by hypothesis always $= 0$, we have for $\eta^j = \delta^j$

$$\sum_{i=1}^{n} \beta^j \delta^j = \sum_{j=1}^{n} (\delta^j)^2 = 0\,,$$

and therefore $d = 0$ and $b = c$.

3. Let T be a linear transformation in R_x^m. Show that T is regular if and only if the corresponding matrix of T is regular with respect to an arbitrary coordinate system (i.e., the matrix is of rank m).

4. Let $x_1^* = L_1, \ldots, x_m^* = L_m$ be elements of the space $R_{x^*}^m$ dual to R_x^m. Show that they are linearly independent precisely when the system of equations

$$L_i\, x = 0 \qquad (i = 1, \ldots, m)$$

has the vector $x = 0$ as its only solution in R_x^m.

It follows from this that $x = 0$ provided $L\,x = 0$ for all operators $x^* = L$ of the dual space.

5. Let

$$A = (\alpha_i^j)\,, \qquad B = (\beta_i^j)$$

be matrices, A with n rows and m columns, B with m rows and p columns. We denote the transposes by A' and B'. Verify that:

a. $(A\,B)' = B'\,A'$.

b. If $n = m$ and A is regular, i.e., of rank m, the matrix A^{-1} having the property

$$A\,A^{-1} = A^{-1}\,A = I$$

exists, where I is the unit matrix (δ_i^j) (δ_i^j is the Kronecker delta). Furthermore, one then has

$$(A^{-1})' = (A')^{-1}\,.$$

6. Let $T\,x$ be a linear transformation of the linear space R_x^m. Prove: If and only if

$$K(T) = K(T^2)$$

are the subspaces $K(T)$ and $T\,R_x^m$ linearly independent and

$$R_x^m = K(T) + T\,R_x^m\,.$$

7. Let U and V be linearly independent complements of the space R_x^m and therefore for every x in R_x^m

$$x = u + v$$

with uniquely determined vectors u in U and v in V.

Show that $u = P\,x$, $v = Q\,x$ are linear transformations with the property

$$P^2 x = P\,x\,, \qquad Q^2 x = Q\,x\,.$$

$P\,x$ is called the projection of x onto U in the "direction" V and $Q\,x = v$ the projection of x onto V in the direction U.

8. Show conversely: If $P\,x$ is a linear transformation of R_x^m with the property $P^2 x = P\,x$, there exist uniquely determined linearly independent complements U and V of R_x^m such that $u = P\,x$ is the projection of x onto U in the direction V and $v = x - P\,x = Q\,x$ is the projection of x onto V in the direction U.

9. Let T be a linear transformation in R_x^m and λ a real number.

Show that the set of all solutions to the equation

$$T\,x = \lambda\,x$$

is a subspace of R_x^m.

If the dimension of this subspace is $d > 0$, λ is called an *eigenvalue* of T of multiplicity d; the solutions x are the associated *eigenvectors* and the subspace the *eigenspace* belonging to λ.

10. Show that the transformation T, taking into account the multiplicity of the eigenvalues, can have at most m eigenvalues.

Hint. The eigenspaces belonging to the various eigenvalues are linearly independent.

11. Let $T\,x \equiv P\,x$ be the projection onto U in the direction V. Determine the eigenvalues and the eigenspaces of P.

§ 4. Bilinear and Quadratic Functions

4.1. Real bilinear and quadratic functions. Let R_x^m, R_y^n and R_z^p be three linear spaces and B a mapping that assigns to each ordered pair (x, y) an element z:

$$(x, y) \overset{B}{\to} z\,.$$

The mapping is called *bilinear* provided it is linear in both arguments x, y, which is to be indicated by the notation

$$z = B\,x\,y\,.$$

In the following we shall restrict ourselves to the case $p = 1$ and discuss, which is then no further restriction, only *real-valued* bilinear functions $B\,x\,y$.

We additionally assume $n = m$. The argument spaces R^m_x and R^n_y are then linearly isomorphic. If vectors which correspond to one another in some isomorphic mapping of these spaces are identified, we are in what follows treating real bilinear functions of the vectors x and y, which vary independently in the m-dimensional linear space R^m.

If a coordinate system is fixed in R^m in which

$$x = \sum_{i=1}^{m} \xi^i a_i \quad \text{and} \quad y = \sum_{j=1}^{m} \eta^j a_j ,$$

it follows from the bilinearity of B that

$$B\,x\,y = \sum_{i,j=1}^{m} \xi^i \eta^j B\,a_i\,a_j = \sum_{i,j=1}^{m} \beta_{ij}\,\xi^i \eta^j$$

becomes a bilinear form of the coordinates ξ^i and η^j with real coefficients. Conversely, every such form with arbitrary real coefficients defines, if ξ^i and η^j are interpreted as coordinates with respect to a linear coordinate system, a real bilinear function on R^m. The square matrix of the coefficients

$$\beta_{ij} = B\,a_i\,a_j$$

is called the matrix of the bilinear form with respect to the fixed coordinate system.

For $y = x$ the bilinear function $B\,x\,y$ becomes the associated *quadratic* function or *form*

$$B\,x\,x = B\,x^2 = \sum_{i,j=1}^{m} \beta_{ij}\,\xi^i \xi^j ,$$

which is related to the generating bilinear function through the *polarization identity*

$$B\,(x + y)^2 - B\,(x - y)^2 = 2\,(B\,x\,y + B\,y\,x) .$$

The bilinear function is *symmetric* if

$$B\,y\,x = B\,x\,y$$

and *alternating* if

$$B\,y\,x = -\,B\,x\,y .$$

Any bilinear function B can be represented in a unique way as the sum of a symmetric bilinear function S and an alternating function A:

$$B\,x\,y = S\,x\,y + A\,x\,y ,$$

where

$$S\,x\,y = \frac{1}{2}\,(B\,x\,y + B\,y\,x) , \qquad A\,x\,y = \frac{1}{2}\,(B\,x\,y - B\,y\,x) .$$

The quadratic function generated by an alternating bilinear function vanishes identically, and it follows from the polarization identity that the converse is also true. The symmetric part S of a bilinear function B generates the same quadratic function as B itself and, according to the polarization identity,

$$B\,(x+y)^2 - B\,(x-y)^2 = 4\,S\,x\,y\,.$$

The symmetric part S is uniquely determined by the quadratic function $B\,x^2$.

A quadratic function is called positive or negative *definite* if it vanishes only for $x = 0$ and otherwise assumes only positive or negative values, respectively. It is *semidefinite* if it also vanishes for certain vectors $x \neq 0$, but is otherwise positive or negative, respectively. It is *indefinite* if it assumes positive as well as negative values. We also use the corresponding terminology for the generating symmetric bilinear function.

4.2. The inertia theorem. Let $B\,x^2$ be a real quadratic function in the space R^m and $B\,x\,y$ the generating symmetric bilinear function; consequently,

$$B\,(x+y)^2 - B\,(x-y)^2 = 4\,B\,x\,y\,.$$

We prove the following

Inertia theorem. *In R^m there exist coordinate systems $e_1, \ldots, e_m$ such that*

$$(*)\quad B\,e_i\,e_j = 0 \quad \textit{for} \quad i \neq j\,, \qquad B\,e_i^2 = +1 \quad \textit{or} \quad = -1 \quad \textit{or} = 0\,.$$

Then if $B\,e_i^2 = +1$ for $i = 1, \ldots, p$, $B\,e_i^2 = -1$ for $i = p+1, \ldots, p+q$ and $B\,e_i^2 = 0$ for $i = p+q+1, \ldots, p+q+r = m$ and if U^p, V^q, W^r are the linearly independent subspaces spanned by these three sets of vectors:

$$R^m = U^p + V^q + W^r\,,$$

the following invariances hold:

The dimensions p, q, r, of the subspaces named are invariant numbers characteristic of the function B for every coordinate system with the properties (*), and indeed the subspace W^r is itself invariant.

Before we go on to the proof, with a view to later metric concepts, we introduce the following terminology.

Two vectors x and y are said to be *orthogonal* to one another with respect to the given symmetric bilinear function B if

$$B\,x\,y = B\,y\,x = 0\,.$$

Further, x is called a positive or negative vector according as $B\,x^2 > 0$ or $B\,x^2 < 0$, and a "unit vector" if $B\,x^2 = \pm 1$, while x is a "null vector" provided $B\,x^2 = 0$; all with respect to B. A positive or negative vector can be normalized to a unit vector by multiplying with

$$\lambda = \frac{1}{\sqrt{|Bx^2|}}\,.$$

By this, a coordinate system with the properties (*) is orthogonal and normalized, for short *orthonormal*, with respect to B.

In such a coordinate system a symmetric bilinear function has an especially simple form. Namely, if

$$x = \sum_{i=1}^{m} \xi^i e_i\,, \qquad y = \sum_{i=1}^{m} \eta^i e_i\,,$$

then

$$B\,x\,y = \sum_{i=1}^{p} \xi^i \eta^i - \sum_{j=p+1}^{p+q} \xi^j \eta^j\,.$$

The subspaces U^p, V^q, W^r with vectors

$$u = \sum_{i=1}^{p} \xi^i e_i\,, \qquad v = \sum_{j=p+1}^{p+q} \xi^j e_j\,, \qquad w = \sum_{k=p+q+1}^{p+q+r} \xi^k e_k$$

are pairwise orthogonal in that

$$B\,u\,v = B\,u\,w = B\,v\,w = 0\,,$$

and for every x in R^m the representation $x = u + v + w$ is unique. Furthermore,

$$B\,u^2 = \sum_{i=1}^{p} (\xi^i)^2\,, \qquad B\,v^2 = -\sum_{j=p+1}^{p+q} (\xi^j)^2\,, \qquad B\,w^2 = 0\,,$$

and B is therefore positive definite in U^p, negative definite in V^q, while W^r contains nothing but null vectors.

4.3. First proof of the inertia theorem. If R^m contains only null vectors, it follows from the polarization identity that $B\,x\,y \equiv 0$. All vectors are orthogonal to one another with respect to B, and any coordinate system satisfies the inertia theorem. We have $U^p = V^q = 0$ and R^m reduces to the null space $W^r = W^m$.

If this is not the case, then R^m contains positive or negative vectors, consequently also unit vectors. Let e_1, say, be a positive unit vector, so that $B\,e_1^2 = 1$.

Then the set of all vectors x_1 orthogonal to e_1 ($B\,e_1\,x_1 = B\,x_1\,e_1 = 0$) is obviously a subspace. We claim that this subspace and the one-dimensional subspace (e_1) generated by e_1 are linearly independent complements in R^m. In fact, every vector x can be decomposed in a unique way into components

$$x = \xi^1 e_1 + x_1$$

from these subspaces; for from

$$0 = B\,x_1\,e_1 = B\,(x - \xi^1\,e_1)\,e_1 = B\,x\,e_1 - \xi^1$$

ξ^1 is uniquely determined to be

$$\xi^1 = B\,x\,e_1\,.$$

$\xi^1 e_1$ is the orthogonal *projection* of the vector x onto e_1 and x_1 the projecting *normal* with respect to B. The orthogonal and linearly independent complement to (e_1), which consists of these normals, has dimension $m - 1$ and can be denoted by R^{m-1}. One proceeds with R^{m-1} just as above with R^m and continues the procedure until a subspace R^r of dimension $r \geqq 0$ is reached. It contains nothing but null vectors, where hence, by the polarization identity, $B\,x\,y \equiv 0$. One has then found in turn $m - r$ positive or negative pairwise orthogonal unit vectors which are completed, with an arbitrary coordinate system for the null space R^r orthogonal to these unit vectors, to a complete coordinate system which is orthonormal with respect to B. If $p \geqq 0$ of these vectors are positive and $q \geqq 0$ are negative, then $p + q + r = m$; the positive unit vectors generate a p-dimensional subspace U^p, where B is positive definite; in the q-dimensional subspace generated by the negative unit vectors B is negative definite; and $W^r = R^r$ contains nothing but null vectors.

This establishes the existence of a coordinate system of the kind required in the inertia theorem, and it remains to prove the asserted invariances.

First, the invariance of the null space $R^r = W^r$ follows from the fact that this space contains precisely those vectors which are orthogonal to *every* R^m-vector x with respect to B. In fact, according to the above, for an arbitrary $x = u + v + w$ and a w_0 from W^r

$$B\,x\,w_0 = B\,u\,w_0 + B\,v\,w_0 + B\,w\,w_0 = 0\,.$$

Conversely, if the identity $B\,x\,y_0 = 0$ holds in R^m for a $y_0 = u_0 + v_0 + w_0$, then for $x = u_0$ one has

$$0 = B\,u_0\,y_0 = B\,u_0^2 + B\,u_0\,v_0 + B\,u_0\,w_0 = B\,u_0^2\,,$$

and therefore, since B is positive definite in U^p, $u_0 = 0$. In the same way it follows that $v_0 = 0$, and consequently $y_0 = w_0$, which proves the invariance of the null space W^r.[1]

---

[1] The invariant null space contains nothing but null vectors, to be sure, but in general by no means *all* null vectors. In fact, if $x = u + v + w$,

$$B\,x^2 = B\,u^2 + B\,v^2 = 0$$

precisely when $-B\,v^2 = B\,u^2$. Only if $B\,x^2$ vanishes identically or is semidefinite does it follow from the above that $u = v = 0$ and therefore $x = w$.

On the other hand, the positive space U^p and the negative space V^q are in general not invariant as subspaces, however their dimensions p and q are.

To see this, we consider a second decomposition of the required kind,

$$\overline{R}^m = \overline{U}^{\overline{p}} + \overline{V}^{\overline{q}} + \overline{W}^r ,$$

where by the above $\overline{W}^r = W^r$. Let the dimensions of $\overline{U}^{\overline{p}}$ and $\overline{V}^{\overline{q}}$ be $\overline{p}$ and $\overline{q}$; the claim is that $\overline{p} = p$ and $\overline{q} = q$.

In fact, for an arbitrary u in U^p one has uniquely

$$u = \overline{u} + \overline{v} + \overline{w}$$

by virtue of the second decomposition. Here $\overline{u} = A\,u$ is obviously a linear mapping of the space U^p into the space $\overline{U}^{\overline{p}}$, and indeed a *regular* mapping. For from $\overline{u} = A\,u = 0$ it follows that $u = \overline{v} + \overline{w}$ and hence

$$B\,u^2 = B\,(\overline{v} + \overline{w})^2 = B\,\overline{v}^2 + 2\,B\,\overline{v}\,\overline{w} + B\,\overline{w}^2 = B\,\overline{v}^2 \leqq 0;$$

thus because B is positive definite in U^p, $B\,u^2 = 0$ and $u = 0$.

But then by virtue of 3.3 one must have $p \leqq \overline{p}$, and since the converse also holds for reasons of symmetry, one has $\overline{p} = p$ and consequently also $\overline{q} = m - \overline{p} - \overline{r} = m - p - r = q$, with which the invariance claims of the inertia theorem have been established.

4.4. E. Schmidt's orthogonalization process. Second proof of the inertia theorem. We give in addition a second variant of the above proof, which is no shorter, but which does give rise to considerations that are of interest in themselves.

Consider first the case where B is *definite* in R^m, e.g. positive definite. Under this hypothesis we shall construct an orthonormal system with respect to B, starting from an arbitrary coordinate system

$$a_1, \ldots, a_m .$$

Since $a_1 \neq 0$, $B\,a_1^2 > 0$. If the real number $\pm\lambda_{11}$ is defined by

$$\lambda_{11}^2 = B\,a_1^2 ,$$

the equation

$$a_1 = \lambda_{11}\,e_1 ,$$

because $\lambda_{11} \neq 0$, yields a positive unit vector e_1.

Then project a_2 onto e_1 and thus determine the number λ_{21} so that

$$B\,(a_2 - \lambda_{21}\,e_1)\,e_1 = 0 ,$$

from which

$$\lambda_{21} = B\,a_2\,e_1$$

follows. Since a_1 and a_2 are linearly independent, so also are e_1 and a_2; consequently, the normal is $a_2 - \lambda_{21}\,e_1 \neq 0$, and $B\,(a_2 - \lambda_{21}\,e_1)^2 > 0$. The roots of the equation

$$\lambda_{22}^2 = B\,(a_2 - \lambda_{21}\,e_1)^2$$

are thus real and $\neq 0$, so that a positive unit vector orthogonal to e_1 is determined by

$$a_2 = \lambda_{21} e_1 + \lambda_{22} e_2 .$$

In the third step we project a_3 onto the subspace $(e_1, e_2) = (a_1, a_2)$ and thus determine the numbers λ_{31} and λ_{32} so that

$$B(a_3 - \lambda_{31} e_1 - \lambda_{32} e_2)\, e_i = 0$$

for $i = 1, 2$; they are

$$\lambda_{3i} = B\, a_3\, e_i .$$

Because of the linear independence of the vectors a_1, a_2 and a_3, e_1, e_2 and a_3 are also linearly independent, and consequently the normal $a_3 - \lambda_{31} e_1 - \lambda_{32} e_2$ of a_3 on $(e_1, e_2) = (a_1, a_2)$ is different from zero. Therefore $B(a_3 - \lambda_{31} e_1 - \lambda_{32} e_2)^2 > 0$. We determine $\lambda_{33} \neq 0$ from

$$\lambda_{33}^2 = B(a_3 - \lambda_{31} e_1 - \lambda_{32} e_2)^2$$

and define by means of

$$a_3 = \lambda_{31} e_1 + \lambda_{32} e_2 + \lambda_{33} e_3$$

a third unit vector e_3 which is orthogonal to e_1 and e_2.

Continuing in this way, one obtains a system of equations for determining the orthonormal system $e_1, \ldots, e_m$

$$\begin{aligned}
a_1 &= \lambda_{11} e_1 , \\
a_2 &= \lambda_{21} e_1 + \lambda_{22} e_2 , \\
a_3 &= \lambda_{31} e_1 + \lambda_{32} e_2 + \lambda_{33} e_3 , \\
&\vdots \\
a_m &= \lambda_{m1} e_1 + \lambda_{m2} e_2 + \lambda_{m3} e_3 + \cdots + \lambda_{mm} e_m ,
\end{aligned}$$

where for $j < i \leqq m$

$$\lambda_{ij} = B\, a_i\, e_j$$

and for $i = 1, \ldots, m$

$$\lambda_{ii}^2 = B(a_i - \lambda_{i1} e_1 - \cdots - \lambda_{i(i-1)} e_{i-1})^2 .$$

For every i, $(e_1, \ldots, e_i) = (a_1, \ldots, a_i)$, and by successive solution of the above system of equations one obtains

$$\begin{aligned}
e_1 &= \mu_{11} a_1 , \\
e_2 &= \mu_{21} a_1 + \mu_{22} a_2 , \\
e_3 &= \mu_{31} a_1 + \mu_{32} a_2 + \mu_{33} a_3 , \\
&\vdots \\
e_m &= \mu_{m1} a_1 + \mu_{m2} a_2 + \mu_{m3} a_3 + \cdots + \mu_{mm} a_m .
\end{aligned}$$

That completes the orthogonalization procedure of E. Schmidt.

Now let B be an arbitrary real symmetric bilinear function on R^m which does not vanish identically. Then if Bx^2 assumes positive values, for example, let U^p be a maximal positive space, that is, a subspace of the highest possible dimension p where B is positive definite. In this subspace we can by means of the Schmidt orthogonalization procedure construct a coordinate system

$$e_1, \ldots, e_p$$

which is orthonormal with respect to B. We thus have $B\,e_i\,e_j = 0$ for $i \neq j$, $B\,e_i^2 = +1$ and $U^p = (e_1, \ldots, e_p)$.

We are now going to project the R^m-vector x onto U^p and determine the corresponding normal of x to U^p.

It is thus a question of decomposing x into two components

$$x = \sum_{i=1}^{p} \xi^i\, e_i + n = u_0 + n$$

in such a way that the normal n is orthogonal to all vectors u and consequently $B\,n\,u = 0$ for every u in U^p. For this it is necessary and sufficient for n to be orthogonal to all of the generators e_j of U^p:

$$B\,n\,e_j = B\,(x - \sum_{i=1}^{p} \xi^i\, e_i)\,e_j = B\,x\,e_j - \xi^j = 0\,.$$

Accordingly, the projection is

$$u_0 = \sum_{i=1}^{p} e_i\, B\,x\,e_i$$

and $n = x - u_0$. Note that this projection as well as the corresponding normal is uniquely determined by x and U^p and is thus independent of the choice of the orthonormal system e_i in U^p. For if

$$x = u_0' + n'\,,$$

is a second decomposition of the required kind then the U^p-vector $u_0' - u_0 = n - n'$ is orthogonal to U^p and in particular to itself, and consequently $B\,(u_0' - u_0)^2 = 0$. Since B is definite in U^p, we must have $u_0' = u_0$ and as a consequence also $n' = n$.

The set of all U^p-normals n obviously form a subspace N^{m-p}, the linearly independent *orthogonal complement* of U^p in R^m with respect to B. Because of the maximality of U^p, N^{m-p} can include no positive vectors; for if $B\,n^2 > 0$, then for every u in U^p we would have

$$B\,(u + n)^2 = B\,u^2 + 2\,B\,u\,n + B\,n^2 = B\,u^2 + B\,n^2 \geqq B\,n^2 > 0\,,$$

and B would therefore be positive definite in the $(p + 1)$-dimensional space generated by U^p and n.

If N^{m-p} contains negative vectors, let V^q be a maximal negative subspace of dimension q in N^{m-p} and

$$e_{p+1}, \ldots, e_{p+q}$$

an orthonormal coordinate system constructed by means of the Schmidt orthogonalization process, so that hence $B\,e_i\,e_j = 0$ for $i \neq j$ and $B\,e_i^2 = -1$.

Then if W^r is the orthogonal complement of V^q in N^{m-p} constructed by the above method, W^r contains nothing but null vectors and is of dimension $r = m - p - q$. One has

$$R^m = U^p + V^q + W^r,$$

and if one adds an arbitrary basis

$$e_{p+q+1}, \ldots, e_{p+q+r} = e_m$$

for the subspace W^r to the above $p + q$ vectors, a coordinate system has been constructed in R^m that satisfies the requirements of the inertia theorem. The invariance of the space W^r and of the dimensions p and q is proved in the preceding section.

4.5. Orthogonal transformations. In connection with the inertia theorem we add a few supplementary remarks.

The first concerns the determination of *all* coordinate systems for the space R^m that are orthonormal with respect to the symmetric bilinear function B.

Thus let $\bar{e}_i$ be a second basis besides e_i which is orthonormal with respect to B. According to the inertia theorem this basis likewise contains p positive and q negative unit vectors and r null vectors $\bar{e}_i$. The null vectors $\bar{e}_i$ span the same null space $\overline{W}^r = W^r$ as the r null vectors e_i. If one further orders the vectors in both systems, for example so that the positive vectors are written first, then the negative ones and the null vectors last, one has for all indices $i, j = 1, \ldots, m$

$$B\,\bar{e}_i\,\bar{e}_j = B\,e_i\,e_j\,.$$

For any ordering of this kind, according to 3.8 there exists a uniquely determined linear transformation

$$\bar{x} = T\,x$$

which maps e_i onto $\bar{e}_i$. Since both systems of vectors are linearly independent, this transformation is regular. Then if

$$x = \sum_{i=1}^{m} \xi^i\,e_i\,, \qquad y = \sum_{j=1}^{m} \eta^j\,e_j\,,$$

we have

$$\bar{x} = \sum_{i=1}^{m} \xi^i\,T\,e_i = \sum_{i=1}^{m} \xi^i\,\bar{e}_i\,,$$

$$\bar{y} = \sum_{j=1}^{m} \eta^j\,T\,e_j = \sum_{j=1}^{m} \eta^j\,\bar{e}_j\,,$$

and

$$B\,\bar{x}\,\bar{y} = \sum_{i,j=1}^{m} \xi^i \eta^j B\,\bar{e}_i\,\bar{e}_j = \sum_{i,j=1}^{m} \xi^i \eta^j B\,e_i\,e_j = B\,x\,y\,,$$

and thus

$$B(T\,x)\,(T\,y) = B\,x\,y\,.$$

The symmetric bilinear function $B\,x\,y$ is thus invariant with respect to the regular linear transformation T. Such a linear transformation of the space R^m is said to be *orthogonal* with respect to B.

If, conversely, T is an arbitrary regular transformation that leaves $B\,x\,y$ invariant and maps e_i onto $T\,e_i = \bar{e}_i$, then the m vectors $\bar{e}_i$, because of the regularity of T, are linearly independent and moreover, as a consequence of the equations

$$B\,\bar{e}_i\,\bar{e}_j = B(T\,e_i)\,(T\,e_j) = B\,e_i\,e_j\,,$$

orthonormal relative to B. They hence form a basis for the space R^m that is orthonormal relative to B. We see:

If we understand a linear transformation T to be orthogonal with respect to the symmetric bilinear function $B\,x\,y$ whenever it is, first, regular and, second, leaves the latter bilinear function invariant, then all ordered coordinate systems for R^m that are orthonormal relative to B, on the one hand, and all transformations of this space which are orthogonal relative to B, on the other, are in one-to-one correspondence.

The linear transformations T of the space R^m which are orthogonal for a symmetric bilinear function B obviously form a group of transformations which are orthogonal relative to B. It is a subgroup of the group mentioned in 3.7 of all regular linear transformations of the space.

4.6. Degenerate bilinear functions. Let $B\,x\,y$ be a real not necessarily symmetric, bilinear function on the linear space R^m. Those vectors y for which

$$B\,x\,y = 0$$

holds identically in x obviously form a subspace in R^m. Provided the dimension r of this space is positive we say that $B\,x\,y$ is r-fold *degenerate* with respect to y. Thus if $B\,x\,y$ is not degenerate with respect to y, it follows from the above identity in x that $y = 0$.

If in an arbitrary coordinate system a_i

$$x = \sum_{i=1}^{m} \xi^i a_i\,, \qquad y = \sum_{j=1}^{m} \eta^j a_j\,,$$

and hence

$$B\,x\,y = \sum_{i,j=1}^{m} \xi^i \eta^j B\,a_i\,a_j = \sum_{i,j=1}^{m} \beta_{ij}\,\xi^i \eta^j\,,$$

then the fact that B is r-fold degenerate with respect to y is equivalent with the linear homogeneous system of equations

$$\sum_{j=1}^{m} \beta_{ij} \eta^j = 0 \qquad (i = 1, \ldots, m)$$

having precisely r linearly independent solution vectors y. Since the transposed system of equations

$$\sum_{i=1}^{m} \beta_{ij} \xi^i = 0 \qquad (j = 1, \ldots, m)$$

then likewise has precisely r linearly independent vectors x (cf. 3.10, exercises 1 and 2), we see: The bilinear function $B\,x\,y$ is r-fold degenerate with respect to y as well as with respect to x, and hence simply r-fold degenerate if the rank of the matrix (β_{ij}) is $m - r$. In particular, B is not degenerate precisely when this rank is m and the matrix is therefore regular.

If B is symmetric, then B is obviously r-fold degenerate precisely when the dimension of the null space W^r mentioned in the inertia theorem is equal to r.

Note in addition that the polarized, symmetric, bilinear function corresponding to a semidefinite quadratic function $B\,x^2$ is obviously always degenerate. A nondegenerate symmetric bilinear function always generates a quadratic function that is definite or indefinite, never semidefinite.

4.7. Theorem of Fréchet-Riesz. Let $B\,x\,y$ be a nondegenerate real bilinear function in R^m; symmetry is not required. For a fixed y

$$B\,x\,y = L\,x$$

is a real linear function of x, thus an element of the space dual to R^m. When y runs through the space R^m, we obtain in this way all of the elements of the dual space, each once. Namely, if $a_1, \ldots, a_m$ is a coordinate system in R^m, where

$$y = \sum_{i=1}^{m} \eta^i a_i \,,$$

then

$$L\,x = B\,x\,y = \sum_{i=1}^{m} \eta^i B\,x\,a_i = \sum_{i=1}^{m} \eta^i L_i\,x \,,$$

and here the operators L_i on the right are linearly independent and hence form a basis for the dual space if B is not degenerate.

For each L in the space dual to R^m there exists, therefore, a unique y in R^m such that for all x in R^m

$$L\,x = B\,x\,y \,.$$

This is the *theorem of Fréchet-Riesz* in the present elementary case where the dimension m is finite[1].

4.8. Adjoint linear transformations. Retaining the above hypotheses, let T now be an arbitrary linear transformation. Then for a fixed y, $B(T\,x)\,y$ is a linear function of x, and according to the theorem of Fréchet-Riesz, for each y in R^m there exists, therefore, a unique y^* in R^m such that

$$B(T\,x)\,y = B\,x\,y^*$$

identically in x. One verifies at once that

$$y^* = T^*\,y$$

is a linear transformation of the space R^m, the transformation *adjoint* to T relative to B. We thus have

$$B(T\,x)\,y = B\,x(T^*\,y)$$

identically in x and y.

Provided the nondegenerate bilinear function $B\,x\,y$ is in addition symmetric, this identity can be written $B\,y(T\,x) = B(T^*\,y)\,x$ or, after switching x and y,

$$B(T^*\,x)\,y = B\,x(T\,y)\,.$$

Thus T is conversely the linear transformation adjoint to T^*, and consequently

$$(T^*)^* = T^{**} = T\,.$$

The relation of adjunction is thus involutory.

The linear transformations that are *self-adjoint* (or *symmetric*) relative to a given symmetric, nondegenerate, bilinear function B, those for which $T^* = T$ deserve special attention.

If the linear transformation T is orthogonal with respect to B and therefore regular, $B(T\,x)\,(T\,y) = B\,x\,y$, and thus

$$B(T\,x)\,y = B\,x(T^{-1}\,y)\,.$$

According to this

$$T^* = T^{-1}\,,$$

a relation which is obviously equivalent with the original definition of a transformation which is orthogonal with respect to B, provided B is symmetric and nondegenerate.

If a linear transformation commutes with its adjoint,

$$T\,T^* = T^*\,T\,,$$

then T is said to be *normal* with respect to B. Self-adjoint orthogonal transformations are special normal transformations.

---

[1] Actually the theorem of Fréchet-Riesz is understood to be the deeper corresponding theorem in infinite dimensional Hilbert space. We have retained the same name for the almost trivial case of a finite dimensional space.

4.9. Exercises. 1. Let $B = (\beta_i^j)$ be a symmetric square matrix with m rows and columns. Prove that there are regular square matrices M such that

$$M' B M = D = (\varepsilon_i^j),$$

where M' is the transpose of M and D stands for a diagonal matrix with $\varepsilon_i^j = 0$ for $j \neq i$ and $\varepsilon_i^i = +1$ for $i = 1, \ldots, p$, $\varepsilon_i^i = -1$ for $i = p+1, \ldots, p+q$, $\varepsilon_i^i = 0$ for $i = p+q+1, \ldots, p+q+r = m$.

Show further that provided $M = M_0$ accomplishes the above, the remaining M can be obtained from

$$M = T M_0,$$

where T stands for an arbitrary regular matrix for which

$$T' B T = B.$$

2. Let $B\,x\,y$ and $C\,x\,y$ be two bilinear functions on the linear space R^n such that $B\,x\,y$ implies $C\,x\,y = 0$. Prove that

$$C\,x\,y \equiv \varkappa\, B\,x\,y,$$

where $\varkappa$ stands for a real constant.

Hint. If $B\,x\,y \equiv 0$, then $C\,x\,y \equiv 0$, and there is nothing to prove. Otherwise, let $B\,x_0\,y_0 \neq 0$. For an arbitrary pair of vectors x, y set $x = \xi\,x_0 + x_1$, $y = \eta\,y_0 + y_1$ and determine the coefficients ξ, η so that $B\,x_1\,y_0 = B\,x_0\,y_1 = 0$, which is the case if $B\,x\,y_0 = \xi\,B\,x_0\,y_0$, $B\,x_0\,y = \eta\,B\,x_0\,y_0$. Then by hypothesis $C\,x_1\,y_0 = C\,x_0\,y_1 = 0$ too, and $B\,x\,y = \xi\,\eta\,B\,x_0\,y_0 + B\,x_1\,y_1$, $C\,x\,y = \xi\,\eta\,C\,x_0\,y_0 + C\,x_1\,y_1$, and therefore

$$B\,x_0\,y_0\,C\,x\,y - C\,x_0\,y_0\,B\,x\,y = B\,x_0\,y_0\,C\,x_1\,y_1 - C\,x_0\,y_0\,B\,x_1\,y_1.$$

Since the right hand side of this equation is independent of ξ and η, the equation holds for every pair of vectors $x' = \xi'\,x_0 + x_1$, $y' = \eta'\,y_0 + y_1$ with an arbitrary choice of the coefficients ξ' and η'. If the latter are chosen so that $B\,x'\,y' = \xi'\,\eta'\,B\,x_0\,y_0 + B\,x_1\,y_1 = 0$, then by hypothesis $C\,x'\,y' = 0$ too, and the right hand side of the above equation is therefore $= 0$, from which the assertion follows with $\varkappa = C\,x_0\,y_0/B\,x_0\,y_0$.

3. Let $B\,x\,y$ be a symmetric bilinear function in R^m and U a subspace of R^m.

Set up the necessary and sufficient condition for the existence of a normal with respect to B at a given point x of U.

Give the general expression for these normals and show in particular that the normal is uniquely determined precisely when B is not degenerate in U.

4. Prove, provided $B\,x\,y$ is positive definite in R^m, the so-called *Bessel inequality*

$$B\,p^2 \leqq B\,x^2 ,$$

where p stands for the orthogonal projection of x on U, and show further that the so-called *Parseval equation*

$$B\,p^2 = B\,x^2$$

holds only for $x = p$.

5. Let $B\,x\,y$ be a nondegenerate bilinear function in R^m and $T\,x$ a linear transformation of R^m, $T^*\,x$ its adjoint transformation relative to B. Prove that the kernel of these transformations have the same dimension.

6. With the hypotheses and notations of the preceding exercise we fix in R^m an arbitrary coordinate system. With respect to this coordinate system the bilinear function B and the linear transformations T and T^* have well-defined square matrices which we denote by the same letters

$$B = (\beta_i^j) , \qquad T = (\tau_i^j) , \qquad T^* = (\tau^{*j}_i) .$$

Prove that

$$T^* = (B^{-1})'\,T'\,B' , \qquad T = B^{-1}(T^*)'\,B ,$$

and more particularly show:

a. T is self-adjoint relative to B if

$$T\,B^{-1} = B^{-1}\,T' .$$

b. T is orthogonal relative to B if

$$T\,B^{-1} = B^{-1}(T^{-1})' .$$

c. T is normal relative to B if

$$T'\,B\,T\,B^{-1} = B\,T\,B^{-1}\,T' .$$

Finally show that provided B is symmetric and definite, the coordinate system can be chosen so that

$$T^* = T' .$$

7. Let R_x^m and R_y^n be linear spaces in each of which a real bilinear function is given which we denote by $(x_1, x_2)_x$ and $(y_1, y_2)_y$ for short. Show:

a. Provided the bilinear functions are not degenerate, the linear mappings $y = A\,x$ from R_x^m into R_y^n and the linear mappings $x = A^*\,y$ from R_y^n into R_x^m are pairwise *adjoint* to each other so that

$$(A\,x, y)_y = (x, A^*\,y)_x$$

holds identically in x and y.

b. If the bilinear functions are in addition definite, then

$$R_x^m = K(A) + A^* R_y^n , \qquad R_y^n = K(A^*) + A R_x^m ,$$

where $K(A)$ and $K(A^*)$ stand for the kernels of the mappings A and A^*.

Hint. a is a direct consequence of the theorem of Fréchet-Riesz. To prove b, show that the subspaces $K(A)$ and $A^* R_y^n$ have no vectors in common other than the zero vector and are therefore linearly independent. Namely, if $x = A^* y$ is a zero vector, then because $A x = 0$

$$(x, x)_x = (x, A^* y)_x = (A x, y)_y = 0 ,$$

and therefore $x = A^* y = 0$.

Remark. The above clearly repeats in a shorter formulation what has already been said in exercises 1 and 2 of 3.10.

8. Let $A x y$ be a real bilinear alternating function that is defined in the space R^m and that does not vanish identically; thus for arbitrary vectors in the space

$$A x y = - A y x$$

and $A x x = 0$. Prove:

There exist coordinate systems $e_1, \ldots, e_m$ in R^m and a number n ($\leq m/2$) such that

$$A e_{2i-1} e_{2i} = - A e_{2i} e_{2i-1} = 1$$

for $i = 1, \ldots, n$, while $A e_h e_k = 0$ for every other index pair h, k and consequently

$$A x y = \sum_{i=1}^{n} (\xi^{2i-1} \eta^{2i} - \xi^{2i} \eta^{2i-1}) ,$$

where $\xi^1, \ldots, \xi^m$ and $\eta^1, \ldots, \eta^m$ stand for the coordinates of x and y in such a distinguished coordinate system.

Hint. Since $A x y$ does not vanish identically, there exist two vectors a_1 and a_2 such that $A a_1 a_2 = - A a_2 a_1 > 0$. These vectors are obviously linearly independent, and the same holds for the normalized vectors

$$e_1 = \frac{a_1}{\sqrt{A a_1 a_2}} , \qquad e_2 = \frac{a_2}{\sqrt{A a_1 a_2}} ,$$

for which $A e_1 e_2 = - A e_2 e_1 = 1$ and which span a two-dimensional subspace U_1^2 in R^m.

Each vector x in the space R^m can now be decomposed in a unique way into two components

$$x = p_1 + x_1$$

so that $p_1 = \xi^1 e_1 + \eta^1 e_2$ lies in U_1^2, while x_1 stands perpendicular to U_1^2, relative to A. For

$$A e_1 (x - p_1) = A e_2 (x - p_1) = 0$$

implies that $\xi^1 = - A e_2 x$, $\eta^1 = A e_1 x$.

The normals x_1 form in R^m a linearly independent complement R_1^{m-2} to U_1^2, and if $y = q_1 + y_1 = \xi^2 e_1 + \eta^2 e_2 + y_1$ is the above decomposition for y,

$$A\,x\,y = A\,p_1\,q_1 + A\,x_1\,y_1 = \xi^1 \eta^2 - \xi^2 \eta^1 + A\,x_1\,y_1 .$$

If $A\,x_1\,y_1 \equiv 0$ in R_1^{m-2}, then we are done. Otherwise, one continues the above technique in R_1^{m-2}, and in this way, step by step, one finally reaches a subspace R_n^{m-2n} whose vectors x_n, y_n are orthogonal to the pairwise perpendicular subspaces U_i^2 $(i = 1, \ldots, n)$ and where in addition $A\,x_n\,y_n \equiv 0$. Then

$$x = \sum_{i=1}^{n} p_i + x_n , \qquad y = \sum_{i=1}^{n} q_i + y_n$$

and

$$A\,x\,y = \sum_{i=1}^{n} A\,p_i\,q_i = \sum_{i=1}^{n} (\xi^{2i-1}\,\eta^{2i} - \xi^{2i}\,\eta^{2i-1}) ,$$

q.e.d.

Remark. The vectors in the space R_n^{m-2n} are, as is easily seen, not only mutually perpendicular, but are perpendicular, relative to A, to all of the vectors in the space R^m, and the space R_n^{m-2n} is therefore through this property *uniquely* determined by A. In particular, the number n is according to this an invariant which is uniquely determined by A.

9. With the hypotheses and notations of the previous exercise show: All coordinate systems of the kind mentioned there are obtained from one by means of the group of regular transformations T that leave A invariant, so that

$$A\,T\,x\,T\,y \equiv A\,x\,y .$$

§ 5. Multilinear Functions

5.1. Real n-linear functions. A real function defined for the vectors $x_1, \ldots, x_n$ of the space R^m

$$M\,x_1 \ldots x_n$$

is said to be n-linear if it is linear in each of its arguments. For $n = 1$, M is a linear, for $n = 2$, a bilinear function.

In an arbitrary coordinate system for the space R^m let

$$x_j = \sum_{i,j=1}^{m} \xi_j^{i_j} a_{i_j} \qquad (j = 1, \ldots, n) .$$

Then

$$M\,x_1 \ldots x_n = \sum_{i_1, \ldots, i_n = 1}^{m} \mu_{i_1 \ldots i_n} \xi_1^{i_1} \ldots \xi_n^{i_n} ,$$

with

$$\mu_{i_1 \ldots i_n} = M\,a_{i_1} \ldots a_{i_n} ,$$

is a real homogeneous form of degree n in the coordinates of the vectors x_j. Conversely, such a form with arbitrary real coefficients and a given coordinate system a_i in R^m defines a real n-linear function.

The n vectors x_j admit the $n!$ permutations of the symmetric permutation group. These permutations are even or odd according as they can be broken up into an even or an odd number of transpositions (x_i, x_j). The $n!/2$ even permutations form the alternating subgroup of the symmetric permutation group.

An n-linear vector function M which remains unchanged under permutations from the symmetric permutation group, and thus for any transposition of the vectors x_i, is called *symmetric*.

If, on the other hand, it has the value M_1 for precisely those permutations of the alternating permutation group, then it has a value M_2 $(\neq M_1)$ for all of the odd permutations. $M_1 - M_2$ is then an *alternating* n-linear function, which for any transposition of the vectors x_i only changes its sign.

If all of the permutations of the alternating permutation group are applied to an arbitrary n-linear function $M\, x_1 \ldots x_n$, the sum of the functions obtained is either symmetric or has at most two values M_1^* and M_2^*. The n-linear alternating function

$$\frac{1}{n!}(M_1^* - M_2^*)$$

is then called the *alternating part* of the n-linear function M and is denoted by

$$\wedge\, M\, x_1 \ldots x_n\,.^1$$

In what follows the real alternating multilinear functions of several vectors will be of particular interest to us.

5.2. Alternating functions and determinants. Thus let

$$D\, x_1 \ldots x_n$$

be a real, multilinear, alternating function defined in R^m.

For $n = 1$, $D\, x_1$ is a real linear function. It is convenient to also think of such a function as being "alternating", because all theorems valid for in fact alternating multilinear functions $(n > 1)$ also then hold for $n = 1$, as is easily verified in each individual case.

[1] In Cartan's alternating calculus the transition from a multilinear form to its alternating part is usually indicated by separating its arguments by the mark $\wedge$. We prefer to write this symbol *in front of* the operator of the multilinear form. The symbol becomes a *linear operator*, which presents advantages for the techniques of the alternating calculus.

Switching two vectors x_i and x_j $(i \neq j)$ changes the sign of D, and for $x_i = x_j$, D must therefore vanish. From this we see more generally:

If the vectors $x_1, \ldots, x_n$ are *linearly dependent,* then

$$D\,x_1 \ldots x_n = 0\,.$$

For $n = 1$ this says that any simply linear function $D\,x_1$ vanishes for $x_1 = 0$. If $n > 1$, then one of the vectors, for example

$$x_n = \sum_{i=1}^{n-1} \lambda_i\,x_i\,,$$

is a linear combination of the others, and therefore because of the linearity of D in x_n

$$D\,x_1 \ldots x_n = \sum_{i=1}^{n-1} \lambda_i\,D\,x_1 \ldots x_{n-1}\,x_i\,,$$

and here all of the terms on the right vanish.

If the number n of arguments is greater than the dimension of the space R^m, the argument vectors are always linearly dependent, and therefore $D = 0$. Alternating n-linear functions that do not vanish identically thus exist in R^m only for $n \leqq m$. In what follows we consider in particular the case $n = m$.

Then if the value of D is given for *one* linearly independent vector system $a_1, \ldots, a_m$, D is *uniquely* determined in R^m. In fact, for an arbitrary system of vectors $x_1, \ldots, x_m$

$$x_j = \sum_{i_j=1}^{m} \xi_j^{i_j}\,a_{i_j}\,,$$

and since

$$D\,a_{i_1} \ldots a_{i_m} = \pm\,D\,a_1 \ldots a_m\,,$$

with the sign $+$ or $-$ according as the permutation $j \to i_j$ $(j = 1, \ldots, m)$ is even or odd,

$$D\,x_1 \ldots x_m = \delta\,D\,a_1 \ldots a_m\,,$$

where the real number

$$\delta = \sum_{i_1,\ldots,i_m=1}^{m} \pm\,\xi_1^{i_1} \ldots \xi_m^{i_m} = \sum_{j_1,\ldots,j_m=1}^{m} \pm\,\xi_{j_1}^1 \ldots \xi_{j_m}^m = \det\,(\xi_j^i) = \det\,(\xi_i^j)$$

is the m-rowed determinant of the coordinates ξ_j^i. The value $D\,x_1 \ldots x_m$ is therefore, according to the above equation, uniquely determined if $D\,a_1 \ldots a_m$ is given.

From this it follows that D vanishes in R^m identically if it is equal to zero for one single linearly independent system of vectors a_i. If this case is excluded, then D vanishes only if the m argument vectors are linearly dependent. There then exists precisely one m-linear alternating function, namely

$$D\, x_1 \ldots x_m = \alpha \det (\xi^i_j) ,$$

which is uniquely determined up to the arbitrarily normalizable factor $D\, a_1 \ldots a_m = \alpha$.

However, we shall not make use of the ordinary theory of determinants. To the contrary, this theory can be derived from the concept of an m-linear alternating vector function on the space R^m. In order to prove the multiplication rule for determinants, say, using this approach, start with two arbitrary m-rowed determinants $\det (\xi^i_j)$ and $\det (\eta^j_k)$ and, taking $a_1, \ldots, a_m$ to be a basis for R^m, set

$$x_j = \sum_{i=1}^{m} \xi^i_j\, a_i , \qquad y_k = \sum_{j=1}^{m} \eta^j_k\, x_j ,$$

from which

$$y_k = \sum_{i=1}^{m} \left(\sum_{j=1}^{m} \eta^j_k\, \xi^i_j \right) a_i$$

follows. Then on the one hand

$$D\, y_1 \ldots y_m = \det \left(\sum_{j=1}^{m} \eta^j_k\, \xi^k_i \right) D\, a_1 \ldots a_m ,$$

and on the other hand

$$D\, y_1 \ldots y_m = \det (\eta^j_k)\, D\, x_1 \ldots x_m = \det (\eta^j_k) \det (\xi^i_j)\, D\, a_1 \ldots a_m ,$$

and consequently, since $D\, a_1 \ldots a_m \neq 0$,

$$\det (\eta^j_k) \det (\xi^i_j) = \det \left(\sum_{j=1}^{m} \eta^j_k\, \xi^i_j \right) = \det \left(\sum_{j=1}^{m} \xi^j_i\, \eta^j_k \right) = \det \left(\sum_{j=1}^{m} \xi^i_j\, \eta^k_j \right) .$$

5.3. Orientation of a simplex. Referring to what was said in 2.4 and 2.5, we consider an m-dimensional simplex $s^m(x_0, \ldots, x_m)$ in a linear space R^n $(m \leqq n)$ with the vertices $x_0, \ldots, x_m$ and linearly independent edges $x_1 - x_0, \ldots, x_m - x_0$. These edges generate an m-dimensional subspace U^m, and the simplex lies in a hyperplane E^m parallel to the space U^m whose points

$$x = \sum_{i=0}^{m} \mu^i\, x_i$$

have unique barycentric coordinates μ^i relative to the simplex with

$$\sum_{i=0}^{m} \mu^i = 1 .$$

In order to *orient* all simplexes lying in E^m or in planes parallel to it (observe that the edges of such simplexes thus determine the same subspace U^m), we take on the space U^m an m-linear real alternating function D, which is uniquely determined up to an arbitrary factor, and for each of the simplexes $s^m(x_0, \ldots, x_m)$ mentioned we form the expression

$$D\,(x_1 - x_0) \ldots (x_m - x_0) \equiv \varDelta(x_0, \ldots, x_m)\,.$$

Since the edges $x_i - x_0$ are linearly independent vectors in the space U^m, this real number is different from zero and therefore positive or negative. We define:

For a given ordering of the vertices of the simplex, $s^m(x_0, \ldots, x_m)$ is positively or negatively oriented with respect to $\varDelta$ according as the above expression is positive or negative.

The function $\varDelta$ is, to be sure, not linear, but still it is alternating, and therefore changes its sign for any transposition (x_i, x_j). For $i, j \neq 0$ this is evident. For a transposition (x_0, x_j) set

$$x_i - x_j = (x_i - x_0) + (x_0 - x_j)$$

for each i, from which it follows that

$$\begin{aligned}
&\varDelta(x_j, x_1, \ldots, x_{j-1}, x_0, x_{j+1}, \ldots, x_m)\\
&= D\,(x_1 - x_j) \ldots (x_{j-1} - x_j)\,(x_0 - x_j)\,(x_{j+1} - x_j) \ldots (x_m - x_j)\\
&= D\,(x_1 - x_0) \ldots (x_{j-1} - x_0)\,(x_0 - x_j)\,(x_{j+1} - x_0) \ldots (x_m - x_0)\\
&= -\,D\,(x_1 - x_0) \ldots (x_{j-1} - x_0)\,(x_j - x_0)\,(x_{j+1} - x_0) \ldots (x_m - x_0)\\
&= -\,\varDelta(x_0, x_1, \ldots, x_{j-1}, x_j, x_{j+1}, \ldots, x_m)\,.
\end{aligned}$$

Thus $\varDelta$ also changes its sign for this transposition and is therefore alternating. Note that this also holds for $m = 1$.

From this it follows by virtue of the above definition that the orientation of a simplex remains unchanged under even permutations of its vertices and changes its sign under odd ones[1].

The m-dimensional simplex $s^m(x_0, \ldots, x_m)$ has the $(m - 1)$-dimensional side simplexes

$$s_i^{m-1}(x_0, \ldots, \hat{x}_i, \ldots, x_m) \qquad (i = 0, \ldots, m)\,,$$

where $\hat{}$ indicates the omission of the point thus designated. If s^m is oriented in the above way by $\varDelta(x_0, \ldots, x_m)$, then we define the orientation of the side simplexes s_i^{m-1} *induced* by this orientation by means of the signs of the alternating functions

$$\varDelta_i(x_0, \ldots, \hat{x}_i, \ldots, x_m) \equiv (-1)^i\,\varDelta(x_0, \ldots, x_i, \ldots, x_m)\,.$$

[1] This is the usual definition of orientation. We have given preference to the one above because the function $\varDelta(x_0, \ldots, x_m)$ not only decides the orientation of the simplex $s^m(x_0, \ldots, x_m)$, but also has a significance for the simplex which becomes evident in the theorem that follows in 5.6.

Observe that Δ_i has the same meaning with respect to the side simplex s_i^{m-1} and the space U_i^{m-1} spanned by its edges as

$$\Delta(x_0, \ldots, x_m) = D\, h_1 \ldots h_m \qquad (h_i = x_i - x_0)$$

has for the space U^m. For, taking at first $i \neq 0$,

$$\Delta_i(x_0, \ldots, \hat{x}_i, \ldots, x_m) = (-1)^i D\, h_1 \ldots h_{i-1} (x_i - x_0)\, h_{i+1} \ldots h_m$$
$$\equiv D_i\, h_1 \ldots \hat{h}_i \ldots h_m$$

becomes for a fixed x_i a nonzero $(m-1)$-linear alternating function of the edges $h_1, \ldots, h_{i-1}, h_{i+1}, \ldots, h_m$ of the side simplex s_i^{m-1}, which can be taken to be the fundamental form of the space U_i^{m-1}. For $i = 0$ write, for example, $x_i - x_0 = (x_i - x_1) + (x_1 - x_0)$, and hence

$$\Delta_0(\hat{x}_0, x_1, \ldots, x_m) = D\,(x_1 - x_0)\, k_2 \ldots k_m \qquad (k_i = x_i - x_1)$$
$$= D_0\, k_2 \ldots k_m\,.$$

For a fixed x_0, D_0 is here a nonzero $(m-1)$-linear alternating function of the edges $k_2, \ldots, k_m$ of the side simplex s_0^{m-1} which can be used as the fundamental form of the space U_0^{m-1}.

5.4. Simplicial subdivisions. Let s^m be a closed m-dimensional simplex (i.e. the closed convex hull of the points $x_0, \ldots, x_m$). We consider a *subdivision* D of s^m, i.e. a finite set of closed m-dimensional subsimplexes s_i^m $(i = 1, 2, \ldots, N)$ with the properties:

1. *The simplex s^m is the union of the subsimplexes s_i^m.*

2. *Any two subsimplexes s_i^m and s_j^m have no common interior points.*

The subdivision D is *simplicial*, if in addition:

3. *Each $(m-1)$-dimensional face s^{m-1} of the subsimplexes s_i^m that contains interior points of s^m belongs as a face to precisely two subsimplexes s_i^m and s_j^m.*

Following an idea due to H. Whitney [1] we will construct a simplicial subdivision D of the simplex s^m.

5.5. Construction of a simplicial subdivision. With the vertices $x_0, x_1, \ldots, x_m$ of the given simplex s^m we form the points

$$x_{ij} = \frac{1}{2}(x_i + x_j) \qquad (0 \leqq i \leqq j \leqq m;\ x_{ii} = x_i)$$

and partially order them by a relation $\leqq$ such that

$$x_{ij} \leqq x_{hk} \quad \text{if } h \leqq i \quad \text{and} \quad j \leqq k\,.$$

Consider now all increasing sequences σ of $m+1$ points x_{hk} (beginning with one of the points x_i $(i = 0, 1, \ldots, m)$ and ending with x_{0m}). For each x_{hk} $(h > 0,\ k < m)$ there are two possible successors, namely $x_{(h-1)k}$ and $x_{h(k+1)}$. The number of sequences σ is 2^m, and they correspond to 2^m one-dimensional oriented polygonal paths. The Figure illustrates the case $m = 4$.

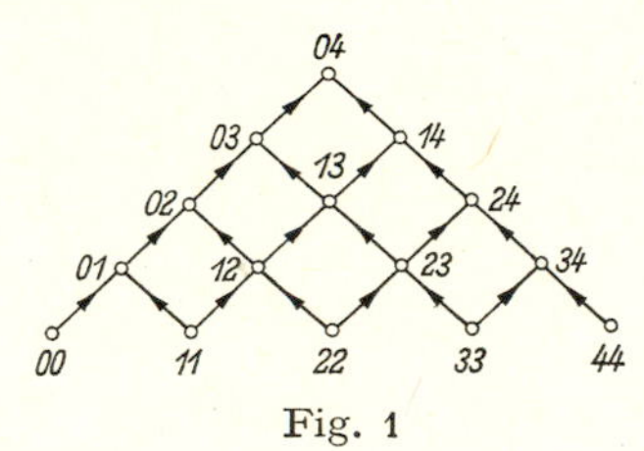

Fig. 1

The simplicial subdivision D is now defined by the 2^m simplexes s_σ^m which have the points of σ as vertices. In order to show that D is a simplicial subdivision, we first prove the property 3.

Let $s_\sigma^{m-1}(x_{ii}, \ldots, \hat{x}_{rt}, \ldots, x_{0m})$ be a $(m-1)$-dimensional face of $s_\sigma^m(x_{ii}, \ldots, x_{rt}, \ldots, x_{0m})$. One has to distinguish three cases:

(α) The neighbors of x_{rt} in the sequence σ are

$$x_{(r+1)t}, x_{(r-1)t} \qquad \text{or} \qquad x_{r(t-1)}, x_{r(t+1)} .$$

Then the face s_σ^{m-1} consists of points of the boundary ∂s^m of the given simplex s^m, since all points $x \in s_\sigma^{m-1}$ have the barycentric coordinate 0 with respect to the vertex x_{rr} in the first case and x_{tt}, in the second.

(β) Similarly, if $r = t = i$ or if $r = 0$ and $t = m$, then the face $s_\sigma^{m-1} \subset \partial s^m$.

(γ) The neighbors of x_{rt} are

$$x_{(r+1)t}, x_{r(t+1)} \qquad \text{or} \qquad x_{r(t-1)}, x_{(r-1)t} .$$

In these cases there are precisely two subsimplexes of the subdivision D which have $s_\sigma^{m-1}(x_{ii}, \ldots, \hat{x}_{rt}, \ldots, x_{0m})$ as a common face, namely the given subsimplex s_σ^m and the subsimplex $s_{\sigma'}^m$ in which the point x_{rt} has been replaced by $x_{r(t+1)}$ in the first case and by $x_{(r-1)t}$ in the second.

For the proof of properties 1 and 2, let the sequences σ be numbered as follows:

$$\begin{array}{llllll}
\sigma_1: & x_{00}, & x_{01}, & x_{02}, & \ldots, & x_{0m}, \\
\sigma_2: & x_{11}, & x_{01}, & x_{02}, & \ldots, & x_{0m}, \\
\sigma_3: & x_{11}, & x_{12}, & x_{02}, & \ldots, & x_{0m}, \\
\cdot & \cdot & \cdot & \cdot & \cdots & \cdot \\
\sigma_{2^m}: & x_{mm}, & x_{(m-1)m}, & x_{(m-2)m}, & \ldots, & x_{0m}.
\end{array}$$

Then the corresponding consecutive simplexes $s_{\sigma_i}^m = s_i^m$ and $s_{\sigma_{i+1}}^m = s_{i+1}^m$ $(i = 1, 2, \ldots, 2^m - 1)$ have presicely one common face s_i^{m-1}. The vertices of s_i^{m-1} lie on the boundary ∂s^m of the given simplex s^m and have the same order in both simplexes s_i^m and s_{i+1}^m. Hence the orientations of these simplexes are opposite (cf. 5.3). By Lemma A, which will be formulated in 5.7, one concludes that the simplexes s_i^m and s_{i+1}^m contain no common interior points.

Because of the convexity of s^m, the $(m-1)$-dimensional simplex s_i^{m-1} $(1 \leqq i \leqq 2^m - 1)$ cuts s^m in two closed convex polyhedrons with no common interior point. One of them, P_i^m, has the point x_0 as a vertex. Denote by Q_i^m the other, complementary polyhedron.

We prove now: The interiors of the simplexes $s_1^m, \ldots, s_i^m$ are disjoint, and the union

$$\bigcup_{j=1}^{i} s_j = P_i^m .$$

These statements are evident for $i = 1$, $P_1^m = s_1^m$. Assuming that they hold for $i = 1, \ldots, k$ $(\leqq 2^m - 1)$, we will prove their validity for $i = k + 1$.

By assumption the polyhedron P_k^m is the union of the simplexes $s_1^m, \ldots, s_k^m$, any two of which have no common interior point, and the face s_k^{m-1} separates P_k^m from the complement Q_k^m. As we have seen, the face s_k^{m-1} also separates the simplexes s_k^m and s_{k+1}^m, and therefore $s_{k+1}^m \subset Q_k^m$. Hence, s_{k+1}^m has no interior points in common with any of the simplexes $s_1^m, \ldots, s_k^m$. The union $P_k^m \cup s_{k+1}^m$ is a polyhedron cut by the face s_{k+1}^{m-1} from the given simplex s^m and having the point x_0 as a vertex. Thus, $P_k^m \cup s_{k+1}^m = P_{k+1}^m$, and P_{k+1}^m is, as stated, the union of the simplexes $s_1^m, \ldots, s_{k+1}^m$.

Repeating this reasoning for $k = 1, \ldots, 2^m - 1$, one concludes that the properties 1 and 2 are valid.

5.6. Additivity of the function Δ. We now consider a decomposition of the closed simplex $s^m(x_0, \ldots, x_m)$ (i.e., of the closed convex hull of the points $x_0, \ldots, x_m$) into a finite number of m-dimensional subsimplexes:

$$s^m(x_0, \ldots, x_m) = \sum_k s_k^m(x_0^k, \ldots, x_m^k) ,$$

with the two first properties 1, 2 of 5.4.

1. *s^m is the union of the closed subsimplexes s_k^m.*
2. *Any two subsimplexes have no common interior points.*

For such a decomposition one has the following

Theorem. *If s^m and the subsimplexes s_k^m have the same orientation with respect to*

$$\Delta(x_0, \ldots, x_m) = D\,(x_1 - x_0) \ldots (x_m - x_0) ,$$

then

$$\Delta(x_0, \ldots, x_m) = \sum_k \Delta(x_0^k, \ldots, x_m^k) ,$$

and the sum on the right is therefore independent of the decomposition.

In this sense the function Δ is an *additive* set function.

5.7. Lemmas. In order to be able to establish the proof without disturbing interruptions, we wish to present a few preparatory considerations.

Let

$$x = \sum_{j=0}^{m} \mu^j x_j$$

be an arbitrary point in the plane of the simplex $s^m(x_0, \ldots, x_m)$. Because $\sum_{j=0}^{m} \mu^j = 1$ one has

$$x - x_0 = \sum_{j=1}^{m} \mu^j (x_j - x_0) ,$$

and consequently if we replace the point x_i in $\Delta(x_0, \ldots, x_m)$ by x, first for $i \neq 0$,

$$\begin{aligned} &\Delta(x_0, \ldots, x_{i-1}, x, x_{i+1}, \ldots, x_m) \\ &= D\,(x_1 - x_0) \ldots (x_{i-1} - x_0)\,(x - x_0)\,(x_{i+1} - x_0) \ldots (x_m - x_0) \\ &= \sum_{j=1}^{m} \mu^j D\,(x_1 - x_0) \ldots (x_{i-1} - x_0)\,(x_j - x_0)\,(x_{i+1} - x_0) \ldots (x_m - x_0) \end{aligned}$$

and thus, since all terms vanish for $j \neq i$,

$$\Delta(x_0, \ldots, x_{i-1}, x, x_{i+1}, \ldots, x_m) = \mu^i \Delta(x_0, \ldots, x_{i-1}, x_i, x_{i+1}, \ldots, x_m) .$$

For $i = 0$ write $x_j - x_0 = (x_j - x_m) + (x_m - x_0)$, for example, from which it follows that

$$\Delta(x_0, x_1, \ldots, x_m) = D\,(x_1 - x_m) \ldots (x_{m-1} - x_m)\,(x_m - x_0) .$$

If instead of x_0 one substitutes

$$x = \sum_{j=0}^{m} \mu^j x_j = x_m + \sum_{j=0}^{m-1} \mu^j (x_j - x_m) ,$$

then

$$\begin{aligned} \Delta(x, x_1, \ldots, x_m) &= -\mu^0 D(x_1 - x_m) \ldots (x_{m-1} - x_m)\,(x_0 - x_m) \\ &= \mu^0 \Delta(x_0, x_1, \ldots, x_m) , \end{aligned}$$

from which it can be seen that the above equation also holds for $i = 0$. If in addition exercise 3 in 2.6 is taken into account, one has

Lemma A. *If*

$$s^m(x_0, \ldots, x_{i-1}, x_i, x_{i+1}, \ldots, x_m)$$

and

$$\bar{s}^m(x_0, \ldots, x_{i-1}, x, x_{i+1}, \ldots, x_m)$$

are simplexes of the same m-dimensional hyperplane with the common side simplex $s_i^{m-1}(x_0, \ldots, \hat{x}_i, \ldots, x_m)$ *and if in the barycentric representation of* x *with respect to* s^m *the coefficient of* x_i *is equal to* μ^i*, then for* $i = 0, \ldots, m$

$$\Delta(x_0, \ldots, x_{i-1}, x, x_{i+1}, \ldots, x_m) = \mu^i \Delta(x_0, \ldots, x_{i-1}, x_i, x_{i+1}, \ldots, x_m) .$$

Since $\mu^i < 0$ is the necessary and sufficient condition for the simplexes to have no common interior points, this is the case precisely when for the given ordering of the vertices the simplexes are oppositely oriented.

If in this lemma i is set equal to $0, \ldots, m$ successively, addition yields

Lemma B. *For every x in the plane of the simplex s^m*

$$\sum_{i=0}^{m} \Delta(x_0, \ldots, x_{i-1}, x, x_{i+1}, \ldots, x_m) = \Delta(x_0, \ldots, x_m) .$$

Observe that this equation already implies the additivity to be proved in the special case of a "star" decomposition. For if x lies in the interior or on the boundary of s^m,

$$s^m(x_0, \ldots, x_m) = \sum_{i=0}^{m} s^m(x_0, \ldots, x_{i-1}, x, x_{i+1}, \ldots, x_m)$$

is obviously a simplicial star decomposition of s^m centered at x, where because $\mu^i \geqq 0$ the subsimplexes on the right are according to Lemma A all oriented like s^m.

5.8. Proof of the theorem. After these preparations we now give a general proof of the asserted additivity of Δ, whereby we restrict ourselves to simplicial decompositions

$$s^m(x_0, \ldots, x_m) = \sum_k s_k^m(x_0^k, \ldots, x_m^k)$$

of the simplex s^m.

Letting x stand for a temporarily arbitrary point in the plane of s^m, we have, according to Lemma B, for every k

$$\Delta(x_0^k, \ldots, x_m^k) = \sum_{i=0}^{m} \Delta(x_0^k, \ldots, x_{i-1}^k, x, x_{i+1}^k, \ldots, x_m^k) ,$$

and therefore

$$\sum_k \Delta(x_0^k, \ldots, x_m^k) = \sum_k \sum_{i=0}^{m} \Delta(x_0^k, \ldots, x_{i-1}^k, x, x_{i+1}^k, \ldots, x_m^k) .$$

The problem is the evaluation of the double sum on the right.

For this let s^{m-1} be an $(m-1)$-dimensional side simplex of the subsimplexes that contain interior points of s^m. Since the decomposition is simplicial, there exist precisely two subsimplexes,

$$s_h^m(x_0^h, \ldots, x_{p-1}^h, x_p^h, x_{p+1}^h \ldots, x_m^h)$$

and

$$s_k^m(x_0^k, \ldots, x_{q-1}^k, x_q^k, x_{q+1}^k, \ldots, x_m^k) ,$$

that have the side simplex s^{m-1} in common, where x_p^h and x_q^k are the only vertices of the neighboring simplexes s_h^m and s_k^m not common to both. Now since these simplexes have no common interior points and are for the above sequence of vertices by hypothesis oriented in the same way, it follows from Lemma A that the ordering

$$x_0^h, \ldots, x_{p-1}^h, x_q^k, x_{p+1}^h, \ldots, x_m^h$$

must be an *odd* permutation of the ordering

$$x_0^k, \ldots, x_{q-1}^k, x_q^k, x_{q+1}^k, \ldots, x_m^k$$

of the vertices of s_k^m. But then

$$x_0^h, \ldots, x_{p-1}^h, x, x_{p+1}^h, \ldots, x_m^h$$

is likewise an odd permutation of the ordering

$$x_0^k, \ldots, x_{q-1}^k, x, x_{q+1}^k, \ldots, x_m^k,$$

and the corresponding Δ-terms in the above double sum consequently cancel each other out.

Thus in this double sum only those terms remain which come from side simplexes s^{m-1} of the subsimplexes that contain no interior and thus only boundary points of the decomposed simplex s^m. If in addition the point x is now shifted to a vertex of the simplex, to x_0, for example, then only the terms which correspond to the side simplex s_0^{m-1} of s^m opposite to x_0 remain. If, finally, in these remaining terms x_0 is brought into the first place by means of an *even* permutation of the vertices, which is obviously possible for $m > 1$ and which leaves the orientations unchanged, then

$$\sum_k \Delta(x_0^k, \ldots, x_m^k) = \sum_h \Delta(x_0, y_1^h, \ldots, y_m^h),$$

where as always the sum on the left is taken over all subsimplexes of the original simplicial decomposition of s^m and the sum on the right over the induced simplicial decomposition

$$s_0^{m-1}(x_1, \ldots, x_m) = \sum_h s_h^{m-1}(y_1^h, \ldots, y_m^h)$$

of the side simplex s_0^{m-1}.

Now if $y_1, \ldots, y_m$ denote arbitrary points in the plane of the side simplex s_0^{m-1}, according to the remark in 5.3,

$$\Delta(x_0, y_1, \ldots, y_m) = D_0(y_2 - y_1) \ldots (y_m - y_1)$$

is the (up to a real factor) uniquely determined $(m-1)$-linear alternating fundamental form for the space U_0^{m-1} parallel to s_0^{m-1}. We assert that the expressions

$$\Delta(x_0, y_1^h, \ldots, y_m^h)$$

all have the sign of $\Delta(x_0, \ldots, x_m)$ and consequently are oriented in the same way.

In fact, there is a subsimplex in the original decomposition that up to an even permutation of the vertices is equal to

$$s_h^m(x_0^h, y_1^h, \ldots, y_m^h)\,,$$

where x_0^h is an interior vertex of the decomposition. This simplex is therefore oriented like

$$s^m(x_0, x_1, \ldots, x_m)\,,$$

but on the other hand also like

$$s^m(x_0, y_1^h, \ldots, y_m^h)\,.$$

For in the barycentric representation

$$x_0^h = \mu^0\, x_0 + \sum_{i=1}^{m} \mu^i\, x_i$$

the coefficients μ^i, thus in particular μ^0, are positive. If the barycentric representations of $x_1, \ldots, x_m$ with respect to $y_1^h, \ldots, y_m^h$ are substituted here, the barycentric representation of x_0^h with respect to $x_0, y_1^h, \ldots, y_m^h$ is obtained, where the coefficient of x_0 is unchanged and equal to μ^0 and therefore positive. Now by virtue of Lemma A

$$\Delta(x_0^h, y_1^h, \ldots, y_m^h) = \mu^0\, \Delta(x_0, y_1^h, \ldots, y_m^h)\,,$$

from which it can be seen that $\Delta(x_0, y_1^h, \ldots, y_m^h)$ in fact has the sign of $\Delta(x_0^h, y_1^h, \ldots, y_m^h)$ and thus for every h the sign of $\Delta(x_0, x_1, \ldots, x_m)$. The induced simplicial decomposition of s_0^{m-1} is therefore oriented in the same way.

Assuming the theorem to have been proved for dimensions $< m$, it follows from the above that

$$\sum_h \Delta(x_0, y_1^h, \ldots, y_m^h) = \Delta(x_0, x_1, \ldots, x_m)\,.$$

But then also

$$\sum_k \Delta(x_0^k, \ldots, x_m^k) = \sum_h \Delta(x_0, y_1^h, \ldots, y_m^h) = \Delta(x_0, \ldots, x_m)\,,$$

and the theorem is also true for dimension m.

For the dimension $m = 1$, the theorem is trivial, and the proof is therefore complete.

5.9. Exercises. 1. Let

$$D\, x_1 \ldots x_n$$

be a real n-linear alternating function in the linear space R^m, $n \leqq m$, and further

$$x_j = \sum_{i=1}^{m} \xi_j^i\, a_i \quad (j = 1, \ldots, n)$$

in a coordinate system $a_1, \ldots, a_m$ for this space. Show that

$$D\, x_1 \ldots x_n = \sum_{1 \leqq i_1 < \cdots < i_n \leqq m} \delta^{i_1 \ldots i_n} D\, a_{i_1} \ldots a_{i_n},$$

where $\delta^{i_1 \ldots i_n}$ stands for the determinant formed by the i_1-th, $\ldots$, i_n-th rows of the matrix

$$(\xi_j^i) = \begin{pmatrix} \xi_1^1 \ldots \xi_n^1 \\ \vdots \quad\quad \vdots \\ \xi_1^m \ldots \xi_n^m \end{pmatrix}.$$

2. Show that the above $\binom{m}{n}$ determinants $\delta^{i_1 \ldots i_n}$ are linearly independent, n-linear alternating functions that span the linear space of all such functions of $x_1, \ldots, x_n$. Note in particular the extreme cases $n = 1$ and $n = m$.

3. Let T be a linear transformation of the space R^m and D the space's real, m-linear alternating fundamental form, which is uniquely determined up to a real factor.

Show: The quotient

$$\frac{D\, T\, x_1 \ldots T\, x_m}{D\, x_1 \ldots x_m}$$

is independent of the vectors $x_1, \ldots, x_m$ and in every coordinate system equal to the determinant

$$\det T = \det (\tau_i^j)$$

of the transformation T; this determinant is consequently an invariant (independent of the coordinate system).

4. Show generally that for each k $(1 \leqq k \leqq m)$ the quotient

$$q_k = \frac{1}{D\, x_1 \ldots x_m} \sum_{1 \leqq i_1 < \cdots < i_k \leqq m} D\, x_1 \ldots T\, x_{i_1} \ldots T\, x_{i_k} \ldots x_m$$

is independent of the vectors $x_1, \ldots, x_m$ and in any coordinate system is equal to $(-1)^{m-k}$ times the coefficient of λ^{m-k} in the mth degree polynomial

$$\det (T - \lambda\, I) = \det (\tau_i^j - \lambda\, \delta_i^j),$$

where I denotes the identity transformation.

Remark. In particular,

$$q_1 = \frac{1}{D\, x_1 \ldots x_m} \sum_{i=1}^{m} D\, x_1 \ldots x_{i-1}\, T\, x_i\, x_{i+1} \ldots x_m$$

is equal to the *trace*

$$\operatorname{Tr} T \equiv \sum_{i=1}^{m} \tau_i^i$$

of the transformation T (or of the matrix (τ_i^j)).

5. Prove: In order that λ_0 be an eigenvalue of the transformation T, it is necessary and sufficient that λ_0 satisfy the *secular equation*

$$\det(T - \lambda I) = \det(\tau_i^j - \lambda \delta_i^j) = 0$$

(cf. 3.10, exercises 9—10).

Further show: If λ_0 is an n-fold ($1 \leqq n \leqq m$) eigenvalue of the transformation, then it is at least an n-fold root of the secular equation and is an n-fold root of this equation precisely when the kernels of $T - \lambda_0 I$ and $(T - \lambda_0 I)^2$ coincide.

Hint. With no loss of generality assume $\lambda_0 = 0$. Since in the preceding exercise the expression for the coefficients q_k of $(-1)^{m-k}\lambda^{m-k}$ is independent of the vectors $x_1, \ldots, x_m$, it is possible to take for $x_1, \ldots, x_n$ a basis of the eigenspace corresponding to the n-fold eigenvalue $\lambda_0 = 0$, i.e., of the kernel $K(T)$ of the transformation T. Then q_k certainly vanishes for $k = m, m-1, \ldots, m-n+1$, so that $\lambda = 0$ is at least an n-fold root of the secular equation. The coefficient for λ^n is

$$(-1)^{m-n} q_{m-n} = \frac{1}{D x_1 \ldots x_m} D x_1 \ldots x_n \, T x_{n+1} \ldots T x_m \, .$$

This coefficient is $\neq 0$ precisely when $x_1, \ldots, x_n, T x_{n+1}, \ldots, T x_m$ are linearly independent, which according to exercise 3.10.6 is the case if and only if $K(T) = K(T^2)$, so that $K(T)$ and $T(R^m)$ are linearly independent complements in the space R^m. This is, for example, *not* the case for the linear transformation T defined by

$$T x_1 = \cdots = T x_n = 0 \, , \quad T x_{n+1} = x_n, \ldots, T x_m = x_{m-1} \, .$$

In fact, as is at once apparent, the secular equation here is $\lambda^m = 0$, while $\lambda = 0$ is only an n-fold eigenvalue of the transformation T.

6. We consider an m-dimensional simplex

$$s^m = s^m(x_0, \ldots, x_m)$$

and order the vertices into $(m+1)!$ different sequences

$$x_{i_0}, \ldots, x_{i_m} \, .$$

Prove that the closed simplexes $s^m(p)$ corresponding to these $(m+1)!$ permutations p with vertices

$$y_{i_k} = \frac{1}{k+1} \sum_{j=0}^{k} x_{i_j} \qquad (k = 0, \ldots, m)$$

at the centers of gravity of the side simplexes of $0^{\text{th}}, \ldots, m^{\text{th}}$ dimension of s^m decompose this closed simplex simplicially.

Show further: If D is the m-linear alternating fundamental form (which is uniquely determined up to a real factor) of the space $R^m =$

$= (h_1, \ldots, h_m)$ spanned by the edges $h_i = x_i - x_0$ of the simplex s^m, or using the earlier notation,

$$D\, h_1 \ldots h_m = \Delta(x_0, \ldots, x_m)\,,$$

then for each of the above subsimplexes $s^m(p)$

$$|\Delta(y_{i_0}, \ldots, y_{i_m})| = \frac{1}{(m+1)!}\, |\Delta(x_0, \ldots, x_m)|\,.$$

Finally, show: If

$$\bar{y}(p) = \frac{1}{m+1} \sum_{k=0}^{m} y_{i_k}$$

stands for the center of gravity of the subsimplex $s^m(p)$ corresponding to the permutation p, then

$$\frac{1}{(m+1)!} \sum_{p} \bar{y}(p) = \frac{1}{m+1} \sum_{i=0}^{m} x_i\,.$$

In other words, the center of gravity of these $(m+1)!$ centers of gravity is equal to the center of gravity of the simplex s^m.

Remark. The above simplicial decomposition of s^m is called the *barycentric* decomposition of first order. If each subsimplex $s^m(p)$ is again barycentrically decomposed, one obtains the barycentric decomposition of second order of s^m, etc.

7. Let $h_i = x_i - x_0$ $(i = 1, \ldots, m)$ be linearly independent vectors of a linear space and D the m-linear alternating fundamental form of the space R^m spanned by these vectors. Show:

a. The m simplexes

$$s_i^m(x_0 + h_m, \ldots, x_{i-1} + h_m, x_i, \ldots, x_{m-1}, x_i + h_m)\,,$$

where $i = 0, \ldots, m-1$, are oriented in the same way and have the same "volume" in that for each i

$$\Delta(x_0 + h_m, \ldots, x_{i-1} + h_m, x_i, \ldots, x_{m-1}, x_i + h_m) = D\, h_1 \ldots h_m\,.$$

b. The simplexes named decompose the *prism*

$$x = \mu\, h_m + \sum_{i=0}^{m-1} \mu^i x_i\,,$$

where $0 \leqq \mu \leqq 1$, $\mu^i \geqq 0$, $\sum_{i=0}^{m-1} \mu^i = 1$, simplicially.

Remark. This simplicial decomposition of the m-dimensional prism is a generalization of the decomposition which for $m = 2$ and $m = 3$ had already been given by Euclid.

8. Prove that the m-dimensional parallelopiped

$$x = x_0 + \sum_{i=1}^{m} \mu^i h_i \qquad (0 \leqq \mu^i \leqq 1)$$

spanned at the point x_0 by the linearly independent vectors $h_1, \ldots, h_m$ can be simplicially decomposed into $m!$ "equal volume" like-oriented simplexes.

Hint. The proof follows from problem 7 by means of induction on the dimension m.

§ 6. Metrization of Affine Spaces

6.1. The natural topology of linear spaces of finite dimension. A linear space of *finite* dimension possesses a "natural" topology which goes back to the topology of the real multiplier domain. If $a_1, \ldots, a_m$ is a basis of the space in which

$$x = \sum_{i=1}^{m} \xi^i a_i, \qquad x_0 = \sum_{i=1}^{m} \xi_0^i a_i,$$

then, for example, the limit $x \to x_0$ can be defined by the m limits $\xi^i \to \xi_0^i$ $(i = 1, \ldots, m)$, which are meaningful in the domain of the real numbers, and, indeed, this can be done in a way that is independent of the choice of a coordinate system: If according to this definition $x \to x_0$ in one coordinate system, then it follows from the finite, linear formulas for the transformation of the coordinates that the same is the case in all coordinate systems. The corresponding holds for the remaining basic topological concepts and relations such as the accumulation point of a point set, interior point of a region, etc.

It is because of the existence of this natural topology that the basic notions and relations in the "absolute analysis" developed in the following chapters will be for the greatest part of a purely affine character in finite dimensional spaces. We shall, however, partly in order to be able to formulate the concepts and theorems conveniently and partly for technical reasons having to do with the proofs, almost everywhere use the metric concepts introduced in this section. But the notions and theorems of absolute analysis in themselves will mostly be independent of the particular choice of the auxiliary metrics introduced.

6.2. The Minkowski-Banach metric. The in many ways simplest d-dimensional point set of an affine space is the closed d-dimensional simplex

$$s^d(x_0, \ldots, x_d)$$

with the vertices x_i and the d linearly independent edges $x_1 - x_0, \ldots, x_d - x_0$. This configuration is the most elementary possible principally because the point set

$$x = \sum_{i=0}^{d} \mu^i x_i, \qquad \sum_{i=0}^{d} \mu^i = 1, \qquad \mu^i \geqq 0,$$

is defined in an *affine* way, without any kind of metric, and invariantly retains its character as a d-dimensional simplex for arbitrary regular linear transformations of the space.

In establishing a general measure theory for point sets of the affine space it is thus expedient first to define the notion of measure accordingly for simplexes.

For $d = 0$ the simplex shrinks to one single point whose "measure" we set $= 0$.

For $d = 1$ we are dealing with a closed line segment

$$x = \mu^0 x_0 + \mu^1 x_1 , \quad \mu^0 + \mu_1 = 1 , \quad \mu^0 \geqq 0 , \quad \mu^1 \geqq 0 .$$

We start by introducing a suitable definition for the measure or length of such a segment $x_0 x_1$.

If this notion of length is to conform to our usual ideas, then it will be a real number $|x_0, x_1|$, defined for any segment $x_0 x_1$, that satisfies the following postulates:

A. *The length $|x_0, x_1|$ is independent of the orientation of the simplex $s^1(x_0, x_1)$ and invariant with respect to parallel translations of the latter.*

According to this,

$$|x_0, x_1| = |x_1, x_0| = |0, x_1 - x_0| = |0, x_0 - x_1| .$$

We therefore denote this length more briefly with

$$|x_1 - x_0| = |x_0 - x_1| .$$

B. *The length is to be additive in the following sense: For an interior point*

$$x = \mu^0 x_0 + \mu^1 x_1 , \quad \mu^0 + \mu^1 = 1 , \quad \mu^0 > 0 , \quad \mu^1 > 0$$

of the segment we have

$$|x - x_0| + |x_1 - x| = |x_1 - x_0| .$$

C. *We have*

$$|x_1 - x_0| \geqq 0 ,$$

and $= 0$ *only if* $x_1 = x_0$ *and the segment degenerates to one single point.*

These postulates are equivalent with the following ones (cf. 6.11, problem 1).

Associated with each vector x of the affine space is a real number $|x|$, the *length* or the *norm* of the vector, which has the following properties:

1. *For every real* λ, $|\lambda x| = |\lambda| \, |x|$.

2. $|x| \geqq 0$, *and* $= 0$ *only for* $x = 0$.

When in addition the *triangle inequality*

3. $|x_2 - x_0| \leqq |x_2 - x_1| + |x_1 - x_0|$

holds for each triangle in the space, one has a *Minkowski-Banach metric* for the affine space, and, indeed, a Minkowski or Banach one according as the dimension of the space is finite or infinite.

It is easily shown that the "unit sphere"

$$|x| \leqq 1$$

of such a metric is a *convex* point set.

Conversely, keeping the remaining postulates, the triangle inequality can be replaced by requiring the unit sphere to be convex (cf. 6.11, problem 2).

In what follows, the case, considered by Minkowski, of a finite dimensional space will be treated almost exclusively. Regarding such a Minkowski metric the following should be observed.

In some coordinate system $a_1, \ldots, a_m$, let

$$x = \sum_{i=1}^{m} \xi^i a_i .$$

Then each Minkowskian length $|x|$ is a continous function of the coordinates, and, conversely, the latter, as functions of x, are continuous with respect to the given metric.

It obviously suffices to show this for $\xi^1 = \cdots = \xi^m = 0$ or $x = 0$.

For this we set

$$\sum_{i=1}^{m} (\xi^i)^2 = \varrho^2 ,$$

and thus claim that $|x| \to 0$ for $\varrho \to 0$ and conversely.

In fact, as a consequence of the triangle inequality and 1.

$$|x| \leqq \sum_{i=1}^{m} |\xi^i| \, |a_i| ,$$

and therefore

$$|x|^2 \leqq \left(\sum_{i=1}^{m} |\xi^i| \, |a_i|\right)^2 \leqq \sum_{i=1}^{m} |a_i|^2 \sum_{i=1}^{m} (\xi^i)^2 = K^2 \varrho^2 ;$$

consequently $|x| \leqq K \varrho$ and $|x| \to 0$ as $\varrho \to 0$.

Since according to this $|x|$ is a continuous function of the m coordinates ξ^i, $|x|$ has on the surface

$$\sum_{i=1}^{m} (\xi^i)^2 = 1$$

a nonnegative lower bound k which is reached for at least one system ξ_0^i. Because

$$x_0 = \sum_{i=1}^{m} \xi_0^i a_i \neq 0 ,$$

$k = |x_0|$ is by 2 positive and, as a consequence of 1, for an arbitrary x

$$|x| = \left|\varrho \frac{x}{\varrho}\right| = \varrho \left|\frac{x}{\varrho}\right| \geqq k \varrho ,$$

from which it follows, conversely, that $\varrho \to 0$ for $|x| \to 0$.

Hence if $x \to 0$ in the natural topology of the space, then $|x| \to 0$ also in every Minkowski metric, and conversely.

6.3. Norms of linear operators. Let

$$y = A x$$

be a linear mapping of the Minkowski space R_x^m into the Minkowski space R_y^n. Since in the natural topologies of these spaces obviously $A x \to 0$ as $x \to 0$, this is also the case in the Minkowski metrics: $|A x| \to 0$ as $|x| \to 0$. By the triangle inequality, for arbitrary x and h in R_x^m

$$||A (x + h)| - |A x|| \leqq |A (x + h) - A x| = |A h| ,$$

and it follows from this that $A x$ as well as $|A x|$ are continuous functions of x. Consequently, on the m-dimensional sphere $|x| = 1$, $|y| = |A x|$ has a finite upper bound

$$\sup_{|x|=1} |A x| = |A| ,$$

which is even reached at at least one point on the sphere. This is the *norm* of the linear operator A with respect to the metrics introduced in the spaces R_x^m and R_y^n.

For an arbitrary $x \neq 0$ one has

$$|A x| = |x| \left|A \left(\frac{x}{|x|}\right)\right| \leqq |A| \, |x| .$$

If $A x$ vanishes identically, then $|A| = 0$; it can be seen from the above inequality that, conversely, the identical vanishing of $A x$ follows from $|A| = 0$. Therefore $|A| = 0$ precisely when A is the null operator of the mn-dimensional operator space introduced in 3.5.

Further, for every real λ

$$|\lambda A| = |\lambda| \, |A| .$$

Finally, it follows from the triangle inequality that when B is a second linear operator in this space that

$$|(A + B) x| = |A x + B x| \leqq |A x| + |B x| \leqq (|A| + |B|) \, |x| ;$$

hence the inequality

$$|A + B| \leqq |A| + |B|$$

is valid for the norms.

We see: If the operator space mentioned is metrized by the introduction of Minkowski metrics in the spaces R_x^m and R_y^n, then this norm metric is likewise a Minkowski one.

Moreover, notice the following. If R_z^p stands for a third Minkowski space and

$$z = C\,y$$

is a linear mapping from R_y^n into R_z^p with the norm

$$|C| = \sup_{|y|=1} |C\,y|\,,$$

then one obtains for the composite linear mapping $C\,B$

$$|z| = |C\,B\,x| \leqq |C|\,|B\,x| \leqq |C|\,|B|\,|x|\,,$$

and hence for the norms

$$|C\,B| \leqq |C|\,|B|\,.$$

The norm of a multilinear mapping

$$y = M\,x_1 \ldots x_p$$

between two Minkowski spaces R_x^m and R_y^n is defined in a corresponding fashion as the least upper bound

$$|M| = \sup_{|x_1|=\cdots=|x_p|=1} |M\,x_1 \ldots x_p|\,.$$

For arbitrary vectors $x_1, \ldots, x_p$ in R_x^m one thus has

$$|M\,x_1 \ldots x_p| \leqq |M|\,|x_1| \ldots |x_p|\,.$$

6.4. The euclidean metric. In a general Minkowski metric there is no measurement of angles. On the other hand, in elementary euclidean geometry, besides the measurement of lengths, a measure for the angle ϑ formed by two vectors x and y is introduced which is tied to the lengths of these vectors by the "law of cosines"

$$|x + y|^2 = |x|^2 + |y|^2 + 2\,|x|\,|y| \cos\vartheta\,.$$

If y is replaced in this formula by $-y$, the addition of both equations yields the *parallelogram identity*

4. $|x + y|^2 + |x - y|^2 = 2\,|x|^2 + 2\,|y|^2\,,$

which now only contains lengths of vectors and according to which the sum of the squares of the diagonals is equal to the sum of the squares of the four sides of the parallelogram with the vertices 0, x, y, $x + y$.

If one adds this equation, which makes sense in any Minkowski metric, as a fourth postulate to the three already mentioned, the thus specialized metric is, as is known, a *euclidean* one, i.e.:

There exists in the affine space a uniquely determined bilinear, symmetric, positive definite function $G\,x\,y$ such that for every x

$$|x| = +\sqrt{G\,x^2}\,.$$

If, conversely, a norm of this kind is introduced in an affine space, whereby the fundamental metric form G has the properties mentioned above, but may otherwise be chosen arbitrarily, then this metric satisfies the three postulates 1, 2, 3 of a Minkowski metric and in addition the parallelogram identity 4.

In both of the following sections we shall for the sake of completeness briefly prove these assertions, which are fundamental for the euclidean metric.

6.5. Derivation of G from the four postulates. In an affine space (here the dimension makes no difference) suppose the length of vectors is defined in a way which satisfies the four postulates 1—4 mentioned above. The claim is that a bilinear, symmetric, positive definite function $G\,x\,y$ then exists so that $|x| = +\sqrt{G\,x^2}$.

Assuming the correctness of the assertion, it follows from the polarization formula (cf. 4.1) for G that

$$4\,G\,x\,y = G\,(x+y)^2 - G\,(x-y)^2 = |x+y|^2 - |x-y|^2\,,$$

and thus it must be shown that this expression is actually a bilinear, symmetric, positive definite function which stands in the asserted relationship to $|x|$.

First, the above definition for $y = x$ yields by 1, according to which $|0|^2 = 0$ and $|2\,x|^2 = 4\,|x|^2$, that

$$G\,x^2 = |x|^2$$

and therefore according to 2 is a positive definite function of x which stands in the proper relationship to $|x|$.

Since further $|y - x| = |x - y|$, $G\,x\,y$ is symmetric, and it remains to demonstrate the linearity. For the most part this will be a consequence of the parallelogram identity 4.

Thus the equations

$$G\,(x+y)\,z = G\,x\,z + G\,y\,z\,, \quad G(\lambda\,x)\,y = \lambda\,G\,x\,y$$

must be verified.

The first identity is according to the definition of G equivalent with

$$|x+y+z|^2 - |x+y-z|^2 = |x+z|^2 - |x-z|^2 + |y+z|^2 - \\ - |y-z|^2\,.$$

Now by 4

$$2\,|x+z|^2 + 2\,|y+z|^2 = |x+y+z+z|^2 + |x-y|^2;$$

furthermore,

$$|x + y + z + z|^2 + |x + y|^2 = 2\,|x + y + z|^2 + 2\,|z|^2\,,$$

and therefore

$$2\,|x + z|^2 + 2\,|y + z|^2 = 2\,|x + y + z|^2 + 2\,|z|^2 - |x + y|^2 + |x - y|^2.$$

With $-z$ in place of z, since $|-z| = |z|$, this becomes

$$2\,|x - z|^2 + 2\,|y - z|^2 = 2\,|x + y - z|^2 + 2\,|z|^2 - |x + y|^2 + |x - y|^2$$

and the subtraction of these two identities yields the desired identity for G.

In order to prove the second equation, $G(\lambda\, x)\, y = \lambda\, G\, x\, y$, for real λ, observe that

$$4\, G(\lambda\, x)\, y = |\lambda\, x + y|^2 - |\lambda\, x - y|^2$$

is as a consequence of the triangle inequality 3 and 1 a continuous function of λ.

For one has

$$|\lambda'\, x + y| = |(\lambda' - \lambda'')\, x + \lambda''\, x + y| \leq |(\lambda' - \lambda'')\, x| + |\lambda''\, x + y|$$
$$= |\lambda' - \lambda''|\, |x| + |\lambda''\, x + y|\,,$$

and hence, since λ' and λ'' can be exchanged,

$$\big||\lambda'\, x + y| - |\lambda''\, x + y|\,\big| \leqq |\lambda' - \lambda''|\, |x|\,.$$

It is therefore possible to restrict the multiplier λ to rational and even to positive rational values. But then the asserted homogeneity in λ follows from the above proved additivity. In the first place, for a positive integral $\lambda = p$, because $p\, x = \underbrace{x + \cdots + x}_{p}$,

$$G(p\, x)\, y = p\, G\, x\, y\,,$$

and thus, when $p\, x$ is replaced by x,

$$G\left(\frac{1}{p}\, x\right) y = \frac{1}{p}\, G\, x\, y\,,$$

and consequently for a positive rational $\lambda = p/q$

$$G\left(\frac{p}{q}\, x\right) y = \frac{p}{q}\, G\, x\, y\,,$$

which completes the proof.

6.6. Derivations of the four postulates from G. Suppose, conversely, that $G\, x\, y$ is an arbitrary, bilinear, symmetric, positive definite function in the finite or infinite dimensional affine space under consideration. We shall briefly present the well-known proof for the fact that the norm

$$|x| = +\sqrt{G\,x^2}$$

satisfies postulates 1—4.

Properties 1 and 2 are clear without further ado. Moreover,

$$|x+y|^2 = G\,(x+y)^2 = G\,x^2 + 2\,G\,x\,y + G\,y^2 = |x|^2 + G\,x\,y + |y|^2\,,$$

and with $-y$ in place of y, we have an equation whose addition to the above relation yields the parallelogram identity.

And, finally, to prove the triangle inequality

$$|x+y|^2 \leqq |x|^2 + 2\,|x|\,|y| + |y|^2\,,$$

as a glance at the preceding identity shows, the inequality $G\,x\,y \leqq \leqq |x|\,|y|$, that is, Schwarz's inequality

$$(G\,x\,y)^2 \leqq G\,x^2\,G\,y^2\,,$$

must hold. In the trivial cases $x = 0$ or $y = 0$ it together with the triangle inequality clearly hold with the equality sign. In the general case these inequalities result from the identity

$$G\,x^2\,G\,(y-\lambda\,x)^2 = G\,x^2\,G\,y^2 - (G\,x\,y)^2 + (\lambda\,G\,x^2 - G\,x\,y)^2$$

for

$$\lambda = \frac{G\,x\,y}{G\,x^2} = \frac{G\,x\,y}{|x|^2}\,,$$

from which the relation

$$G\,x^2\,G\,y^2 - (G\,x\,y)^2 = G\,x^2\,G\,(y-\lambda\,x)^2 \geqq 0$$

follows. Here equality holds for $y = \lambda\,x$, so that x and y are linearly dependent. In order for equality to then also hold in the triangle inequality one must furthermore have $G\,x\,y = |x|\,|y|$, and hence $\lambda = |y|/|x| > 0$; the linearly dependent vectors x and y must therefore have the same direction.

In summary, a euclidean space can, as is customarily done, be defined as an affine space with an arbitrary, bilinear, symmetric, positive definite *fundamental metric form*

$$G\,x\,y = (x, y)$$

called the *inner product* or the *scalar product* of the vectors x and y, where $|x|^2 = (x, x)$; or as an affine space in which a metric is introduced that satisfies the four often mentioned postulates.

6.7. Angles and orthogonality in the euclidean space. By virtue of Schwarz's inequality one can define the angle ϑ formed by the vectors x and y using the equation

$$G\,x\,y = (x, y) = |x|\,|y|\cos\vartheta\,.$$

Accordingly, these vectors are orthogonal (with respect to G) when $G\,x\,y = 0$, in agreement with the general terminology already used in 4.2.

All that has been said about euclidean spaces up to this point is independent of the dimension. If this dimension is now finite and $= m$, it then follows from the general inertia theorem in 4.2 that coordinate systems $e_1, \ldots, e_m$ exist which are orthonormal with respect to the inner product, and $G\,e_i\,e_j = \delta_i^j$. Thus if

$$x = \sum_{i=1}^{m} \xi^i e_i , \qquad y = \sum_{j=1}^{m} \eta^j e_j ,$$

then in such an orthonormal system $\xi^i = G\,x\,e_i$, $\eta^j = G\,y\,e_j$, and therefore

$$G\,x\,y = (x, y) = \sum_{i=1}^{m} \xi^i \eta^i \quad \text{and} \quad G\,x^2 = |x|^2 = \sum_{i=1}^{m} (\xi^i)^2 .$$

Orthonormal coordinate systems can be constructed from arbitrary bases by means of the Schmidt orthogonalization process. Once such a system has been found, all other orthonormal systems are obtained by means of the group of linear transformations T which are orthogonal with respect to the fundamental metric form, which thus preserve the inner product:

$$(T\,x, T\,y) = (x, y) .$$

6.8. The principal axis problem. Let an arbitrary real *symmetric* bilinear function $B\,x\,y$ be given in the euclidean space R^m which has been metrized with the fundamental form $G\,x\,y = (x, y)$.

The principal axis problem consists in finding a coordinate system $e_1, \ldots, e_m$ (the principal axes) orthonormal with respect to G such that

$$B\,x\,y = \sum_{i=1}^{m} \lambda_i \xi^i \eta^i ,$$

that is, $B\,e_i\,e_j = \lambda_i\,\delta_i^j$.

Since the treatment of this classical problem of linear algebra may be assumed as known here, only the line of thought behind one of the many solutions to this problem is to be briefly sketched. We present the details as problems in 6.11.

According to the Fréchet-Riesz theorem a unique linear transformation $T\,x$ of R^m exists such that

$$B\,x\,y = (x, T\,y) = G\,x\,T\,y .$$

Because of the symmetry of G and B

$$(x, T\,y) = (T\,x, y) ,$$

and T is therefore self-adjoint with respect to G. Conversely, each such transformation defines by means of the above equation a bilinear symmetric function B.

Assuming the principal axis problem to have been solved for a given B, it follows from the above representation of B that

$$(T e_i, e_j) = B e_i e_j = \lambda_i \delta_i^j = (\lambda_i e_i, e_j),$$

and thus, since G is not degenerate,

$$T e_i = \lambda_i e_i \qquad (i = 1, \ldots, m).$$

The self-adjoint linear transformation T therefore has m eigenvalues λ_i with the associated eigenvectors e_i (cf. 3.10, problems 9—10). Conversely, the principal axis problem is obviously solved when it can be shown that every linear transformation of R^m which is self-adjoint with respect to G has precisely m eigenvalues λ_i whose associated eigenvectors e_i form an orthonormal coordinate system. This is proved in a coordinate-free way in exercises 7—12 of 6.11.

6.9. Measure of a simplex. Assume that $1 \leqq p \leqq m$ and that $s^p(x_0, \ldots, x_p)$ is a p-dimensional closed simplex of the affine space R^m. The p linearly independent edges $x_i - x_0$ span a subspace U^p in R^m parallel to the plane of the simplex, and for the time being we restrict ourselves to p-dimensional simplexes that are parallel to a given subspace U^p.

If a measure is to be defined for these p-dimensional simplexes in a natural way, the simplest would seem to be to retain the first three postulates $A - C$ in 6.2 in an appropriately generalized form. Accordingly we postulate:

A. *The measure of simplexes parallel to U^p remains invariant under reorientation and parallel translation.*

This postulate permits us to take the vertices $x_0, \ldots, x_p$ in an arbitrary sequence and to shift the origin of the space to one of the vertices, e. g., to x_0. Then $x_1, \ldots, x_p$ are the linearly independent edges of the simplex, and, in agreement with the notation in the case $p = 1$, we denote the measure of the simplex $s^p(0, x_1, \ldots, x_p)$ which is to be defined, by

$$|x_1, \ldots, x_p| .$$

B. *For a decomposition of the simplex in the sense of* 5.6 *the measure is an additive function.*

Observe that for a decomposition of the simplex s^p the subsimplexes lie in the same subspace U^p determined by s^p and in this sense are parallel; for $p = 1$ one has precisely postulate B of 6.2.

C. *The measure is a positive definite function of the edges x_i; thus*

$$|x_1, \ldots, x_p| \geqq 0 ,$$

and $= 0$ *precisely when the edges are linearly dependent and the simplex consequently degenerates to a simplex of lower dimension.*

For $p = 1$ this condition coincides with postulate C of 6.2. Taking the theorem in 5.6 into account, these postulates are easily satisfied if one defines

$$|x_1, \ldots, x_p| = |D\, x_1 \ldots x_p|$$

for the p-dimensional simplexes "in the direction U^p", which are thus parallel to the fixed subspace U^p. Here D is the not identically vanishing, real, p-fold linear alternating function of the edges x_i, which is uniquely determined in U^p up to a normalizing factor.

6.10. Connection with the fundamental metric form. Final definition of $|x_1, \ldots, x_p|$. Now assume, that R^m is a euclidean space with inner product $G\, x\, y = (x, y)$, which until now was not relevant. In order to find the connection between the above definition with the one-dimensional metric determined by G, it is convenient to orthogonalize the subspace U^p with respect to G.

Therefore, let $e_1, \ldots, e_p$ be an arbitrary orthonormal coordinate system in U^p and set

$$x_i = \sum_{j=1}^{p} \xi_i^j e_j \qquad (i = 1, \ldots, p)\,.$$

By 5.2

$$D\, x_1 \ldots x_p = \det(\xi_i^j)\, D\, e_1 \ldots e_p\,.$$

Using the theorem on the multiplication of determinants deduced there, according to which $\left(\det(\xi_i^j)\right)^2 = \det\left(\sum_{j=1}^{p} \xi_h^j \xi_k^j\right) = \det(G\, x_h\, x_k)$,

$$(D\, x_1 \ldots x_p)^2 = \det(G\, x_h\, x_k)\, (D\, e_1 \ldots e_p)^2\,.$$

This formula shows, first, that

$$\det(G\, x_h\, x_k) \geqq 0\,,$$

and $= 0$ precisely when $x_1, \ldots, x_p$ are linearly dependent. This is a generalization of Schwarz's inequality, which is included here for $p = 2$.

This formula further shows that $|D\, x_1 \ldots x_p|$ is invariant under orthogonal transformations of the subspace U^p and that $|D\, e_1 \ldots e_p|$ is independent of the choice of the orthonormal system e_1. We can therefore fix this positive quantity arbitrarily for each p-dimensional subspace U^p.

Now if the measure being defined for the simplex s^p is to be "isotropic", that is, independent of the direction U^p of the simplex (as is the case, of course, for $p = 1$), then the factor

$$|D\, e_1 \ldots e_p| = |D(U^p)|$$

must be assigned the same value for *every* p-dimensional subspace. It is now only a question of doing this in an appropriate way. It seems natural, moreover, to make this assignment so that the unit cube receives measure 1. According to problem 8 in 5.9 one gets then, for every p,

$$p!|D\, e_1 \ldots e_p| = 1 .$$

Thus in

$$|x_1, \ldots, x_p| = \frac{1}{p!}\sqrt{\det\,(G\, x_h\, x_k)}$$

we have a definition for the euclidean measure or the volume of a p-dimensional simplex that satisfies all postulates proposed. This definition is also satisfactory insofar as the volume remains invariant under orthogonal transformations of the euclidean space.

6.11. Exercises. 1. Show that postulates A, B, C and 1, 2 in 6.2 are equivalent.

Hint. One shows first, keeping in mind the final remark in 1.1, that

$$\left|\frac{p}{q}\, x\right| = \frac{p}{q}\,|x|$$

follows from B for every positive, rational p/q. Then if λ is a positive irrational number, let (α) be the class of positive rational numbers $< \lambda$ and (β) the class of rational numbers $> \lambda$. It then follows from the above and B that

$$\alpha|x| < |\lambda\, x| < \beta|x| ,$$

on the one hand, while, on the other hand,

$$\alpha|x| < \lambda|x| < \beta|x| .$$

Consequently, for every pair α, β

$$||\lambda\, x| - \lambda|x|| < (\beta - \alpha)|x| ,$$

from which it follows that $|\lambda\, x| = \lambda|x|$. Hence because of A, $|\lambda\, x| = |\lambda|\; |x|$ for every real λ.

Conversely, A and B are immediate consequences of 1.

2. Show that the unit sphere $|x| \leqq 1$ in a Minkowski-Banach metric is a convex point set and, conversely, that the triangle inequality is a consequence of the convexity together with 1.

3. Let $x_1, x_2, \ldots$ be an infinite sequence of points in the space R^m. Prove the *Cauchy convergence criterion*:

In order for there to exist a limit point

$$\bar{x} = \lim_{n \to \infty} x_n$$

in the sense of the natural topology of the space, it is necessary and sufficient that as $p, q \to \infty$

$$|x_q - x_p| \to 0$$

in an arbitrary Minkowski metric. In a Hilbert or Banach space the validity of the Cauchy criterion is an axiom, independent of the other axioms defining these spaces.

4. Let $s^m(x_0, \ldots, x_m)$ be a simplex in the m-dimensional linear space R^m. Show that there exists precisely one euclidean metric so that the lengths of all the edges $|x_j - x_i|$ have unit length in this metric.

Further determine the angle between two adjoining edges and the height of such a regular m-dimensional simplex.

5. Prove Schwarz's inequality as a consequence of the Pythagorean theorem, according to which in a euclidean space

$$|y|^2 = |p|^2 + |n|^2 .$$

Here p stands for the projection of the arbitrary vector y onto a vector $x \neq 0$, and $n = y - p$ is the projecting normal.

6. Let $e_1, \ldots, e_m$ be a coordinate system which is orthonormal with respect to the inner product

$$G\,x\,y = (x, y)$$

of a euclidean space R^m. Establish for an arbitrary vector x of the space the *orthogonal expansion* with respect to G:

$$x = \sum_{i=1}^{m} e_i\, G\, x\, e_i = \sum_{i=1}^{m} (x, e_i)\, e_i .$$

Develop in a corresponding way the vector x into an orthogonal series with respect to a nondegenerate symmetric bilinear function $B\,x\,y$.

7. In a euclidean space with the inner product $G\,x\,y = (x, y)$ let

$$p = P\,x$$

be the orthogonal projection of x onto a subspace U and

$$q = x - P\,x = Q\,x$$

be the projecting normal.

Prove that P and Q are self-adjoint linear transformations with respect to G which are characterized as orthogonal projections by this property and by the (purely affine) property $P^2 = P$, $Q^2 = Q$ (cf. 3.10, exercise 7), which is valid for any projection.

8. Let T be a self-adjoint linear transformation of the euclidean space R^m and set

$$\sup_{|x|=|y|=1} |(T\,x, y)| = |B|\,,$$
$$\sup_{|x|=1} |(T\,x, x)| = |Q|\,,$$
$$\sup_{|x|=1} |T\,x| = |T|\,.$$

Prove that

$$|B| = |Q| = |T|\,.$$

Hint. For arbitrary x and y

$$|(T\,x, y)| \leqq |B|\,|x|\,|y|\,, \quad |(T\,x, x)| \leqq |Q|\,|x|^2\,, \quad |T\,x| \leqq |T|\,|x|\,.$$

From $|T\,x|^2 = (T\,x, T\,x)$ it thus follows that $|T\,x| \leqq |B|\,|x|$, hence $|T| \leqq |B|$. Further, it follows from the polarization formula and the parallelogram identity that

$$4\,|(T\,x, y)| \leqq |Q|\,(|x+y|^2 + |x-y|^2) = 2\,|Q|\,(|x|^2 + |y|^2)\,.$$

Hence $|B| \leqq |Q|$ and therefore

$$|T| \leqq |B| \leqq |Q|\,.$$

On the other hand, it goes without saying that $|Q| \leqq |B|$, and as a consequence of Schwarz's inequality, $|(T\,x, y)| \leqq |T\,x|\,|y|$. Therefore $|B| \leqq |T|$, and hence

$$|Q| \leqq |B| \leqq |T|\,.$$

9. Because of the finite dimension of R^m, the sphere $|x| = 1$ is compact, and the least upper bound $|Q|$ is thus achieved for at least one unit vector e_1.

Prove that

$$T\,e_1 = \lambda_1\,e_1$$

and that any further possible eigenvalues of T are in absolute value $\leqq |\lambda_1|$.

Hint. According to Schwarz's inequality

$$|Q| = |(T\,e_1, e_1)| \leqq |T\,e_1|\,|e_1| \leqq |T|\,,$$

and therefore, since $|Q| = |T|$,

$$|(T\,e_1, e_1)| = |T\,e_1|\,|e_1|\,,$$

and hence $T\,e_1$ and e_1 are linearly dependent.

10. Let $U_1^{d_1}$ be the eigenspace of dimension d_1 corresponding to the eigenvalue λ_1 found above and $P_1\,x$ the orthogonal projection of x onto $U_1^{d_1}$. If one sets

$$x = P_1\,x + (x - P_1\,x) = P_1\,x + Q_1\,x\,,$$

then

$$T\,x = \lambda_1\,P_1\,x + T\,Q_1\,x\,.$$

Show that

$$T_1 = T\,Q_1 = Q_1\,T\,Q_1 = Q_1\,T$$

and that T_1 is consequently self-adjoint and, further, that it maps $U_1^{d_1}$ onto $x = 0$ and the $(m - d_1)$-dimensional orthogonal complement $R_1^{m-d_1}$ of $U_1^{d_1}$ into itself and that there $T_1\,x \equiv T\,x$.

11. Treat $T_1 = T$ in $R_1^{m-d_1}$ exactly as T was above in R^m and show by continuing the method followed that T has eigenvalues $\lambda_1, \ldots, \lambda_k$ $(k \leqq m)$ with multiplicity sum $\sum_{i=1}^{k} d_i = m$, where $|\lambda_1| \geqq \cdots \geqq |\lambda_k|$, further that the corresponding eigenspaces $U_i^{d_i}$ as a consequence of the method have to be pairwise orthogonal and yield R^m as direct sum, and finally that

$$T\,x = \sum_{i=1}^{k} \lambda_i\,P_i\,x\,,$$

where $P_i\,x$ stands for the projection of x onto the eigenspace $U_i^{d_i}$.

12. Orthogonalize the eigenspaces $U_i^{d_i}$ and show that in this way there results a principal axis system $e_1, \ldots, e_m$ for the symmetric bilinear function

$$B\,x\,y = (T\,x, y) = (x, T\,y)\,.$$

Remark. The results in the above exercises 7—12 contain a complete solution to the principal axis problem in the euclidean space R^m.

13. Let $A\,x$ be a skewsymmetric linear transformation of the euclidean space R_x^n, that is, $A^* = -A$. Then an m $(\leqq n/2)$ exists and an orthonormal coordinate system $e_1, \ldots, e_n$ such that

$$A\,e_{2i-1} = \varrho_i\,e_{2i}\,, \qquad A\,e_{2i} = -\,\varrho_i\,e_{2i-1}$$

for $i = 1, \ldots, m$, while $A\,e_i = 0$ for $i > 2\,m$.

Hint. The transformation A^2 is self-adjoint and has negative eigenvalues $-\,\varrho_1^2, \ldots, -\,\varrho_m^2$, which are each double, and the $(n - 2\,m)$-fold eigenvalue zero. The orthonormalized vectors

$$e_{2i-1}, e_{2i} = \frac{A\,e_{2i-1}}{|A\,e_{2i-1}|} = \frac{A\,e_{2i-1}}{\varrho_i}$$

are the eigenvectors that correspond to the eigenvalue $-\,\varrho_i^2$.

14. Let R_x^m and R_y^n be euclidean spaces with the inner products $(x_1, x_2)_x$ and $(y_1, y_2)_y$, respectively, and A and A^* adjoint linear mappings (cf. 4.9, exercise 7), so that

$$(A\,x, y)_y = (x, A^*\,y)_x\,.$$

Show:

a. $A^*\,A = T_x$ and $A\,A^* = T_y$ are self-adjoint linear transformations of the spaces R_x^m and R_y^n, respectively.

b. If $K_x^p(A)$ and $K_y^q(A^*)$ are the kernels of the mappings A and A^*, respectively, of dimensions p and q, then zero is a p-fold eigenvalue of the transformation T_x with the eigenspace $K_x^p(A)$ and a q-fold eigenvalue for T_y with the eigenspace $K_y^q(A^*)$.

c. The $r = m - p = n - q$ nonzero eigenvalues of the transformations T_x and T_y are precisely the same and of the same multiplicity.

15. Let $G\,x\,y$ be a bilinear symmetric positive definite function in an affine space of arbitrary dimension.

Give a proof of the generalized Schwarz inequality

$$\det(G\,x_h\,x_k) \geqq 0$$

based on the fact that with arbitrary real coefficients λ_i

$$G(\lambda_1 x_1 + \cdots + \lambda_m x_m)^2 = \sum_{i,j=1}^{m} \lambda_i \lambda_j G\,x_i\,x_j$$

is a positive semidefinite quadratic form of the λ_i which only vanishes for

$$\lambda_1 x_1 + \cdots + \lambda_m x_m = 0\,.$$

16. We close this section with the following consideration of E. Schmidt's orthogonalization procedure, described in 4.4; the details are left to be carried out by the reader.

Let U^m be a subspace of a euclidean space with the inner product $G\,x\,y = (x, y)$. Suppose D is the up to a real factor unique fundamental form of this subspace and that $a_1, \ldots, a_m$ is an arbitrary orthonormal coordinate system in U^m.

Then if $x_1, \ldots, x_m$ and $y_1, \ldots, y_m$ are two arbitrary systems of vectors in U^m, one has

$$x_h = \sum_{i=1}^{m} \xi_h^i a_i\,, \qquad y_k = \sum_{j=1}^{m} \eta_k^j a_j$$

and

$$D\,x_1 \ldots x_m = \det(\xi_h^i)\, D\,a_1 \ldots a_m\,,$$

$$D\,y_1 \ldots y_m = \det(\eta_k^j)\, D\,a_1 \ldots a_m\,.$$

Since the coordinate system a_i is orthonormal with respect to $G\,x\,y = (x, y)$,

$$\sum_{i=1}^{m} \xi_h^i \eta_k^i = (x_h, x_k)\,,$$

and the multiplication law for determinants therefore yields

$$D\,x_1 \ldots x_m\, D\,y_1 \ldots y_m = \det\big((x_h, y_k)\big)\,(D\,a_1 \ldots a_m)^2\,.$$

The formula

$$(D\,x_1 \ldots x_m)^2 = \det\big((x_h, x_k)\big)\,(D\,a_1 \ldots a_m)^2$$

already mentioned follows from here for $y_i = x_i$.

Presuming this, let $z_1, \ldots, z_m$ be a linearly independent system of vectors in U^m. If this system is orthogonalized by means of the Schmidt procedure, one obtains m equations of the form

$$z_i = \lambda_{i1} e_1 + \cdots + \lambda_{ii} e_i \qquad (i = 1, \ldots, m),$$

where $e_1, \ldots, e_m$ is an orthonormal system for the space U^m. This system and the coefficients λ_{hk} $(1 \leqq k \leqq h \leqq m)$ are uniquely determined by the vectors z_i, in the given order, provided at each step of the procedure the sign of λ_{ii} is fixed. We wish, for example, to take $\lambda_{ii} > 0$ for all indices i, which because

$$D\, e_1 \ldots e_{i-1}\, z_i \ldots z_m = \lambda_{ii} \ldots \lambda_{mm}\, D\, e_1 \ldots e_m$$

means that the simplexes $s^m(0, e_1, \ldots, e_{i-1}, z_i, \ldots, z_m)$ are all oriented like the simplex $s^m(0, z_1, \ldots, z_m)$. It is a question of setting up explicit expressions for the coefficients λ_{hk} in the given vectors z_i.

To this end we take for $1 \leqq k \leqq h \leqq m$ in the above formula

$$x_1 = z_1, \ldots, x_{k-1} = z_{k-1}, x_k = z_k, x_{k+1} = e_{k+1}, \ldots, x_m = e_m,$$

$$y_1 = z_1, \ldots, y_{k-1} = z_{k-1}, y_k = z_h, y_{k+1} = e_{k+1}, \ldots, y_m = e_m,$$

and thus obtain

$$D\, z_1 \ldots z_{k-1} z_k e_{k+1} \ldots e_m D\, z_1 \ldots z_{k-1} z_h e_{k+1} \ldots e_m = \Delta_{hk} (D\, a_1 \ldots a_m)^2,$$

where

$$\Delta_{hk} = \begin{vmatrix} (z_1, z_1) & \ldots & (z_1, z_{k-1}) & (z_1, z_h) \\ \vdots & & \vdots & \vdots \\ (z_k, z_1) & \ldots & (z_k, z_{k-1}) & (z_k, z_h) \end{vmatrix}$$

But on the other hand, according to Schmidt's scheme,

$$D\, z_1 \ldots z_{k-1} z_k e_{k+1} \ldots e_m = \lambda_{11} \ldots \lambda_{(k-1)(k-1)} \lambda_{kk} D\, e_1 \ldots e_m,$$

$$D\, z_1 \ldots z_{k-1} z_h e_{k+1} \ldots e_m = \lambda_{11} \ldots \lambda_{(k-1)(k-1)} \lambda_{hk} D\, e_1 \ldots e_m,$$

and, therefore,

$$\begin{aligned} D\, z_1 \ldots z_{k-1} z_k e_{k+1} \ldots e_m\, D\, z_1 \ldots z_{k-1} z_h e_{k+1} \ldots e_m \\ = \lambda_{11}^2 \ldots \lambda_{(k-1)(k-1)}^2 \lambda_{kk} \lambda_{hk} (D\, e_1 \ldots e_m)^2 \\ = \lambda_{11}^2 \ldots \lambda_{(k-1)(k-1)}^2 \lambda_{kk} \lambda_{hk} (D\, a_1 \ldots a_m)^2. \end{aligned}$$

Since $(D\, a_1 \ldots a_m)^2 \neq 0$, comparison of both expressions yields

$$\lambda_{11}^2 \ldots \lambda_{(k-1)(k-1)}^2 \lambda_{kk} \lambda_{hk} = \Delta_{hk}$$

for $1 \leqq k \leqq h \leqq m$. In particular in view of the choice of sign $\lambda_{ii} > 0$,

$$\lambda_{11}^2 \ldots \lambda_{(k-1)(k-1)}^2 \lambda_{kk} = + \sqrt{\Delta_{(k-1)(k-1)} \Delta_{kk}} .$$

In this way we obtain the required representation

$$\lambda_{hk} = \frac{\Delta_{hk}}{\sqrt{\Delta_{(k-1)(k-1)} \Delta_{kk}}},$$

where Δ_{00} is to be set $= 1$.

II. Differential Calculus

§ 1. Derivatives and Differentials

1.1. Vector functions. Let R_x^m and R_y^n be two linear spaces and G_x^m a region in R_x^m, that is, a point set which is open and connected with respect to the natural topology.

In G_x^m let a single-valued *vector function*

$$x \to y = y(x)$$

be given, providing a mapping from G_x^m into R_y^n. The simplest examples of such functions are the linear mappings $y = A\,x$ of the space R_x^m into the space R_y^n. In what follows we shall be dealing with more general mappings and vector functions.

If coordinate systems a_i and b_j are introduced in R_x^m and R_y^n, so that

$$x = \sum_{i=1}^{m} \xi^i a_i\,, \qquad y = \sum_{j=1}^{n} \eta^j b_j\,,$$

corresponding to the relation $y = y(x)$ there is a system of n real functions

$$\eta^j = \eta^j(\xi^1, \ldots, \xi^m) \qquad (j = 1, \ldots, n)$$

of the real variables $\xi^1, \ldots, \xi^m$. Conversely, any such system given in G_x^m defines by means of the coordinates systems a_i and b_j a vector function $y = y(x)$.

For $n = 1$ this function is reduced to one single component, hence to a real function

$$\eta = \eta(\xi^1, \ldots, \xi^m)$$

of the m real variables ξ^i. If in addition $m = 1$, we have the elementary case of a real function of one real variable.

The vector function $y(x)$ and the corresponding mapping $x \to y$ is *continuous* at the point x in the domain G_x^m of definition provided it is continuous at x in the natural topologies in the spaces R_x^m and R_y^n. The same is also true with respect to arbitrary Minkowski metrics, in particular also euclidean metrics, and conversely. Consequently, for an $\varepsilon > 0$ there exists a $\varrho_\varepsilon > 0$ such that in R_y^n

$$|y\,(x + h) - y(x)| < \varepsilon$$

provided $|h| < \varrho_\varepsilon$ in R_x^m. The real components $\eta^j(\xi^1, \ldots, \xi^m)$ are then also continuous, and, conversely, the continuity of $y(x)$ follows from the continuity of these functions.

1.2. The derivative. In the simplest case $m = n = 1$, where $y = y(x)$ can be thought of as a real function of the real variable x, the derivative at the point x is defined in the elements of the differential calculus as the limit

$$\lim_{|h| \to 0} \frac{y(x+h) - y(x)}{h} = \alpha(x) ,$$

provided this limit exists and is finite.

For a general vector function this definition is obviously meaningless. But the above definition can also be written

$$y\,(x+h) - y(x) = \alpha(x)\,h + |h|\,(h; x) ,$$

where $|(h; x)| \to 0$ for $|h| \to 0$, and in this equivalent form it can be generalized in a natural way.

In fact, $y = \alpha(x)\,h$ defines a linear mapping of the real h-axis into the real y-axis and, conversely, each such mapping has the above form. That yields the following definition of differentiability and of the derivative of a general vector function $y = y(x)$:

The mapping

$$y = y(x)$$

defined in the region G_x^m is said to be differentiable at the point x provided a linear mapping $A(x)\,h$ of R_x^m into R_y^n exists such that

$$y\,(x+h) - y(x) = A(x)\,h + |h|\,(h; x) ,$$

where the length (measured in R_y^n) $|(h; x)| \to 0$ when the length (measured in R_x^m) $|h| \to 0$. The linear operator $A(x)$ is called the derivative of the vector function $y(x)$ at the point x.

The derivative is at every point $x \in G_x^m$ where it exists a *linear operator*, which only in the simplest case $m = n = 1$ can be characterized by a single real number $\alpha(x)$. However, even for this generalized derivative, we retain the usual Lagrange notation $A(x) = y'(x)$. The defining equation

$$y\,(x+h) - y(x) = y'(x)\,h + |h|\,(h; x)$$

states that the mapping $y\,(x+h) - y(x)$ of the neighborhood of the point x can in the first approximation be replaced by the linear mapping $y'(x)\,h$.

The above definition presumes the introduction of some kind of Minkowski metrics in the spaces R_x^m and R_y^n. However, from what was said in I.6.2 it follows that differentiability is independent of this choice, and the existence of the derivative is therefore an affine property of the function $y(x)$ as long as the dimensions m and n are *finite.*

Further observe that the operator $y'(x)$, provided the function is differentiable at all, is uniquely determined. For if one has

$$y(x + h) - y(x) = A(x)\,h + |h|\,(h;x)_1 = B(x)\,h + |h|\,(h;x)_2\,,$$

then

$$A(x)\,h - B(x)\,h = \bigl(A(x) - B(x)\bigr)\,h = |h|\,\bigl((h;x)_2 - (h;x)_1\bigr) = |h|\,(h;x)\,.$$

If h is replaced by $\lambda\,h$, where h is an arbitrary vector in the space R_x^m and λ is a sufficiently small real multiplier, then

$$\bigl(A(x) - B(x)\bigr)\,h = |h|\,(\lambda\,h;x) \to 0$$

for $\lambda \to 0$. Consequently, $\bigl(A(x) - B(x)\bigr)\,h = 0$, and hence for every h in R_x^m

$$A(x)\,h = B(x)\,h = y'(x)\,h\,.$$

1.3. The differential. Following the example of elementary calculus we call the vector

$$dy = y'(x)\,h$$

the *differential* at the point x of the differentiable function $y(x)$, corresponding to the argument differential h.

If in particular $y = A\,x$ is a linear mapping, then

$$A\,(x + h) - A\,x = A\,h\,,$$

and the derivative A' thus is identical with the linear operator A. For $y = x$, $dy = h = dx$, and accordingly we can write

$$dy = y'(x)\,dx\,,$$

whereby the argument differential dx is an arbitrary vector in the space R_x^m.

This equation suggests introducing the Leibniz notation

$$y'(x) = \frac{dy}{dx}\,,$$

where, of course, the right hand side is not a quotient, but a symbol for the linear operator $y'(x)$. We shall in what follows often operate, instead of with the derivatives, with the in many respects more convenient differentials

$$dy = y'(x)\,dx = \frac{dy}{dx}\,dx\,.$$

1.4. Coordinate representation of the derivative. If one introduces coordinate systems a_i and b_j in the linear spaces R_x^m and R_y^n, so that

$$x = \sum_{i=1}^{m} \xi^i\,a_i, \qquad y = \sum_{j=1}^{n} \eta^j\,b_j\,,$$

then, as already mentioned, the vector function $y(x)$ can be represented by the system

$$\eta^j = \eta^j(\xi^1, \ldots, \xi^m) \qquad (j = 1, \ldots, n)$$

of real functions.

If this vector function is differentiable at the point x, then for every

$$\Delta x = dx = \sum_{i=1}^{m} d\xi^i a_i ,$$

provided $x + dx$ lies in G_x^m, one has

$$\Delta y = \sum_{j=1}^{n} \Delta\eta^j b_j = y(x + \Delta x) - y(x) = y'(x)\, dx + |dx|\, (dx; x) ,$$

where $|(dx; x)| \to 0$ as $|dx| \to 0$. Corresponding to the linear mapping $y'(x)$ (with respect to the fixed coordinate sytems) is a matrix

$$y'(x) \to (\alpha_i^j(x)) ,$$

and one has for $j = 1, \ldots, n$

$$\Delta\eta^j = \sum_{i=1}^{m} \alpha_i^j(x)\, d\xi^i + |dx|\, (dx; x)^j ,$$

where $|(dx; x)^j| \to 0$ for $|dx| \to 0$.

This means that all of the components η^j are differentiable in the sense of ordinary differential calculus. The partial derivatives

$$\frac{\partial\eta^j}{\partial\xi^i} = \alpha_i^j(x)$$

exist, and one has for $j = 1, \ldots, n$

$$\Delta\eta^j = \sum_{i=1}^{m} \frac{\partial\eta^j}{\partial\xi^i} d\xi^i + |dx|\, (dx; x)^j = d\eta^j + |dx|\, (dx; x)^j .$$

Conversely, it follows from these relations that the vector function $y(x)$ is differentiable in our sense with the linear derivative operator $y'(x)$, which with respect to the fixed coordinate systems is uniquely determined by the functional matrix

$$\left(\frac{\partial\eta^j}{\partial\xi^i}\right).$$

1.5. The differential rules. The definition given above of the differential dy of a vector function $y(x)$,

$$dy = y'(x)\, dx , \qquad \Delta y = y(x + dx) - y(x) = dy + |dx|\, (dx; x) ,$$

is formally the same as in the elementary case of a real function $y = y(x)$ of one real variable x. Since the derivation of the differential rules for such a function does not depend on the differential $dy = y'(x)\, dx$ being a product, but only on dy being linearly dependent on dx, it is clear that the differential rules known from elementary analysis also hold for vector functions.

Thus, if $y_1(x)$ and $y_2(x)$ are two differentiable mappings into R^n_y defined in G^m_x, then with arbitrary real coefficients λ_1 and λ_2 one has

$$d(\lambda_1 y_1 + \lambda_2 y_2) = \lambda_1 dy_1 + \lambda_2 dy_2 .$$

Further, if $\lambda(x)$ is a *real* function differentiable in G^m_x and therefore

$$d\lambda = \lambda'(x)\, dx = \sum_{i=1}^{m} \frac{\partial\lambda}{\partial\xi^i} d\xi^i ,$$

and if $y(x)$ is a differentiable mapping of G^m_x into R^n_y, then one has

$$d(\lambda y) = \lambda\, dy + d\lambda\, y ,$$

and provided $\lambda \neq 0$

$$d\left(\frac{y}{\lambda}\right) = \frac{\lambda\, dy - d\lambda\, y}{\lambda^2} .$$

Finally, the chain rule holds for differentials of composite differentiable mappings. Let R^p_z be a third linear space, which we endow with an arbitrary Minkowski auxiliary metric. Assuming that the vector function $y = y(x)$, which is differentiable at the point x of the region G^m_x, maps this region onto a region G^n_y in R^n_y, where a mapping $z = z(y)$ into R^p_z which is differentiable at the point $y = y(x)$ of G^n_y is given, then

$$z = z(y(x)) = \bar{z}(x)$$

is a mapping of G^m_x into R^p_z. The chain rule states that this mapping is differentiable at the point x with the differential

$$dz = z'(y)\, dy = z'(y)\, y'(x)\, dx .$$

We wish to prove this rule as an example.

Therefore, let $\Delta x = dx$ and $y(x + \Delta x) = y + \Delta y$. Since $y(x)$ is differentiable at the point x, one has

$$\Delta y = y(x + dx) - y(x) = y'(x)\, dx + |dx|\, (dx; x) = dy + |dx|\, (dx; x) ,$$

and, since $z(y)$ is differentiable at the point $y = y(x)$,

$$\Delta z = z(y + \Delta y) - z(y) = z'(y)\, \Delta y + |\Delta y|\, (\Delta y; y) .$$

Hence,

$$\begin{aligned} \Delta z &= z'(y)\, (dy + |dx|\, (dx; x)) + |\Delta y|\, (\Delta y; y(x)) \\ &= z'(y)\, dy + |dx|\, z'(y)\, (dx; x) + |\Delta y|\, (\Delta y; y(x)) . \end{aligned}$$

If $|z'(y)|$ stands for the norm of the linear operator $z'(y)$, then one has

$$|dx|\, |z'(y)\, (dx; x)| \leqq |z'(y)|\, |dx|\, |(dx; x)| ,$$

and hence $|dx|\, z'(y)\, (dx; x)$ is a vector (in R_z^p) that can be denoted by $|dx|\, (dx; x)$. Further,

$$|\Delta y| \leqq |dy| + |dx|\, |(dx; x)| \leqq |dx|\, (|y'(x)| + |(dx; x)|)\,,$$

where $|y'(x)|$ stands for the norm of $y'(x)$. Consequently, $|\Delta y|\, \big(\Delta y; y(x)\big)$ is also of the form $|dx|\, (dx; x)$, and altogether we have

$$\Delta z = z'(y)\, dy + |dx|\, (dx; x) = z'(y)\, y'(x)\, dx + |dx|\, (dx; x)\,,$$

which proves the chain rule.

1.6. The mean value theorem. We wish to derive an analogue of the elementary mean value theorem for the vector function $y(x)$ given in G_x^m.

Let x_1 and x_2 be two points in the region G_x^m such that the connecting line segment

$$x = x(\tau) = x_1 + \tau\, (x_2 - x_1) \qquad (0 \leqq \tau \leqq 1)$$

lies in G_x^m. We make the following assumptions:

1. $y(x)$ is continuous on the closed segment $0 \leqq \tau \leqq 1$.
2. $y(x)$ is differentiable at each interior point $0 < \tau < 1$.

Now let $L\, y$ be an arbitrary real linear function in the space R_y^n; thus, L is an element of the space dual to R_y^n. If using this function the composite function

$$f(\tau) = L\, y\big(x(\tau)\big) = L\, y\, \big(x_1 + \tau\, (x_2 - x_1)\big)$$

is formed, we obtain a continuous real function on the interval $0 \leqq \tau \leqq 1$ which by the chain rule is differentiable and has the derivative

$$f'(\tau) = L\, y'\big(x(\tau)\big)\, (x_2 - x_1)$$

at each interior point of this interval; for $dL = L\, dy$, $dy = y'(x)\, dx$ and $dx = (x_2 - x_1)\, d\tau$.

The real function $f(\tau)$ therefore satisfies the hypotheses of the elementary mean value theorem on the interval $0 \leqq \tau \leqq 1$, according to which

$$f(1) - f(0) = f'(\vartheta)\,,$$

with $0 < \vartheta < 1$. Since, on the other hand, $f(1) - f(0) = L\, \big(y(x_2) - y(x_1)\big)$,

$$L\, \big(y(x_2) - y(x_1)\big) = L\, y'\big(x(\vartheta)\big)\, (x_2 - x_1)\,,$$

where $x(\vartheta) = x_1 + \vartheta\, (x_2 - x_1)$ and ϑ denotes a value (dependent on the choice of L) in the interval $0 < \vartheta < 1$. That is the general formulation of the mean value theorem for a vector function satisfying conditions 1 and 2.

If one introduces a euclidean metric with inner product (y_1, y_2) in R_y^n, it becomes convenient to in particular take

$$L\,y = (y, e) = (e, y)$$

with an arbitrary vector e from R_y^n. If one then sets

$$y(x_1) = y_1\,, \qquad y(x_2) = y_2\,, \qquad x_2 - x_1 = \Delta x\,, \qquad y_2 - y_1 = \Delta y\,,$$

the above mean value theorem yields

$$(\Delta y, e) = (y'(x_e)\,\Delta x, e)\,,$$

with $x_e = x_1 + \vartheta_e\,\Delta x$ and $0 < \vartheta_e < 1$.

In this form the mean value theorem plays the same role in vector analysis as the ordinary mean value theorem in elementary differential calculus. We therefore wish to give it a complete formulation:

Mean value theorem. *Let R_x^m and R_y^n be linear spaces and G_x^m a region of the space R_x^m where a mapping*

$$y = y(x)$$

of G_x^m into R_y^n is given. Suppose further that this vector function satisfies the following conditions on the segment $x = x_1 + \tau\,(x_2 - x_1)$ $(0 \leqq \tau \leqq 1)$, which is contained in G_x^m:

1. *$y(x)$ is continuous on the closed segment $x_1\,x_2$.*
2. *$y(x)$ is differentiable at each interior point of the segment.*

If in R_y^n a euclidean metric is given, there then exists for each vector e in R_y^n at least one mean value $x_e = x_1 + \vartheta_e\,(x_2 - x_1)$ $(0 < \vartheta_e < 1)$, such that

$$(\Delta y, e) = (y'(x_e)\,\Delta x, e)\,,$$

where $\Delta x = x_2 - x_1$, $\Delta y = y(x_2) - y(x_1)$.

When this equation is written

$$(\Delta y - y'(x_e)\,\Delta x, e) = 0\,,$$

it states, interpreted geometrically, that for each vector e in R_y^n there is an interior point x_e of the segment $x_1\,x_2$ such that the difference $\Delta y - y'(x_e)\,\Delta x$ is perpendicular to e.

In particular, if $n = 1$ and hence $y(x) = \eta(x)\,e = \eta(\xi^1, \ldots, \xi^m)e$, one obtains the traditional differential calculus mean value theorem

$$\Delta\eta = \eta'(x)\,\Delta x = \sum_{i=1}^{m} \frac{\partial}{\partial\xi^i}\,\eta\,\left(x_1 + \vartheta\,(x_2 - x_1)\right)\,\Delta\xi^i \qquad (0 < \vartheta < 1)$$

and for $m = 1$ the elementary mean value theorem, on which, of course, the above generalization is based.

1.7. Fundamental theorem of integral calculus. As the first application of the mean value theorem we are going to prove the so-called fundamental theorem of integral calculus.

The proof results from the following simple inequality. If in the mean value theorem e is specially fixed so that $|e| = 1$ and $\Delta y = |\Delta y|\, e$, the result is

$$|\Delta y| = (y'(x_e)\, \Delta x, e)\ ,$$

and as a consequence of Schwarz's inequality

$$|\Delta y| \leqq |y'(x_e)\, \Delta x|\ ,$$

an in itself important estimate which we shall also use later.

Now if the derivative operator $y'(x) = 0$ in the region G_x^m, it follows from here that $|\Delta y| = |y(x) - y(x_0)| = 0$, and therefore $y(x) = y(x_0)$, at first in the largest sphere $|x - x_0| \leqq \varrho$ contained in G_x^m, from which one then deduces in a well-known fashion that $y(x)$ is constant in the entire region G_x^m.

Since, conversely, a constant vector function obviously has the zero operator as its derivative, this completes the proof of the *fundamental theorem of integral calculus* for vector functions:

In order for a function which is defined and differentiable in G_x^m to be constant it is necessary and sufficient that its derivative be the zero operator.

1.8. We wish to prove this theorem again by means of a classical method whose central idea was employed by Goursat in his proof of Cauchy's integral theorem in complex function theory. As we shall later see, this principle can be vastly generalized (cf. III. 2.7—2.9 and IV. 3.10).

Let G_x^m, as above, contain the entire segment

$$x = x_1 + \tau\,(x_2 - x_1) \qquad (0 \leqq \tau \leqq 1)$$

of length $|x_2 - x_1| = l$. If this segment is bisected and x_1^1, x_2^1 stand for the end points of that half for which the difference $|y(x_2) - y(x_1)| = \Delta_1$ is the greatest, it follows from the triangle inequality that

$$\Delta_0 = |y(x_2) - y(x_1)| \leqq 2\,\Delta_1\ .$$

By treating the segment

$$x = x_1^1 + \tau\,(x_2^1 - x_1^1) \qquad (0 \leqq \tau \leqq 1)$$

in the same fashion, one obtains

$$\Delta_1 \leqq 2\,\Delta_2 = 2\,|y(x_2^2) - y(x_1^2)|\ ,$$

and after n steps one has

$$\Delta_0 \leqq 2^n\,\Delta_n = 2^n\,|y(x_2^n) - y(x_1^n)|\ .$$

The segments $x_1\, x_2, x_1^1\, x_2^1, \ldots, x_1^n\, x_2^n, \ldots$ are nested, and the length of the segment $x_1^n\, x_2^n$ is equal to

$$|x_2^n - x_1^n| = 2^{-n}\,|x_2 - x_1| = \frac{l}{2^n}\,.$$

Consequently, one uniquely determined point $\bar{x}$ which is common to all of the segments lies on the closed segment $x_1 x_2$.

At this point $y(x)$ has by hypothesis the zero operator for its derivative operator, and therefore

$$y(x) - y(\bar{x}) = |x - \bar{x}| \, (x - \bar{x}; \bar{x}) ,$$

where $|(x - \bar{x}; \bar{x})| \to 0$ for $|x - \bar{x}| \to 0$. Thus

$$\begin{aligned} \varDelta_n = |y(x_2^n) - y(x_1^n)| &\leqq |y(x_2^n) - y(\bar{x})| + |y(x_1^n) - \bar{x})| \\ &\leqq 2^{1-n} \, l\left(\frac{1}{n}\right) \end{aligned}$$

with $l\ (1/n) \to 0$ for $n \to \infty$, and hence $\varDelta_0 = 0$ and

$$y(x_2) = y(x_1) ,$$

which was to be proved.

1.9. Linear operator functions. The derivative $y'(x)$ of a vector function differentiable in G_x^m is a linear operator function defined in G_x^m.

Generally in what follows we consider a *linear operator* function $A(x)$ defined in a region G_x^m of the linear space R_x^m, which for each fixed x in G_x^m thus maps the space R_x^m linearly into a space R_y^n. Hence for a fixed x

$$y = A(x)\, k$$

is a linear mapping of the space R_k^m into R_y^n.

If, after the introduction of arbitrary Minkowski metrices, the norm

$$|A\ (x + h) - A(x)| = \sup_{|k|=1} |A\ (x + h)\, k - A(x)\, k| \to 0$$

for $|h| \to 0$, the operator $A(x)$ is said to be *continuous* at the point x. Because

$$|A\ (x + h)\, k - A(x)\, k| \leqq |A\ (x + h) - A(x)|\, |k| ,$$

it follows from here that for a fixed k the mapping of G_x^m into R_y^n defined in the former region by $y = A(x)\, k$ is continuous for each k.

The operator $A(x)$ is said to be *differentiable* at the point x if the vector function $A(x)\, k$ is differentiable for each fixed k, that is, if

$$A\ (x + h)\, k - A(x)\, k = B(x, k)\, h + |h|\, (h, k; x) ,$$

where $B(x, k)$ is a linear operator and $|(h, k; x)| \to 0$ for $|h| \to 0$.

To investigate the dependence of the operator $B(x, k)$ on the parameter k, we replace k, successively, by k_1, k_2 and an arbitrary linear combination $\lambda_1 k_1 + \lambda_2 k_2$. If the first two equations multiplied by λ_1 and λ_2, respectively, are added, subtraction of the third yields, because of the linearity of the left hand side in k,

$$\begin{aligned} &B(x, \lambda_1 k_1 + \lambda_2 k_2)\, h - \lambda_1 B(x, k_1)\, h - \lambda_2 B(x, k_2)\, h \\ &= - |h| \{(h, \lambda_1 k_1 + \lambda_2 k_2; x) - \lambda_1 (h, k_1; x) - \lambda_2 (h, k_2, x)\} , \end{aligned}$$

where

$$|\{\ \}| \leqq |(h, \lambda_1 k_1 + \lambda_2 k_2; x)| + |\lambda_1| \, |(h, k_1; x)| + |\lambda_2| \, |(h, k_2; x)| \to 0$$

as $|h| \to 0$. If h is replaced by λh, a factor λ comes out on both sides, and if one then lets λ, for arbitrarily fixed h, converge to zero, the result is

$$B(x, \lambda_1 k_1 + \lambda_2 k_2) h - \lambda_1 B(x, k_1) h - \lambda_2 B(x, k_2) h = 0 .$$

Thus $B(x, k) h$ is not only linear in h, but also in k. We can therefore write

$$B(x, k) h = A'(x) h k ,$$

where for $x \in G_x^m$ $A'(x)$ is a *bilinear* operator on the space R^m with range in R^n. But then

$$(h, k; x) = (h; x) k$$

is also linear in k, and altogether we have

$$A(x + h) k - A(x) k = A'(x) h k + |h| (h; x) k ,$$

where the norm $|(h; x)|$ of the linear operator $(h; x)$ converges to zero for $|h| \to 0$.

The bilinear operator $A'(x)$ is called the *derivative* of the operator $A(x)$ at the point x.

From the inequality

$$|A(x + h) k - A(x) k| \leqq (|A'(x)| + |(h; x)|) \, |h| \, |k|$$

it can be seen that

$$|A(x + h) - A(x)| \to 0$$

for $|h| \to 0$. The continuity of the operator $A(x)$ at the point x therefore follows from its differentiability.

The above argument can be carried out without modification for multilinear operators. Let $A(x)$ be a *p-linear operator function* defined in G_x^m; thus

$$y = A(x) h_1 \ldots h_p$$

is for each $x \in G_x^m$ a multilinear function of the R^m-vectors $h_1, \ldots, h_p$ with a range in R_y^n. The operator $A(x)$ is said to be *continuous* at the point x if the norm

$$|A(x + h) - A(x)| = \sup_{|h_1| = \cdots = |h_p| = 1} |A(x + h) h_1 \ldots h_p - A(x) h_1 \ldots h_p|$$

converges to zero with h. It is *differentiable* at the point x if the vector function $A(x) h_1 \ldots h_p$ is differentiable for each fixed system $h_1, \ldots, h_p$. One then has

$$\begin{aligned} &A(x + h) h_1 \ldots h_p - A(x) h_1 \ldots h_p \\ &= A'(x) h h_1 \ldots h_p + |h| (h; x) h_1 \ldots h_p , \end{aligned}$$

where $A'(x)$, the *derivative* of the operator $A(x)$, is a $(p+1)$-linear operator and the norm $|(h;x)|$ of the p-fold multilinear operator $(h;x)$ converges to zero with $|h|$.

1.10. The second derivative. Now let $y(x)$ be a vector function which is differentiable in the region G_x^m; $y'(x)$ is therefore a linear operator defined in this space. If this operator is differentiable at the point x in the sense of the preceding paragraph, then the derivative is a bilinear operator which we shall denote by $y''(x)$. For a sufficiently small h and an arbitrary $k \in R_x^m$ one has

$$y'(x+h)\,k - y'(x)\,k = y''(x)\,h\,k + |h|\,(h;x)\,k\,,$$

where the norm $|(h;x)| \to 0$ if $|h| \to 0$. The bilinear operator $y''(x)$ is the *second derivative* of the given vector function at the point x.

If coordinate systems are introduced in R_x^m and R_y^n in which

$$x = \sum_{i=1}^{m} \xi^i a_i\,, \qquad y = \sum_{j=1}^{n} \eta^j b_j\,,$$

it follows from the existence of the bilinear operator $y''(x)$, first of all, that the n matrices

$$\left(\frac{\partial^2 \eta^j}{\partial \xi^s\,\partial \xi^t}\right) \quad (j = 1, \ldots, n)$$

exist. If

$$d\eta^j = \sum_{t=1}^{m} \frac{\partial \eta^j}{\partial \xi^t}\, d\xi_k^t$$

is the differential of the component η^j corresponding to the vector differential

$$k = \sum_{t=1}^{m} d\xi_k^t\, a_t\,,$$

one further has

$$\Delta\, d\eta^j = \sum_{s,t=1}^{m} \frac{\partial^2 \eta^j}{\partial \xi^s\,\partial \xi^t}\, d\xi_h^s\, d\xi_k^t + |h|\,((h;x)\,k)^j\,,$$

for the increase in $d\eta^j$ that corresponds to the vector differential

$$h = \sum_{s=1}^{m} d\xi_h^s\, a_s\,.$$

Because of

$$|(h;x)\,k)^j| \leqq |(h;x)\,k| \leqq |(h;x)|\,|k|\,,$$

this increase can also be written

$$\Delta\, d\eta^j = \sum_{s,t=1}^{m} \frac{\partial^2 \eta^j}{\partial \xi^s\,\partial \xi^t}\, d\xi_h^s\, d\xi_k^t + (h;x)^j\,|h|\,|k|\,,$$

with $|(h;x)^j| \to 0$ for $|h| \to 0$.

Conversely, it obviously follows from n such component equations that $y(x)$ is twice differentiable with the second differential

$$d^2y = y''(x)\,h\,k = \sum_{j=1}^{n} \left(\sum_{s,t=1}^{m} \frac{\partial^2 \eta^j}{\partial \xi^s\,\partial \xi^t}\,d\xi^s\,d\xi^t_k \right) b_j\,.$$

1.11. Symmetry of the second derivative. The *fundamental symmetry*

$$y''(x)\,h\,k = y''(x)\,k\,h$$

of the second derivative operator of a vector function follows from the above coordinate representation of the second differential under the well-known sufficient conditions for commutativity of partial differentiation.

However, we wish to go in the opposite direction and give a direct, coordinate-free proof of this symmetry.

We start out from the following, abundantly sufficient assumption which, as we shall later see, can be restricted in various directions:

The second derivative operator y'' exists in a neighborhood of the point x and is continuous at this point, so that the norm

$$|y''(x + \Delta x) - y''(x)| \to 0$$

for $|\Delta x| \to 0$.

For the proof, take h and k so small that the parallelogram with vertices x, $x + h$, $x + k$, $x + h + k$ lies in the mentioned neighborhood of the point x, and form the second difference

$$\Delta^2 y = y(x + h + k) - y(x + h) - y(x + k) + y(x)\,,$$

which is symmetric in h and k. If for brevity we denote the first differences by

$$y(x + h) - y(x) = \varphi(x)\,, \qquad y(x + k) - y(x) = \psi(x)\,,$$

then

$$\Delta^2 y = \varphi(x + k) - \varphi(x) = \psi(x + h) - \psi(x)\,.$$

Now let L be an arbitrary operator from the space dual to R^n_y, $L\,y$ thus being a real linear function of y. From the existence of the derivatives y' and y'' in the neighborhood of the point x it follows, in view of the mean value theorem in 1.6, that, on the one hand,

$$\begin{aligned} L\,\Delta^2 y &= L\big(\varphi(x + k) - \varphi(x)\big) = L\,\varphi'(x + \vartheta_2 k)\,k \\ &= L\big(y'(x + h + \vartheta_2 k) - y'(x + \vartheta_2 k)\big)\,k \\ &= L\,y''(x + \vartheta_1 h + \vartheta_2 k)\,h\,k\,, \end{aligned}$$

where $0 < \vartheta_1, \vartheta_2 < 1$. On the other hand, one obtains in the same way

$$\begin{aligned} L\,\Delta^2 y &= L\big(\psi(x + h) - \psi(x)\big) = L\,\psi'(x + \vartheta_3 h)\,h \\ &= L\big(y'(x + \vartheta_3 h + k) - y'(x + \vartheta_3 h)\big)\,h \\ &= L\,y''(x + \vartheta_3 h + \vartheta_4 k)\,k\,h\,, \end{aligned}$$

with $0 < \vartheta_3, \vartheta_4 < 1$, and consequently

$$L\left(y''(x + \vartheta_1 h + \vartheta_2 k) h k - y''(x + \vartheta_3 h + \vartheta_4 k) k h\right) = 0 .$$

If the original vectors h and k are replaced here by λh and λk, with arbitrarily fixed vectors h and k and sufficiently small real λ, then λ^2 comes out on the left. Because of the continuity of the second derivative at the point x, for $\lambda \to 0$ and for each pair h, k from R^m,

$$L\left(y''(x) h k - y''(x) k h\right) = 0 .$$

Since this holds for every operator L in the space dual to R^n_y, it follows that (cf. I. 3.10, exercise 4)

$$y''(x) h k = y''(x) k h ,$$

with which the asserted symmetry of the bilinear operator $y''(x)$ is proved.

1.12. Higher derivatives. Provided the bilinear, symmetric operator $y''(x)$ is again differentiable in the sense of 1.9, one defines

$$(y''(x))' = y'''(x)$$

to be the third derivative of the vector function $y(x)$. In the same way one has generally the definition

$$\frac{d^{p+1}y}{dx^{p+1}} = y^{(p+1)}(x) = \left(y^{(p)}(x)\right)' .$$

Thus if the derivative operators exist up to the order $p + 1$ this means that for $q = 0, 1, \ldots, p$

$$\begin{aligned} & y^{(q)}(x + h) h_1 \ldots h_q - y^{(q)}(x) h_1 \ldots h_q \\ &= y^{(q+1)}(x) h h_1 \ldots h_q + |h| (h; x) h_1 \ldots h_q , \end{aligned}$$

where the norm $|(h; x)|$ converges to zero with $|h|$ and $y^{(0)}(x) = y(x)$.

From this definition it follows more generally that for nonnegative integral indices i and j

$$\frac{d^i}{dx^i}\left(\frac{d^j y}{dx^j}\right) = \frac{d^j}{dx^j}\left(\frac{d^i y}{dx^i}\right) = \frac{d^{i+j} y}{dx^{i+j}} ,$$

provided the derivative operators which appear exist. It further results from here by means of induction, observing the symmetry of the second derivative proved in 1.11, that under certain sufficient conditions the pth differential

$$d^p y = y^{(p)}(x) h_1 \ldots h_p = y^{(p)}(x) d_1 x \ldots d_p x$$

is a multilinear and *symmetric* function of the arbitrary R^m-vectors $h_i = d_i x$.

1.13. Exercises. 1. Let R_x be a euclidean space and

$$x = |x|\, e(x)\ .$$

Show that

$$d|x| = (e, dx)\ , \qquad |x|\, de = dx - (e, dx)\, e$$

and

$$|x|^2\, e''(x)\, h\, k = 3\, (e, h)\, (e, k)\, e - (e, h)\, k - (e, k)\, h - (h, k)\, e\ .$$

2. The mapping $y = y(x)$ of the euclidean space R_x^m onto itself is *conformal* at the point x provided the derivative $y'(x) = \lambda(x)\, T(x)$, where $\lambda(x)$ is real and $T(x)$ is an orthogonal transformation of the space R_x^m. Prove that the mapping by means of reciprocal radii

$$y = \frac{x}{|x|^2}$$

is conformal.

3. Let $y = y(x)$, $x = x(y)$ be inverse, twice differentiable vector functions. Then for two pairs of associated differentials d_1x, d_1y and d_2x, d_2y

$$\frac{d^2x}{dy^2}\, d_2y\, d_1y + \frac{dx}{dy}\frac{d^2y}{dx^2}\, d_2x\, d_1x = 0\ .$$

4. If the one-to-one twice differentiable variable transformations $x \leftrightarrow y \leftrightarrow z$ are carried out for the differentiable function $u = u(x)$, then

$$\frac{du}{dx}\frac{d^2x}{dz^2}\, d_2z\, d_1z + \frac{du}{dy}\frac{d^2y}{dx^2}\, d_2x\, d_1x + \frac{du}{dz}\frac{d^2z}{dy^2}\, d_2y\, d_1y = 0\ .$$

5. Let $\mu > 0$ and $\varphi(\varrho)$ be a monotonically increasing continuous function defined for $\varrho > 0$ with $\lim\limits_{\varrho \to 0} \varphi(\varrho) = 0$ such that the least upper bound ϱ_x of those values ϱ for which $\varphi(\varrho) < \mu$ is positive (finite or infinite).

Further suppose R_x^m is a euclidean space, $x = |x|\, e(x)$ and

$$y(x) = \int_0^{|x|} (\mu - \varphi(\varrho))\, d\varrho\, e(x) = \mu\, x - \int_0^{|x|} \varphi(\varrho)\, d\varrho\, e(x)\ .$$

Show:

a. The function $y(x)$ is differentiable in the sphere $|x| < \varrho_x$.

b. The norm $|y'(0)| = \mu$.

c. The least upper bound of the norm $|y'(x) - y'(0)|$ for $|x| \leqq \varrho < \varrho_x$ is precisely $= \varphi(\varrho)$.

d. If one sets

$$\mu\, \varrho_x - \int_0^{\varrho_x - 0} \varphi(\varrho)\, d\varrho = \int_0^{\varrho_x - 0} (\mu - \varphi(\varrho))\, d\varrho = \varrho_y\ ,$$

then $y = y(x)$ defines a one-to-one mapping of the spheres $|x| < \varrho_x$, $|y| < \varrho_y$, and ϱ_x and ϱ_y are the largest radii for which this is the case.

6. Let $y = y(x)$ be a differentiable function in the region G_x^m of the space R_x^m with a range in R_y^n; consequently, for each x in G_x^m

$$y(x+h) - y(x) = y'(x)\, h + |h|\, (h; x)$$

with $|(h; x)| \to 0$ for $|h| \to 0$. Prove:

In order that the derivative $y'(x)$ be continuous in G_x^m it is necessary and sufficient that the equation

$$\lim_{|h| \to 0} |(h; x)| = 0$$

hold *uniformly* on each compact subregion of G_x^m.

Proof. Since the ratio of two Minkowski lengths $|x|'$ and $|x|''$, according to what was said in I. 6.2, lies between two positive bounds which are independent of x, the hypotheses and conclusions of the theorem are independent of the choice of the metrics. We can therefore give the spaces R_x^m and R_y^n euclidean metrics.

If one then sets $(h; x) = |(h; x)|\, e$, the mean value theorem, in view of Schwarz's inequality, yields

$$|(h; x)| \leqq |y'(x + \vartheta h) - y'(x)| \qquad (0 < \vartheta < 1)\,.$$

Now if $y'(x)$ is continuous in G_x^m, it is even uniformly continuous on each compact subregion of G_x^m, from which the necessity of the condition in the theorem follows.

The condition is also sufficient. For if x is an arbitrary point in G_x^m, take $\varrho_0 > 0$ at first so small that $x + k$ lies in G_x^m for $|k| \leqq \varrho_0$. If one further sets $h = |h|\, e = \lambda\, e$ with an arbitrary unit vector e,

$$\begin{aligned} &|y'(x+k)\, e - y'(x)\, e| \\ &\leqq \frac{1}{\lambda} (|y(x + \lambda e + k) - y(x + \lambda e)| + |y(x+k) - y(x)| \\ &\qquad + |(\lambda e; x + k)| + |(\lambda e; x)|\,. \end{aligned}$$

In consequence of the uniform convergence $|(h; x)| \to 0$ for $|h| \to 0$ one can, for preassigned $\eta > 0$, at first fix $\lambda > 0$ so small that the last terms are smaller than $\eta/3$ for $|k| \leqq \varrho_0$, and then $\varrho_\eta \leqq \varrho_0$ so that the first term on the right is also smaller than $\eta/3$ for $|k| < \varrho_\eta$. Since this holds for every unit vector e in the space R_x^m, the norm

$$|y'(x+k) - y'(x)| < \eta$$

for $|k| < \varrho_\eta$, which was to be proved.

In general, the following holds: Provided the p-linear operator $A(x)$ is differentiable in G_x^m, and consequently

$$\begin{aligned} &A(x+h)\, h_1 \ldots h_p - A(x)\, h_1 \ldots h_p \\ &= A'(x)\, h\, h_1 \ldots h_p + |h|\, (h; x)\, h_1 \ldots h_p\,, \end{aligned}$$

$A'(x)$ is continuous in G_x^m precisely when

$$\lim_{|h|\to 0} |(h;x)| = 0\,,$$

uniformly on every compact subregion of G_x^m.

7. Let $y_p(x)$ $(p = 1, 2, \ldots)$ be differentiable functions in the region G_x^m of the space R_x^m with ranges in R_y^n. Prove:

Provided the limit function and the limit operator

$$y_p(x) \to y(x)\,, \qquad y_p'(x) \to A(x)$$

exist in G_x^m for $p \to \infty$ and the second convergence is uniform on each compact subregion of G_x^m, $y(x)$ is differentiable and

$$y'(x) = A(x)$$

at every point where $A(x)$ is continuous.

Hint. If for each p one sets

$$y_p(x+h) - y_p(x) = y_p'(x)\,h + |h|\,(h;x)_p$$

with $|(h;x)_p| \to 0$ for $|h| \to 0$, the convergence $(h;x)_p \to (h;x)$ for $p \to \infty$ follows, and therefore

$$y(x+h) - y(x) = A(x)\,h + |h|\,(h;x)\,.$$

It remains to be shown that $|(h;x)| \to 0$ for $|h| \to 0$.

From the mean value theorem it follows that for every p

$$\begin{aligned}|(h;x)_p| &\leqq |y_p'(x + \vartheta_p h) - y_p'(x)| \\ &\leqq |y_p'(x+\vartheta_p h) - A(x+\vartheta_p h)| + |A(x+\vartheta_p h) - A(x)| \\ &\qquad + |A(x) - y_p'(x)|\,,\end{aligned}$$

with $0 < \vartheta_p < 1$, which because of the uniformity of the convergence $y_p' \to A$ and the asserted continuity of A at the point x implies the conclusion.

If, granting the uniform convergence $y_p' \to A$, the derivatives y_p' are continuous in the region G_x^m, $A(x)$ is also continuous, and the equation $y'(x) = A(x)$ holds for each x in G_x^m.

8. Let A be a bilinear operator that maps the product space $R_x^m \times R_y^n$ into R_y^n. Assuming further that A is symmetric,

$$A\,h\,A\,k\,y = A\,k\,A\,h\,y\,,$$

set

$$\underbrace{A\,x\ldots A\,x}_{p}\,y = (A\,x)^p\,y$$

and prove:

For each y_0 from R_y^n the vector function

$$y(x) = \left(\sum_{p=0}^{\infty} \frac{(A\,x)^p}{p!}\right) y_0 = \sum_{p=0}^{\infty} \frac{(A\,x)^p\, y_0}{p!}$$

is meaningful in the entire space R_x^m and satisfies the linear differential equation

$$dy = A\,dx\,y\,.$$

9. If $y = y(x)$ is a differentiable, homogeneous vector function of degree p of the vector x, i.e., $y(\lambda\,x) = \lambda^p\,y(x)$, then Euler's equation $y'(x)\,x = p\,y(x)$ holds.

§ 2. Taylor's Formula

2.1. Powers and polynomials. In preparation for what follows we consider vector powers and polynomials.

Let R_x^m be a linear space and A a constant, multilinear, *symmetric* operator from the former space into the linear space R_y^n, that is, let

$$y = A\,x_1 \ldots x_p$$

be a p-linear symmetric function of the R_x^m-vectors $x_1, \ldots, x_p$. One such vector function of the differentials $h_i = x_i$, for example, is the pth differential of a sufficiently differentiable vector function $y(x)$ for a fixed x.

If one sets $x_1 = \cdots = x_p = x$, this function transforms into a homogeneous vector function of degree p, which we briefly denote by

$$y = A\,x^p$$

and call a pth "power" of x. For $p = 1$ this is a linear, for $p = 2$ a quadratic vector function.

With respect to the coordinate systems a_i and b_j, in which

$$x = \sum_{i=1}^{m} \xi^i\,a_i\,, \qquad y = \sum_{j=1}^{n} \eta^j\,b_j\,,$$

$$\eta^j = \eta^j(\xi^1, \ldots, \xi^m) = \sum_{i_1, \ldots, i_p=1}^{m} \alpha^j_{i_1 \ldots i_p}\,\xi^{i_1} \ldots \xi^{i_p}\,,$$

whereby

$$\alpha^j_{i_1 \ldots i_p} = A^j\,a_{i_1} \ldots a_{i_p}$$

is symmetric in the indices $i_1, \ldots, i_p$.

We are going to calculate the differential of the power $A\,x^p$.

As a consequence of the linearity and *symmetry* of the function $A\,x_1 \ldots x_p$,

$$A\,(x + h)^p = \sum_{i=0}^{p} \binom{p}{i}\,A\,x^{p-i}\,h^i\,,$$

and consequently

$$A\,(x+h)^p - A\,x^p = p\,A\,x^{p-1}\,h + |h|\,(h;x)\,,$$

where the norm

$$|(h;x)| \leqq |A|\,|h| \sum_{i=2}^{p} \binom{p}{i} |x|^{p-i}\,|h|^{i-2}$$

converges to zero as $|h| \to 0$; $|A|$ is the norm of the operator A. Accordingly,

$$dA\,x^p = p\,A\,x^{p-1}\,h = p\,A\,x^{p-1}\,d_1x\,,$$

where the differential h has been denoted by d_1x.

One finds in the same way, formally exactly as for the elementary power $\alpha\,x^p$ of a real variable,

$$\begin{aligned} &p\,A\,(x+h)^{p-1}\,d_1x - p\,A\,x^{p-1}\,d_1x \\ &= p\,(p-1)\,A\,x^{p-2}\,h\,d_1x + |h|\,(h;x)\,d_1x\,, \end{aligned}$$

where the norm $|(h;x)|$ vanishes with $|h|$. Consequently, the second differential corresponding to the differentials d_1x and $h = d_2x$ is

$$d^2A\,x^p = p\,(p-1)\,A\,x^{p-2}\,d_2x\,d_1x = p\,(p-1)\,A\,x^{p-2}\,d_1x\,d_2x\,.$$

In general one obtains for each positive $i \leqq p$

$$d^iA\,x^p = p\,(p-1)\ldots(p-i+1)\,A\,x^{p-i}\,d_1x\,d_2x\ldots d_ix\,.$$

For $i = p$

$$d^p\,A\,x^p = p!\,A\,d_1x\ldots d_px$$

is a multilinear and symmetric vector function of the differentials d_ix, independent of x, and thus for $i > p$

$$d^iA\,x_p = 0\,.$$

A sum of powers

$$P(x) = \sum_{q=0}^{p} A_{p-q}\,x^q$$

we call a *vector polynomial* of degree p in the vector x, which varies in R_x^m. One has for $i \leqq p$

$$d^iP(x) = \sum_{q=1}^{p} q\,(q-1)\ldots(q-i+1)\,A_{p-q}\,x^{q-i}\,d_1x\,d_2x\ldots d_ix\,,$$

and for $i > p$

$$d^i\,P(x) = 0\,.$$

2.2. The Taylor polynomial. In analogy with the elementary differential calculus, we can now form the *Taylor polynomial* of degree p, $T_p(x, x_0)$, for a p-times differentiable vector function $y(x)$ about the expansion center x_0. Take $x_0 = 0$; we are then treating the *Maclaurin polynomial*

$$T_p(x, 0) = \sum_{q=0}^{p} A_{p-q}\, x^p ,$$

which for $x = 0$ has, up to the order p, the same derivative operators as the given function $y(x)$.

According to 2.1, for $i \leqq p$ and $x = 0$

$$d^i T_p(0, 0) = i!\, A_{p-i}\, d_1 x \ldots d_i x ,$$

and therefore

$$i!\, A_{p-i} = T_p^{(i)}(0, 0) = y^{(i)}(0) ,$$

from which

$$T_p(x, 0) = \sum_{q=0}^{p} \frac{1}{q!}\, y^{(q)}(0)\, x^q$$

follows.

With this we have the Maclaurin formula

$$y(x) = T_p(x, 0) + R_{p+1}(x, 0) .$$

It still remains, as in the elementary differential calculus, to investigate the remainder term R_{p+1} for $p \to \infty$.

2.3. The asymptotic behavior of R_{p+1}. If, as above, one assumes that $y(x)$ is p-times differentiable at the origin, this is also true for $R_{p+1}(x, 0) \equiv R_{p+1}(x)$, and one has

$$R_{p+1}(0) = R'_{p+1}(0) = \cdots = R_{p+1}^{(p)}(0) = 0 .$$

Set (in an arbitrary Minkowski metric) $x = |x|\, e$. With x fixed and an arbitrary operator L from the space R_y^{*n} dual to R_y^n form the real function

$$f(\tau) = L\, R_{p+1}(\tau\, e) \qquad (0 \leqq \tau \leqq |x|) .$$

Then

$$f^{(q)}(0) = L\, R_{p+1}^{(q)}(0)\, e^q = 0$$

for $q = 0, 1, \ldots, p$, and for a sufficiently small $|x|$ and $q = 0, 1, \ldots, p - 1$ the derivatives

$$f^{(q)}(\tau) = L\, R_{p+1}^{(q)}\, (\tau\, e)\, e^q$$

also exist on the interval $0 \leqq \tau \leqq |x|$. They are even continuous if we in addition assume that $y^{(p-1)}(x)$ is continuous in a certain neighborhood of the origin. Then, because $f^{(p)}(0) = 0$, for $0 \leqq \tau \leqq |x|$

$$f^{(p-1)}(\tau) = (\tau)_1\, \tau ,$$

with $|(\tau)_1| \to 0$ for $\tau \to 0$. For $\tau = |x|$ we have, according to this,

$$L\, R_{p+1}(x) = f(|x|) = (x)\, |x|^p ,$$

where $|(x)| \to 0$ for $|x| \to 0$.

This holds for each operator L from the space R_y^{*n} to R_y^n, thus in particular also for the inner product

$$L\,y = (a, y)$$

of a euclidean metric introduced in R_y^n, where $R_{p+1}(x) = |R_{p+1}(x)|a$. Then

$$L\,R_{p+1}\,(x) = |R_{p+1}(x)| = (x)|x|^p\,,$$

with $|(x)| \to 0$ for $|x| \to 0$, for arbitrary Minkowski metrics in the spaces R_x^m and R_y^n.

Under the assumptions made:

1. *$y^{(p-1)}(x)$ exists and is continuous in a certain neighborhood of* $x = 0$;

2. *$y^{(p)}(0)$ exists;*

We see: the Maclaurin polynomial gives an asymptotic representation of the vector function $y(x)$ so that in arbitrary Minkowski metrics

$$|x|^{-p}\,\left|y(x) - \sum_{q=0}^{p} \frac{1}{q!}\,y^{(q)}(0)\,x^q\right| \to 0$$

for $|x| \to 0$.

2.4. The remainder formulas in Taylor's theorem. If one assumes that the function $y(x)$ is p-times *continuously* differentiable in the neighborhood of the origin and that moreover the $(p+1)$st derivative operator $y^{(p+1)}(x)$ exists there, then the real auxiliary function

$$f(\tau) = L\,R_{p+1}(\tau\,e)$$

has corresponding properties on the interval $0 \leqq \tau \leqq |x|$ for a sufficiently small $|x|$: it is p-times continuously differentiable and the derivative

$$f^{(p+1)}(\tau) = L\,R_{p+1}^{(p+1)}\,(\tau\,e)\,e^{p+1} = L\,y^{(p+1)}\,(\tau\,e)\,e^{p+1}$$

exists. Since $f(0) = f'(0) = \cdots = f^{(p)}(0) = 0$, the elementary Maclaurin formula with the Lagrange remainder term

$$f(\tau) = \frac{1}{(p+1)!}\,f^{(p+1)}\,(\vartheta\,\tau)\,\tau^{p+1} \qquad (0 < \vartheta < 1)$$

thus yields for $\tau = |x|$

$$L\,R_{p+1}(x) = \frac{1}{(p+1)!}\,L\,y^{(p+1)}\,(\vartheta\,x)\,x^{p+1}$$

and that again for each L of the space R_y^{*n} dual to R_y^n, where by the number ϑ naturally depends on x as well as on L. Corresponding formulas are obtained if other of the known forms of the remainder (like those of Cauchy or of Schlömilch-Roche) are used for $f(\tau)$.

From here one obtains as above estimates for the remainder term $R_{p+1}(x)$. It follows that $R_{p+1}(x) \to 0$, for a fixed x and $p \to \infty$, and consequently the existence of the Maclaurin expansion

$$y(x) = \sum_{q=0}^{\infty} \frac{1}{q!} y^{(q)}(0)\, x^q$$

can be inferred.

It is worth emphasizing that the point set of the linear space R_x^m for which $R_{p+1}(x)$ possibly converges to zero as $p \to \infty$ is independent of the metric and therefore absolutely determined. This concerns the convergence of the Taylor series just as the other notions introduced and treated in this chapter, such as the continuity of a vector function, the existence of certain derivative operators, etc. They have an "affine character", independent of the auxiliary metrics introduced to facilitate formulation or proof, provided the linear spaces R_x and R_y are of *finite* dimension and therefore possess a natural topology.

If, on the other hand, one goes over to normed or Hilbert spaces of infinite dimension, the concepts of absolute analysis remain meaningful, to be sure, but for the most part only with respect to the metric or the topology which is introduced. Thus, for example, the same vector function can be differentiable in the sense of the definition given in 1.2 with respect to one metric, with respect to another not.[1]

2.5. Exercises. 1. Show with the aid of the fundamental theorem of integral calculus: If $y = y(x)$ is a vector function whose $(p+1)$st differential vanishes identically in a region of the linear space R_x^m, $d^{(p+1)}y(x) \equiv 0$, then $y(x)$ is a vector polynomial of degree p. The set of all such polynomials is the general solution of the above differential equation.

2. Let the sequence of powers $A_j x^j \in R_y^n$ $(j = 0, 1, \ldots)$ be so given that the power series $\sum_{j=0}^{\infty} |A_j|\, \varrho^j$ converges ($|A_j|$ = the norm of A_j relative to a Minkowski metric on the space R_y^n). Then the series $\sum_{j=0}^{\infty} A_j x^j$ is absolutely and uniformly convergent for $|x| \leqq r < \varrho$.

3. If the power series $y(x) = \sum_{j=0}^{\infty} A_j x^j$ converges uniformly for $|x| < r$, one obtains the derivative $y'(x)$ in the same sphere by termwise differentiation.

[1] Boundedness of the linear derivative operator $A(x) = y'(x)$ with respect to the metrics must then be required.

§ 3. Partial Differentiation

3.1. Partial derivatives and differentials. In the definitions of the derivative and the differential of a vector function $y(x)$ given in 1.2 and 1.3 it was assumed that the argument differential $dx = h$ varies freely within the space R_x^m. One can for this reason more accurately describe the differential defined in this way

$$dy = y'(x)\, dx$$

as the total differential of the function $y(x)$ at the point x.

If h is restricted in these definitions to a certain subspace U of the space R_x^m, we have the notion of the *partial derivative* or *partial differential* of the vector function in the direction of the subspace U. Hence the function $y(x)$ being differentiable at the point x in the direction of the subspace U means that a linear derivative operator

$$y_U(x) = \frac{\partial y(x)}{\partial u}$$

from the subspace U into the space R_y^n exists so that

$$y\,(x + h) - y(x) = y_U(x)\, h + |h|(h;\, x)\,,$$

where $|(h;\, x)| \to 0$ when h tends to zero *in the subspace* U. The operator $y_U(x)$ is then called the partial derivative and the vector

$$d_U y = y_U(x)\, h$$

the partial differential of the function $y(x)$ in the direction of the subspace U. One shows as in 1.2 that this partial derivative, provided it exists, is uniquely determined.

If the function is differentiable in the direction of the subspace U, then it is of course differentiable with the same derivative operator in the direction of each subspace of U. In particular, it is partially differentiable in *every* direction provided it is totally differentiable. Further, it follows from the uniqueness of the partial derivative that the function is differentiable in the direction of a nonempty intersection

$$W = [U,\, V]$$

provided the partial derivatives $y_U(x)$ and $y_V(x)$ in the directions of the subspaces U and V exist, whereby for $h \in W$

$$y_W(x)\, h = y_U(x)\, h = y_V(x)\, h\,.$$

However, this does not in general imply the existence of the partial derivative in the direction of the space $(U,\, V)$ generated by U and V.

3.2. Functions of several variables. The case where the vector function y is given as a function of several vector variables $x_1, \ldots, x_p$

each of which varies in certain regions $G_{x_i}^{m_i}$ of the linear spaces $R_{x_i}^{m_i}$ can be reduced to the case of one single variable x. For this purpose one forms the product space (cf. I.1.6, exercises 6 and 7)

$$R_x^m = R_{x_1}^{m_1} \times \cdots \times R_{x_p}^{m_p}$$

of dimension $m = m_1 + \cdots + m_p$, where the spaces $R_{x_i}^{m_i}$ are linearly independent subspaces which generate the product space R_x^m. The vector

$$x = x_1 + \cdots + x_p$$

is in the product space, and conversely each x of the space R_x^m can be represented in a unique way as such a sum. In the product domain

$$G_x^m = G_{x_1}^{m_1} \times \cdots \times G_{x_p}^{m_p}$$

the original function

$$y(x_1, \ldots, x_p) = y(x)$$

is therefore a single-valued function of x.

We say the original function $y(x_1, \ldots, x_p)$ is differentiable for $(x_1, \ldots, x_p)$ provided $y(x)$ is totally differentiable at the point $x = x_1 + + \cdots + x_p$ in the product space. For this it is necessary and sufficient that for arbitrary differentials h_i from $R_{x_i}^{m_i}$

$$y(x_1 + h_1, \ldots, x_p + h_p) - y(x_1, \ldots, x_p)$$
$$= \sum_{i=1}^{p} A_i(x_1, \ldots, x_p)\, h + |h|(h; x_1, \ldots, x_p)\,,$$

where the $A_i(x_1, \ldots, x_p)$ stand for linear mappings of the spaces $R_{x_i}^{m_i}$ into the range space R_y^n and $|(h; x_1, \ldots, x_p)| \to 0$ for

$$|h|^2 = \sum_{i=1}^{p} |h_i|^2 \to 0\,.$$

The operators

$$A_i(x_1, \ldots, x_p) = y_{x_i}(x_1, \ldots, x_p) = \frac{\partial y}{\partial x_i}$$

are the partial derivatives of the function $y(x_1, \ldots, x_p)$ with respect to x_i. Interpreted as a derivative of the function $y(x)$, A_i is the partial derivative in the direction of the space $R_{x_i}^{m_i}$ as a subspace of the product space R_x^m. We leave the simple proofs of these assertions to the reader.

3.3. Partial derivatives of higher order. If the vector function $y(x)$ is partially differentiable in the direction of a subspace V of R_x^m at each point x of a region G_x^m, the partial differential

$$d_V y = y_V(x)\, k$$

defines for each fixed k from V a vector function. If it is partially differentiable in the direction of the subspace U at the point x, one shows exactly as in 1.9 and 1.10 that a bilinear operator

$$y_{UV}(x) = \frac{\partial^2 y}{\partial u \, \partial v}$$

defined for each h from U and each k from V exists such that

$$y_V(x + h)\, k - y_V(x)\, k = y_{UV}(x)\, h\, k + |h| (h; x)\, k .$$

Here, calculated in the subspace V, the norm

$$|(h; x)|_V = \sup_{|k|=1} |(h; x)| \to 0$$

when $|h|$ tends to zero in U. The bilinear operator $y_{UV}(x)$ is the second partial derivative of the function $y(x)$ taken first in the direction V and then in the direction U.

Partial derivatives and differentials of higher order are defined similarly.

3.4. Theorem of H. A. Schwarz. Regarding the sequence of the differentiations there is the following generalization of the classical theorem of H. A. Schwarz:

Let $y(x)$ be a vector function with the following properties:

1. *The function $y(x)$ and its partial derivatives $y_U(x)$ and $y_V(x)$ in the direction of the subspaces U and V of the space R_x^m are continuous in a neighborhood of the point x.*

2. *The partial derivative $y_{UV}(x)$ exists in this neighborhood and is continuous at the point x.*

Then the partial derivative $y_{VU}(x)$ also exists at the point x and for each h in U and each k in V

$$y_{VU}(x)\, k\, h = y_{UV}(x)\, h\, k .$$

For $U = V = R_x^m$ this includes as a special case the symmetry of the second total differential $y''(x)\, h\, k$ proved in 1.11.

The following proof is an easy modification of the well-known proof of Schwarz.

As in 1.11 and with the notation already introduced there, it follows from the existence and continuity of the derivatives y_U and y_V in the vicinity of the point x together with the existence of the derivative y_{UV} in this neighborhood that

$$\begin{aligned} L\, \Delta^2 y &= L\big(y_U(x + \vartheta_3 h + k) - y_U(x + \vartheta_3 h)\big)\, h \\ &= L\big(y_V(x + h + \vartheta_2 k) - y_V(x + \vartheta_2 k)\big)\, k \\ &= L\big(y_{UV}(x + \vartheta_1 h + \vartheta_2 k)\big)\, h\, k , \end{aligned}$$

and thus[1]

$$L(y_U(x+\vartheta_3 h+k) - y_U(x+\vartheta_3 h))\, h = L\, y_{UV}(x+\vartheta_1 h+\vartheta_2 k)\, h\, k\,.$$

Because of the hypothesized continuity of the derivative y_{UV} at the point x,

$$y_{UV}(x+\vartheta_1 h+\vartheta_2 k)\, h\, k = y_{UV}(x)\, h\, k + |h|\, |k|\, \varepsilon(h,k)\,,$$

with $\varepsilon(h,k) \to 0$ for $|h|^2+|k|^2 \to 0$. The above equation can therefore be written

$$L(y_U(x+\vartheta h+k)\, h - y_U(x+\vartheta h)\, h - y_{UV}(x)\, h\, k) = |h|\, |k|\, L\, \varepsilon(h,k)\,,$$

where the number $\vartheta = \vartheta_3$ depends on the choice of the real linear form L from the space dual to R^n_y, but always satisfies the condition $0 < \vartheta < 1$.

If in particular one takes L to be the inner product

$$L\, y = (e, y)$$

of a euclidean metric on the range space R^n_y, whereby e denotes a temporarily arbitrary unit vector, then Schwarz's inequality yields

$$|(e, (y_U(x+\vartheta h+k)\, h - y_U(x+\vartheta h)\, h - y_{UV}(x\, h\, k))|$$
$$\leqq |h|\, |k|\, |\varepsilon(h,k)|\,,$$

where, thus, for an arbitrary small $\varepsilon > 0$ a $\varrho_\varepsilon > 0$ exists such that

$$|\varepsilon(h,k)| < \varepsilon \quad \text{for} \quad |h|^2+|k|^2 < \varrho_\varepsilon^2\,.$$

If h is replaced in this equation by λh, a factor of λ comes out on both sides, and as a result for $\lambda \to 0$, because of the continuity of the derivatives y_U and y_V,

$$|(e, \{y_U(x+k)\, h - y_U(x)\, h - y_{UV}(x)\, h\, k\})| < \varepsilon |h|\, |k|$$

for $|k| < \varrho_\varepsilon$.

If, finally, in this inequality, which holds for every unit vector $e \in R^n_y$, one takes e in the direction of the vector $\{\ \}$, then

$$|y_U(x+k)\, h - y_U(x)\, h - y_{UV}(x)\, h\, k| < \varepsilon |h|\, |k|$$

for $|k| < \varrho_\varepsilon$. Thus one can write

$$y_U(x+k)\, h - y_U(x)\, h = y_{UV}(x)\, h\, k + |k|(k; x)\, h\,,$$

where, calculated in U, the norm

$$|(k;x)|_U = \sup_{|h|=1} |(k;x)\, h| < \varepsilon$$

[1] In formulating the mean value theorem in 1.6, from which the above equation was deduced, the existence of the *total* derivative $y'(x)$ at the interior points of the segment $x = x_1 + \tau(x_2 - x_1)$ $(0 < \tau < 1)$ was, to be sure, assumed. However, it is at once clear from the proof of the mean value theorem that even the existence of the *partial* derivative in the direction of the one-dimensional subspace spanned by the vector $x_2 - x_1$ suffices.

for $|k| < \varrho_\varepsilon$. Accordingly, the partial derivative $y_{VU}(x)$ exists, and for each h in U and each k in V

$$y_{VU}(x)\, k\, h = y_{UV}(x)\, h\, k\,,$$

q.e.d.

§ 4. Implicit Functions

4.1. The problem. We consider in the following a function

$$z = z(x, y)$$

of two vector variables x and y which is defined in certain neighborhoods of the points x_0 and y_0 in the linear spaces R_x^m and R_y^n and which has a range in a third space R_z^p.

The problem then is to solve the equation

$$z(x, y) = 0$$

with respect to x or to y in the vicinity of x_0 and y_0 such that $z(x_0, y_0) = 0$. If coordinate systems are introduced in the three linear spaces in which

$$x = \sum_{i=1}^{m} \xi^i a_i\,, \qquad y = \sum_{j=1}^{n} \eta^j b_j\,, \qquad z = \sum_{k=1}^{p} \zeta^k c_k\,,$$

then the equations

$$\zeta^k(\xi^1, \ldots, \xi^m; \eta^1, \ldots, \eta^n) = 0 \qquad (k = 1, \ldots, p)$$

hold for $\xi^i = \xi_0^i$, $\eta = \eta_0^j$, and the problem in this formulation is to solve this system of real equations for the variables ξ^i or η^j in the neighborhood of the points ξ_0^i and η_0^j.

In the case at hand, where the dimensions m, n and p are finite, the results which we shall obtain will, as far as the content is concerned, contain nothing that is in principle new. It is only the coordinate-free approach used that deviates from the traditional and permits a brief treatment. But it is worth mentioning that, with modifications, the method used also achieves the same end in the more general Hilbert space case, which, however, we shall not discuss[1].

If, more particularly, the function $z(x, y)$ is of the form

$$z(x, y) \equiv y(x) - y\,,$$

where R_z^p is thus the same as R_y^n, the problem is to find the inverse of a mapping

$$y = y(x)$$

from the space R_x^m into the space R_y^n near the points x_0 and $y_0 = y(x_0)$. We shall treat this special case of the problem in detail. The investi-

[1] R. Nevanlinna [1], [3], F. Nevanlinna [1], [2].

gation of the general problem only requires a few easy modifications which are in part left to the reader as exercises.

4.2. Inversion of differentiable mappings. Let

$$y = y(x)$$

be a mapping from the m-dimentional linear space R_x^m into the n-dimensional linear space R_y^n that satisfies the following conditions:

A. In a neighborhood of the point x_0, $y(x)$ is continuous and differentiable.

B. The operator $y'(x)$ is regular for $x = x_0$.

C. The operator $y'(x)$ is continuous at the point x_0.

It follows from B that m can be no greater that n. We first consider the case $m = n$.

Under these conditions we shall prove that in a certain neighborhood of the point $y(x_0) = y_0$ the inverse mapping $x = x(y)$ exists and is differentiable and has the derivative

$$x'(y) = (y'(x))^{-1} \qquad (x = x(y)) .$$

In this formulation the theorem has a purely affine meaning. But in order to be able to use the methods of coordinate-free analysis, it is also expedient here to equip the spaces R_x^m and R_y^n with arbitrary euclidean metrics. With respect to these metrics the assumptions made can be stated in the following equivalent fashion, whereby we set $x_0 = y_0 = 0$:

A. *The function $y(x)$ is differentiable in a spherical neighborhood $|x| < \varrho_x$ of the origin.*

B. *We have*

$$\inf_{|h|=1} |y'(0)\, h| = \mu > 0 .$$

Hypotheses C is equivalent to assuming that the norm

$$|y'(x) - y'(0)| \to 0$$

for $|x| \to 0$. This condition can be weakened somewhat.

For this purpose we consider for $0 \leqq \varrho < \varrho_x$ the nonnegative function

$$\varphi(\varrho) \equiv \sup_{|x| \leqq \varrho} |y'(x) - y'(0)| ,$$

which increases monotonically with ϱ. The above hypothesis C states that the limit

$$\lim_{\varrho \to 0} \varphi(\varrho) = \varphi(0+) = 0 .$$

We replace this with the weaker

C'. $\varphi(0+) < \mu$.

From B and the definition of $\varphi(\varrho)$ it follows for $|x| \leqq \varrho \ (< \varrho_x)$ that
$|y'(x)\,h| \geqq |y'(0)\,h| - |y'(x)\,h - y'(0)\,h| \geqq (\mu - \varphi(\varrho))\,|h|$,
and therefore

$$\inf_{|h|=1} |y'(x)\,h| \geqq \mu - \varphi(\varrho)\,. \tag{4.1}$$

As a consequence of hypothesis C′ the least upper bound of those radii ϱ for which $\mu - \varphi(\varrho) > 0$ is positive, and it is consequently possible to assume from the beginning that the number ϱ_x introduced in A is so small that the inequality $\mu - \varphi(\varrho) > 0$ holds for $\varrho < \varrho_x$. Further, since the dimensions m and n of the spaces R_x^m and R_y^n were assumed to be equal, it follows that *for* $|x| < \varrho_x$ *the linear operator* $y'(x)$ *maps the space* R_x^m *one-to-one onto the entire space* R_y^m.

We shall establish the invertibility of the mapping $y = y(x)$ in the following more precise formulation:

Inverse function theorem. *Under the hypotheses* A, B *and* C′ *the mapping*

$$x = x(y) \qquad (x(0) = 0)$$

inverse to

$$y = y(x) \qquad (y(0) = 0)$$

exists in the neighborhood

$$|y| < \varrho_y = \int_0^{\varrho_x - 0} (\mu - \varphi(\varrho))\,d\varrho$$

of the point $y = 0$. *It is also differentiable here, having as its derivative*

$$x'(y) = (y'(x))^{-1} \qquad (x = x(y))\,.$$

As follows from exercise 5 in 1.13, this theorem is precise in the following sense: Without additional assumptions the spherical neighborhoods $|x| < \varrho_x$, $|y| < \varrho_y$ cannot be enlarged in a generally valid fashion.

4.3. Proof of the inverse function theorem. We go on to the proof of the inverse function theorem and show in this order the following:

1. The range set G_y^m in R_y^m, the image of the sphere $|x| < \varrho_x$ by the function $y = y(x)$, is schlicht: two different points x_1 and x_2 of the sphere are mapped onto different points $y_1 = y(x_1)$ and $y_2 = y(x_2)$. The inverse mapping $x = x(y)$ $(x(0) = 0)$ therefore exists in G_y^m.

2. Interior points of the sphere $|x| < \varrho_x$ are mapped onto interior points of G_y^m. The image set G_y^m is consequently an open region.

3. The region G_y^m contains the spherical neighborhood $|y| < \varrho_y$ of the origin.

4. The inverse mapping $x = x(y)$, which is uniquely defined in G_y^m, is differentiable and has the derivative $x'(y) = (y'(x))^{-1}$ $(x = x(y))$.

4.4. G_y^m is schlicht. Let $x_1 \neq x_2$ be two points of the sphere $|x| < \varrho_x$, $x_2 - x_1 = \Delta x$, $y(x_2) - y(x_1) = \Delta y$. We claim that $\Delta y \neq 0$.

With a for the time being arbitrary unit vector e from the space R_y^m, because of the mean value theorem,

$$(\Delta y, e) = (y'(x_e)\, \Delta x, e)\,,$$

where $x_e = x_1 + \vartheta_e\, \Delta x$ $(0 < \vartheta_e < 1)$, and consequently

$$|(\Delta y, e)| \geqq |(y'(0)\, \Delta x, e)| - |(y'(x_e)\, \Delta x - y'(0)\, \Delta x, e)|\,.$$

It follows from Schwarz's inequality, on the one hand, that

$$|(\Delta y, e)| \leqq |\Delta y|\,,$$

and further that

$$\begin{aligned} |(y'(x_e)\, \Delta x - y'(0)\, \Delta x, e)| &\leqq |(y'(x_e) - y'(0))\, \Delta x| \\ &\leqq |y'(x_e) - y'(0)||\Delta x| \leqq \varphi(\varrho)|\Delta x|\,, \end{aligned}$$

where ϱ $(< \varrho_x)$ stands for the greater of the lengths $|x_1|$ and $|x_2|$. Thus

$$|\Delta y| \geqq |(y'(0)\, \Delta x, e)| - \varphi(\varrho)\, |\Delta x|\,.$$

Now take the unit vector e so that the first term on the right is as large as possible, which by Schwarz's inequality is the case for $e = y'(0)\, \Delta x / |y'(0)\, \Delta x|$. It then follows from hypothesis B that

$$|(y'(0)\, \Delta x, e)| = |y'(0)\, \Delta x| \geqq \mu\, |\Delta x|\,,$$

and we obtain the inequality

$$|\Delta y| \geqq (\mu - \varphi(\varrho))\, |\Delta x|\,, \tag{4.2}$$

which is important for the entire proof. From this in particular, because $\mu - \varphi(\varrho) > 0$ for $\varrho < \varrho_x$, it follows that $\Delta y \neq 0$ for $\Delta x \neq 0$, which was to be proved.

4.5. G_y^m is open. If the range set G_y^m includes the entire space R_y^m, there is nothing to prove. Thus let b stand for a point of the latter space that does not belong to G_y^m, so that $y(x) \neq b$ for $|x| < \varrho_x$. Then if $y(x_0) = y_0$ is an arbitrary point of the range set G_y^m, we wish to show that $|b - y_0|$ lies above a positive bound which is independent of b: corresponding to the interior point x_0 of the sphere $|x| < \varrho_x$ there is then an interior point y_0 of the image set.

For this let $|x_0| < \varrho < \varrho_x$ and δ be the greatest lower bound of $|y(x) - b|$ in the sphere $|x| \leqq \varrho$. Because R_x^m is finite dimensional this lower bound is reached at at least one point $x = a$ of this closed sphere:

$$|y(a) - b| = \delta > 0\,.$$

We first show that a is necessarily a boundary point of the sphere $|x| \leqq \varrho$ and that consequently $|a| = \varrho$.

In fact, because the operator $y'(x)$ for $|x| < \varrho_x$ maps the space R_x^m one-to-one onto the *entire* space R_y^m, the vector $h(\in R_x^m)$ can be determined so that

$$y'(a)\, h = b - y(a)\,.$$

Now if we had $|a| < \varrho$, then one could take $\lambda > 0$ so small that

$$|x| = |a + \lambda\, h| \leqq |a| + \lambda |h| \leqq \varrho\,,$$

and one would have

$$y(x) - b = y(a) - b + \lambda\, y'(a)\, h + \lambda(\lambda) = (1 - \lambda)\,(y(a) - b) + \lambda(\lambda)\,,$$

with $|(\lambda)| \to 0$ for $\lambda \to 0$. Finally for a sufficiently small λ

$$|y(x) - b| \leqq (1 - \lambda)|y(a) - b| + \lambda|(\lambda)| = \delta - \lambda(\delta - |(\lambda)|) < \delta\,,$$

which because $|x| \leqq \varrho$ contradicts the definition of δ. Hence $|a| = \varrho$.

After this preparation we have (cf. Fig. 2)

$$|b - y(x_0)| \geqq |y(a) - y(x_0)| - |y(a) - b| = |y(a) - y(x_0)| - \delta\,,$$

where by virtue of inequality (4.2)

$$|y(a) - y(x_0)| \geqq (\mu - \varphi(\varrho))\,|a - x_0| \geqq (\mu - \varphi(\varrho))\,(\varrho - |x_0|)$$

and, as a consequence of the definition of δ, $\delta \leqq |b - y(x_0)|$.

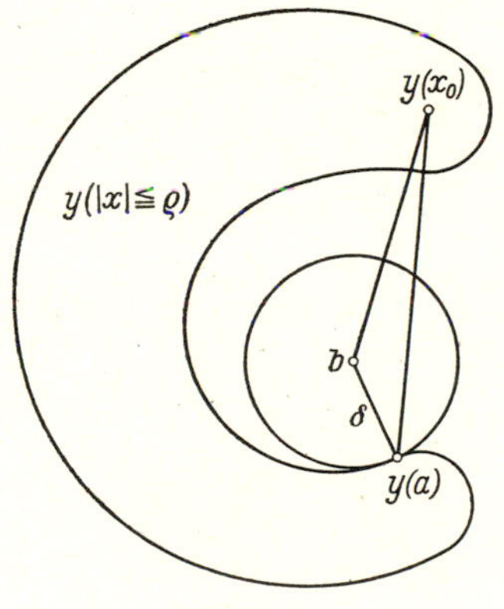

Fig. 2

Thus

$$|b - y(x_0)| \geqq \frac{1}{2}\,(\mu - \varphi(\varrho))\,(\varrho - |x_0|) > 0\,,$$

which proves the assertion: If y_0 belongs to the image point set G_y^m, then for any ϱ in the interval $|x_0| < \varrho < \varrho_x$ the neighborhood

$$|y - y_0| < \frac{1}{2}\,(\mu - \varphi(\varrho))\,(\varrho - |x_0|)\,,$$

lies entirely in G_y^m.

4.6. The region G_y^m contains the sphere $|y| < \varrho_y$. Let e be a unit vector from the space R_y^m. Because the open region G_y^m contains the point $y(0) = 0$, there exists for $y = \lambda e$ ($\lambda \geqq 0$) a unique $\bar{\lambda} > 0$ so that the segment $0 \leqq \lambda \leqq \lambda^*$ lies in G_y^m for $\lambda^* < \bar{\lambda}$, while, provided $\bar{\lambda}$ is finite, this is not the case for $\lambda^* \geqq \bar{\lambda}$. We have to show that for every e

$$\bar{\lambda} \geqq \int_0^{\varrho_x - 0} (\mu - \varphi(\varrho))\, d\varrho\,.$$

For this we take $0 < \lambda^* < \bar{\lambda}$ and set $\lambda^* e = y^*$, $x(y^*) = x^*$, so that the segment $y = \lambda e$ ($0 \leqq \lambda \leqq \lambda^*$) lies in G_y^m and $|x^*| = \varrho^* \to \varrho_x$ for $\lambda^* \to \bar{\lambda}$. Let

$$0 = \lambda_0 < \lambda_1 < \cdots < \lambda_n = \lambda^*$$

be a decomposition of the interval $0 \leqq \lambda \leqq \lambda^*$, $\lambda_i e = y_i$ and $x(y_i) = x_i$. According to (4.2)

$$|y_{i+1} - y_i| \geqq (\mu - \varphi(\bar{\varrho}_i))\, |x_{i+1} - x_i| \geqq (\mu - \varphi(\bar{\varrho}_i))\, |\varrho_{i+1} - \varrho_i|\,,$$

where $\varrho_i = |x_i|$ and $\bar{\varrho}_i = \max\,[\varrho_i, \varrho_{i+1}]$. From here it follows that

$$\lambda^* = \sum_{i=0}^{n-1} (\lambda_{i+1} - \lambda_i) = \sum_{i=0}^{n-1} |y_{i+1} - y_i| \geqq \sum_{i=0}^{n-1} (\mu - \varphi(\bar{\varrho}_i))\, |\varrho_{i+1} - \varrho_i|\,.$$

If the terms on the right with $\varrho_{i+1} < \varrho_i$ are omitted, then endless refinement of the decomposition λ_i yields the estimate

$$\lambda^* \geqq \int_0^{\varrho^*} (\mu - \varphi(\varrho))\, d\varrho\,,$$

from which the desired inequality for $\bar{\lambda}$ follows by taking the limit $\lambda^* \to \bar{\lambda}$.

From

$$\int_0^{\varrho^*} (\mu - \varphi(\varrho))\, d\varrho = \varrho^* (\mu - \varphi(\varrho^*)) + \int_{\varrho=0}^{\varrho^*} \varrho\, d\varphi(\varrho) \geqq \int_{\varrho=0}^{\varrho^*} \varrho\, d\varphi(\varrho)$$

it follows that

$$\varrho_y = \int_0^{\varrho_x - 0} (\mu - \varphi(\varrho))\, d\varrho \geqq \int_{\varrho=0}^{\varrho_x - 0} \varrho\, d\varphi(\varrho)\,,$$

where equality holds on the right provided ϱ_x stands for the least upper bound of those radii ϱ for which $\mu - \varphi(\varrho) > 0$.

4.7. The derivative $x'(y) = (y'(x))^{-1}$ exists in G_y^m. Let y and $y + \Delta y$ be two points of the open image region G_y^m, x and $x + \Delta x$ the uniquely determined preimages of these points in the sphere $|x| < \varrho_x$, so that $y(x) = y$, $y(x + \Delta x) = y + \Delta y$. Then because of the differentiability of the function $y(x)$

$$\Delta y = y'(x)\, \Delta x + |\Delta x|\, (\Delta x; x)\,,$$

with $|(\Delta x; x)| \to 0$ for $|\Delta x| \to 0$.

Since $y'(x)$ is regular in the sphere $|x| < \varrho_x$ and maps the space R_x^m one-to-one onto the entire space R_y^m, the inverse operator $(y'(x))^{-1}$ exists, and by inequality (4.1) its norm is

$$|(y'(x))^{-1}| \leqq (\mu - \varphi(\varrho))^{-1},$$

where $|x| < \varrho < \varrho_x$. The above equation can therefore also be written

$$(y'(x))^{-1} \Delta y = \Delta x + |\Delta x| \, (y'(x))^{-1} (\Delta x; x),$$

where

$$|(y'(x))^{-1} (\Delta x; x)| \leqq (\mu - \varphi(\varrho))^{-1} |(\Delta x; x)|.$$

Further, according to (4.2)

$$|\Delta x| \leqq (\mu - \varphi(\varrho))^{-1} |\Delta y|,$$

and hence

$$\Delta x = (y'(x))^{-1} \Delta y + |\Delta y| (\Delta y; y),$$

where in consequence of the above inequality $|(\Delta y; y)| \to 0$ for $|\Delta y| \to 0$. The inverse mapping $x = x(y)$ is therefore differentiable at the point y with the derivative $x'(y) = (y'(x))^{-1}$ $(x = x(y))$.

This completes the proof of the inverse function theorem of 4.2.

4.8. The case $m < n$. From hypothesis B of the inverse function theorem it follows, as already remarked, that the dimensions of the spaces R_x^m and R_y^n satisfy the inequality $m \leqq n$; above we assumed $m = n$.

If $m < n$, then the equation

$$y = y(x)$$

defines an m-dimensional surface in the space R_y^n which at the point $y(0) = 0$ has the m-dimensional tangent plane

$$E_0 = y'(0) \, R_x^m.$$

Let $P y$ be the orthogonal projection of the vector y onto this tangent plane and

$$\bar{y}(x) \equiv P \, y(x).$$

Now if $y(x)$ satisfies the metric conditions A, B and C', then $\bar{y}(x)$, because $\bar{y}'(x) = P \, y'(x)$, also fulfills hypothesis A and as a consequence of $\bar{y}'(0) \, h = P \, y'(0) \, h = y'(0) \, h$ hypothesis B too. Further, since $|P| = 1$, and consequently

$$\bar{\varphi}(\varrho) = \sup_{|x| \leqq \varrho} |\bar{y}'(x) - \bar{y}'(0)| \leqq \sup_{|x| \leqq \varrho} |y'(x) - y'(0)| = \varphi(\varrho),$$

$\bar{y}(x)$ also satisfies condition C'.

From the inverse function theorem it now follows that the function $\bar{y} = \bar{y}(x)$ maps the sphere $|x| < \varrho_x$ onto an open region $\bar{G}_y$ of the tangent plane E_0 and that the inverse function $x = x(\bar{y}) = x(P\,y)$ exists at the very least in the neighborhood

$$|\bar{y}| < \varrho_y = \int_0^{\varrho_x - 0} (\mu - \varphi(\varrho))\, d\varrho$$

of the origin $\boldsymbol{y} = 0$ on E_0. As a result, we have for the function $y = y(x)$ this

Theorem. *Let $y = y(x)$ be a function defined for $|x| < \varrho_x$ which satisfies conditions* A, B *and* C′ *of* 4.2. *Provided the dimension n of the space R_y^n is greater than the dimension m of R_x^m, then $y = y(x)$ maps the sphere $|x| < \varrho_x$ one-to-one onto a point set $G_y \subset R_y^n$ in such a way that the following holds:*

If $y = y(x)$ is orthogonally projected onto the tangent plane $E_0 = y'(0)\,R_x^m$, then the projections $\bar{y}(x)$ on E_0 cover the sphere

$$|\bar{y}| < \varrho_y = \int_0^{\varrho_x - 0} (\mu - \varphi(\varrho))\, d\varrho$$

without any gaps.

4.9. Solution of the equation $z(x, y) = 0$. Having treated the case $z(x, y) = y(x) - y$ thoroughly in the preceding paragraphs we can now take care of the general equation

$$z(x, y) = 0 \qquad (z(0, 0) = 0)$$

more briefly.

Let R_x^m, R_y^n, R_z^p be three spaces with euclidean metrics, and suppose

$$z = z(x, y) \qquad (z(0, 0) = 0)$$

is a mapping into the space R_z^p defined in the spheres

$$|x| < \varrho_x\,, \qquad |y| < \varrho_y\,.$$

We assume:

The partial derivative $z_x(x, y)$ exists for $|x| < \varrho_x$, $|y| < \varrho_y$, and one has for its greatest lower bound

$$\inf_{\substack{|h|=1 \\ |y|<\varrho_y}} |z_x(0, y)\,h| = \mu > 0\,.$$

From here it follows that $m \leqq p$; as in the inverse function theorem we restrict ourselves to the case $m = p$, so that for each y in the sphere $|y| < \varrho_y$ the operator $z_x(0, y)$ maps the space R_x^m one-to-one onto the entire space R_z^m.

As in the inverse function theorem in 4.2 we require an upper bound for the norm of the operator $z_x(x, y) - z_x(0, y)$. We assume:

For $\varrho < \varrho_x$ let

$$\varphi(\varrho) \equiv \sup_{\substack{|x|\leqq\varrho \\ |y|<\varrho_y}} |z_x(x, y) - z_x(0, y)| < \mu\,.$$

For a fixed y in the sphere $|y| < \varrho_y$ one can then apply the inverse function theorem to the function $z(x, y) - z(0, y)$ of x. It follows that x is defined as a single-valued function $x(z, y)$ of z and y in the spheres

$$|z - z(0, y)| < \varrho_z = \int_0^{\varrho_x - 0} (\mu - \varphi(\varrho))\, d\varrho \tag{4.3}$$

and $|y| < \varrho_y$, so that $x(z(0, y), y) \equiv 0$, $|x(z, y)| < \varrho_x$ and $z \equiv z(x(z, y), y)$.

We now assume that as $\varrho \to 0$ the least upper bound

$$\sup_{|y| \leqq \varrho} |z(0, y)|$$

is smaller than the integral on the right in (4.3). Then if ϱ_0 stands for the least upper bound of those numbers $0 < \varrho < \varrho_y$ for which

$$\sup_{|y| \leqq \varrho} |z(0, y)| < \int_0^{\varrho_x - 0} (\mu - \varphi(\varrho))\, d\varrho\,,$$

then $0 < \varrho_0 \leqq \varrho_y$, and the above result can be applied in particular for $z = 0$. Hence, under this condition, $x = x(0, y) \equiv x(y)$ is defined for $|y| < \varrho_0$ as a single-valued function of y, and $z(x(y), y) \equiv 0$ for $|y| < \varrho_0$ $(\leqq \varrho_y)$.

We therefore have this

Theorem. *Let*

$$z = z(x, y) \qquad (z(0, 0) = 0)$$

be a vector function defined for $|x| < \varrho_x$, $|y| < \varrho_y$ $(x \in R_x^m,\ y \in R_y^n,\ z \in R_z^p)$ *with these properties*:

A. *The partial derivative* $z_x(x, y)$ *exists for* $|x| < \varrho_x$, $|y| < \varrho_y$, *and*

$$\inf_{\substack{|h|=1 \\ |y| < \varrho_y}} |z_x(0, y)\, h| = \mu > 0\,.$$

B. *For* $\varrho < \varrho_x$ *we have*

$$\varphi(\varrho) \equiv \sup_{\substack{|x| \leqq \varrho \\ |y| < \varrho_y}} |z_x(x, y) - z_x(0, y)| < \mu\,.$$

C. $$\sup_{|y| < \varrho_y} |z(0, y)| < \int_0^{\varrho_x - 0} (\mu - \varphi(\varrho))\, d\varrho\,.$$

Under these conditions there exists for $|y| < \varrho_y$ *one and only one function*

$$x = x(y) \qquad (x(0) = 0)\,,$$

such that

$$z(x(y), y) \equiv 0\,.$$

If $z(x, y)$ is totally differentiable, then $x(y)$ is also differentiable, and for $|y| < \varrho_y$

$$x'(y) = -\,(z_x(x, y))^{-1}\, z_y(x, y) \qquad (x = x(y))\,.$$

4.10. The essential part of the above theorem is contained in the following corollary:

Let the vector function $z = z(x, y)$ $\left(x \in R^m_x,\ y \in R^n_y,\ z \in R^p_z,\ z(0, 0) = 0\right)$ which is defined in a neighborhood of the point $x = 0$, $y = 0$ be differentiable with respect to x and let $z(0, y)$, $z_x(0, y)$ furthermore be continuous for $y = 0$.

Then provided

$$\inf_{|h|=1} |z_x(0, 0)\, h| > 0$$

and

$$\lim_{\varrho \to 0} \sup_{|x| \leqq \varrho} |z_x(x, 0) - z_x(0, 0)| < \inf_{|h|=1} |z_x(0, 0)\, h| ,$$

the equation

$$z(x, y) = 0$$

can be solved for x in a neighborhood of the point $x = 0$, $y = 0$: There exists a number $\varrho_y > 0$ and a function $x = x(y)$ which is well-defined for $|y| < \varrho_y$ such that $x(0) = 0$ and

$$z\left(x(y), y\right) \equiv 0 .$$

4.11. Second proof of the inverse function theorem. Because of the great significance of the theory of implicit functions we shall in addition treat the inversion problem using Picard's method of successive approximations. The following discussion can be read directly, independently of the previous investigation. It takes us to our goal under somewhat broader assumptions than those basic to the proof of 4.2—4.7. The following construction also works without modification for the general case of a Banach space.

4.12. The inverse function theorem. We consider a self-mapping $x \to y = y(x)$ $\left(y(0) = 0\right)$ of the Minkowski (or Banach) space R defined for $|x| < r_0 \leqq \infty$ with the following property:

For $0 < r < r_0$ the function $\varphi(x) \equiv y(x) - x$ satisfies the inequality

$$\theta(r) \equiv \sup_{|x| \leqq r} D(x) < 1 , \tag{4.4}$$

where

$$D(x) \equiv \lim_{\Delta x \to 0} \sup \frac{|\varphi(x + \Delta x) - \varphi(x)|}{|\Delta x|} .$$

Under these conditions the mapping $x \to y$ can be inverted: There exists a well-determined mapping $y \to x = x(y)$, uniquely defined in the sphere

$$|y| < \varrho_0 = r_0 - \int_0^{r_0} \theta(r)\, dr ,$$

such that $y \equiv y\left(x(y)\right)$.

We first give two simple lemmas.

Lemma 1. *In the sphere* $|x| < r_0$ *of the space* R *let a single-valued mapping* $y = f(x)$ *into* R *be given. If there exists a number* M $(0 \leqq M \leqq \infty)$ *such that*

$$\limsup_{\Delta x \to 0} \frac{|f(x + \Delta x) - f(x)|}{|\Delta x|} \leqq M ,$$

then for $|a|, |b| < r_0$

$$|f(a) - f(b)| \leqq M\, |a - b| .$$

Proof. Since the assertion is trivial for $a = b$, we assume that $a \neq b$. Let c be an interior point of the line segment $a\,b$. Then

$$|f(a) - f(b)| \leqq |a - c| \frac{|f(a) - f(c)|}{|a - c|} + |c - b| \frac{|f(c) - f(b)|}{|c - b|}$$

and, provided, $a_1\, b_1$ is that one of the subsegments $a\,c$, $c\,b$ corresponding to the greater of the two difference quotients on the right,

$$\frac{|f(a) - f(b)|}{|a - b|} \leqq \frac{|a - c| + |c - b|}{|a - b|} \frac{|f(a_1) - f(b_1)|}{|a_1 - b_1|} = \frac{|f(a_1) - f(b_1)|}{|a_1 - b_1|} .$$

Repetition of this argument yields a sequence of nested line segments $a_n\, b_n$ such that

$$\frac{|f(a) - f(b)|}{|a - b|} \leqq \frac{|f(a_n) - f(b_n)|}{|a_n - b_n|} \quad (n = 1, 2, \ldots) .$$

We now determine this process so that $a_n - b_n \to 0$ for $n \to \infty$. Then the limit point $x = \lim a_n = \lim b_n$ exists $(|x| < r_0)$.

Provided, for a given n, x lies interior to the segment $a_n\, b_n$, then the application of the above argument gives

$$\frac{|f(a) - f(b)|}{|a - b|} \leqq \frac{|f(a_n) - f(b_n)|}{|a_n - b_n|} \leqq \frac{|f(c_n) - f(x)|}{|c_n - x|} ,$$

where c_n is either a_n or b_n. The same inequality holds even if x coincides with either a_n or b_n, where c_n is then taken to be b_n or a_n, respectively.

Taking the limit $n \to \infty$ one finds

$$\frac{|f(a) - f(b)}{|a - b|} \leqq \limsup_{n \to \infty} \frac{|f(c_n) - f(x)|}{|c_n - x|} \leqq M ,$$

which completes the proof of Lemma 1.

Lemma 2. *Assuming that*

$$\limsup_{\Delta x \to 0} \frac{|f(x + \Delta x) - f(x)|}{|\Delta x|} \leqq M(x) ,$$

then for $|a|, |b| < r_0$

$$|f(a) - f(b)| \leqq \int_{ab} M(x)\, |dx| ,$$

where on the right we have the upper (Riemann) integral of $M(x)$ along the segment ab.

Proof. Divide ab into finitely many intervals (Δx). By Lemma 1

$$|f(x+\Delta x) - f(x)| \leq M_\Delta \, |\Delta x| ,$$

where $M_\Delta = \sup M(x)$ on Δx, and the assertion results by summation and then unrestricted refinement of the partition (Δx).

We now come to the proof of the inverse function theorem and show:

1) The mapping $x \to y$ is *univalent* for $|x| < r_0$. As our first result we have

$$|y(a) - y(b)| = |a - b + \varphi(a) - \varphi(b)|$$
$$\geqq |a-b| - |\varphi(a) - \varphi(b)| .$$

If, for example, $|b| \leqq |a|$ $(< r_0)$, then by hypothesis $D(x) \leqq \theta(|x|) \leqq \theta(|a|)$ on the segment ab, and by Lemma 1

$$|\varphi(a) - \varphi(b)| \leqq \theta(|a|) \, |a-b| .$$

Consequently,

$$|y(a) - y(b)| \geqq |a-b| - \theta(|a|) \, |a-b|$$
$$= |a-b| \, \big(1 - \theta(|a|)\big) > 0 .$$

Hence, $y(a) \neq y(b)$ for $a \neq b$.

2) By means of Picard's method of successive approximations we construct the sought-for inverse function $x = x(y)$. To this end one sets

$$x = y - \varphi(x)$$

and for a given y determines the sequence x_n by

$$x_{n+1} = y - \varphi(x_n) \qquad (n = 0, 1, \ldots, x_0 = 0) . \tag{4.5}$$

We must show:

A. There exists a number $\varrho_0 > 0$ such that the points x_n all lie in the sphere $|x| < r_0$ provided $|y| < \varrho_0$.

B. The sequence x_n converges for $|y| < \varrho_0$ and $n \to \infty$ to a point $x = x(y)$, $|x(y)| < r_0$.

It then follows from A and B that $x(y)$ is the sought-for inverse function. For because of hypothesis (4.4) $\varphi(x)$ and thus also $y(x)$ is continuous for $|x| < r_0$. It further results from (4.5) and B, for $n \to \infty$, that

$$x(y) = y - \varphi\big(x(y)\big) , \qquad y - y\big(x(y)\big) = 0 .$$

$x = x(y)$ is thus the sought-for inverse function.

Proof of A. A number $\varrho_0 > 0$ is to be determined so that the points x_n computed by means of (4.5) lie in the sphere $|x| < r_0$. Assuming that y is a point for which the points $x_0, x_1, \ldots, x_n$ have the desired property, one inquires, under which additional conditions is it also true that $|x_{n+1}| < r_0$?

For this one estimates the function $|\varphi(x)|$. Let $0 < |x| = r < r_0$ and $x = r\,e$, where e is a unit vector. If one takes $a = x$, $b = 0$, then as a result of Lemma 2, and in view of (4.4), one has

$$|\varphi(x)| \leqq \int_0^r \theta(r)\,dr\,,$$

and hence

$$|x_{n+1}| = |y - \varphi(x_n)| \leqq |y| + |\varphi(x_n)| \leqq |y| + \int_0^{r_0} \theta(r)\,dr\,.$$

Now if we are to have $|x_{n+1}| < r_0$, it suffices to take $|y|$ smaller than

$$r_0 - \int_0^{r_0} \theta(r)\,dr\,.$$

For this reason we set

$$\varrho_0 = r_0 - \int_0^{r_0} \theta(r)\,dr = \int_0^{r_0} (1 - \theta(r))\,dr \qquad (> 0)\,.$$

With this choice of ϱ_0 all of the points x_n lie in the sphere $|x| < r_0$, provided $|y| < \varrho_0$. Then for $n = 0$, $|x_0| = 0 < r_0$, and the above induction argument shows that $|x_n| < r_0$ holds for every n. Because

$$|x_{n+1}| \leqq |y| + \int_0^{r_0} \theta(r)\,dr = r_0 - (\varrho_0 - |y|)\,,$$

the points x_n even lie in the sphere $|x| \leqq r_0 - (\varrho_0 - |y|)$.

Proof of B. By (4.5) one has for $n > 0$

$$|x_{n+1} - x_n| = |\varphi(x_n) - \varphi(x_{n-1})|\,.$$

If $|y| < \varrho_0$, then the inequality $|x_n| \leqq r_0 - (\varrho_0 - |y|)$ follows from A for every n. The same inequality holds for each point x of the line segment $x_{n-1}x_n$, and by (4.4)

$$D(x) \leqq \theta\,(r_0 - (\varrho_0 - |y|)) \equiv \theta_0 < 1\,.$$

It then follows from Lemma 1 that

$$\begin{aligned} |x_{n+1} - x_n| &= |\varphi(x_n) - \varphi(x_{n-1})| \\ &\leqq \theta_0\,|x_n - x_{n-1}| \\ &\leqq \theta_0^n\,|x_1| = \theta_0^n\,|y|\,. \end{aligned}$$

The series $\sum (x_{n+1} - x_n)$ thus converges uniformly in the sphere $|y| \leqq \varrho_0' < \varrho_0$, from which the existence and even the continuity of the inverse function follows for $|y| < \varrho_0$. The uniqueness of the inverse function results from the single-valuedness of the function $y(x)$.

4.13. Example. The following example shows that under the hypotheses of the inverse function theorem the region in which the mapping $x \to y$ is given cannot be continued outside of the sphere $|y| \leqq \varrho_0 = \int_0^{r_0} (1 - \theta(r))\, dr$.

Define $y(x)$ by

$$y = y(x) = x\,(1 - g(r)) \qquad (|x| = r)\,,$$

where

$$g(r) \equiv \frac{1}{r} \int_0^r \theta(r)\, dr$$

and $\theta(r)$ is for $r \geqq 0$ a continuous, monotonically increasing function such that

$$0 \leqq \theta(r) < 1 \quad \text{for} \quad 0 \leqq r < r_0 \quad \text{and} \quad \theta(r) = 1 \quad \text{for} \quad r \geqq r_0\,.$$

Suppose $|x| = r < r_0$, and further let $\Delta x\ (\neq 0)$ be an increment of x and Δr, Δg and $\Delta\varphi$ the corresponding increments of the functions r, g and $\varphi = y(x) - x = -x\,g(r)$. One has $(g(r) = g)$

$$|\Delta\varphi| = |(x + \Delta x)\,(g + \Delta g) - x\,g| = |x\,\Delta g + \Delta x\,g + \Delta x\,\Delta g|\,.$$

Thus, since $dr \leqq |\Delta x|$,

$$\frac{|\Delta\varphi|}{|\Delta x|} \leqq r\,\frac{|\Delta g|}{|\Delta x|} + g + |\Delta g|\,,$$

and

$$\limsup \frac{|\Delta\varphi|}{|\Delta x|} \leqq r\,g'(r) + g(r) = \theta(r) < 1\,.$$

The hypotheses of the inverse function theorem are thus satisfied, and one concludes that the function $y(x)$ can be uniquely inverted for

$$|y| < \varrho_0 = \int_0^{r_0} (1 - \theta(r))\, dr\,.$$

This also follows at once from the expression

$$y(x) = \frac{x}{r}\,(r - r\,g(r)) = \frac{x}{r} \int_0^r (1 - \theta(r))\, dr\,.$$

The norm $|y(x)|$ increases on the interval $0 \leqq |x| \leqq r_0$ from 0 to ϱ_0 and for $|x| \geqq r_0$ maintains the *constant* value ϱ_0. The sphere $|y| < \varrho_0$ is thus the *precise* domain of existence of the inverse function $x = x(y)$.

4.14. Case where $y(x)$ is differentiable. If $y(x)$ is differentiable for $|x| < r_0$, then $D(x) = |\varphi'(x)| = |y'(x) - I|$[1], where I is the

[1] The norm $|\varphi'|$ is defined as $|\varphi'(x)| = \sup_{|h|=1} |\varphi'(x)\,h|$.

identity mapping. Thus the inverse function theorem holds provided

$$\theta(r) = \sup_{|x| \leqq r_0} |y'(x) - I| < 1 \qquad (r < r_0) .$$

The proof in 4.7 shows that the inverse function $x(y)$ is differentiable for $|y| < \varrho_0$ with the derivative $x'(y) = \left(y'(x)\right)^{-1}$.

In addition, we apply the inverse function theorem to a twice differentiable function $y(x)$. If $y'(0) = I$ and

$$\sup_{|x| \leqq r_0} |y''(x)| = M < \infty ,$$

then it follows from Lemma 1, for an arbitrary constant $k \in R$, that

$$|\left(y'(x) - I\right) k| = |\int_{0x} y''(x)\, dx\, k| \leqq M|k| \int |dx| = M\, r|k| .$$

Hence

$$\theta(r) = \sup_{|x| \leqq r} |y'(x) - I| \leqq M\, r < 1$$

for $r < r_0 = 1/M$.

According to the inverse function theorem, the inverse function $x = x(y)$ exists in the sphere

$$|y| < r_0 - \int_0^{r_0} \theta(r)\, dr = \varrho_0 .$$

If one sets $\theta(r) \equiv M\, r$ for $r \leqq 2/M$, then the example

$$y(x) = x - \frac{x}{r} \int_0^r \theta(r)\ dr = x\left(1 - \frac{M\, r}{2}\right)$$

shows that the radius $\varrho_0 = 1/2M$ cannot be increased. For the mapping $x \to y$ is no longer schlicht for $|x| \leqq r$ $(1/M < r < 2/M)$, and for $|x| < 1/M$ the image region is precisely the sphere $|y| < 1/2M = \varrho_0$.

4.15. Exercises. 1. Let $y = y(x)$ be a mapping from the euclidean space R_x^m into the euclidean space R_y^n which is continuously differentiable in a neighborhood of the point x_0. The point x_0 is to be called a regular point of the function $y(x)$ provided the kernel of the operator $y'(x_0)$ is of the lowest possible dimension. Now let $m > n$. Then the kernel of the operator $y'(x_0)$ is of dimension $p \geqq m - n$, and the point x_0 is regular provided this dimension p is exactly $m - n$. Prove:

Let $m > n$ and suppose x_0 is a regular point of the function $y(x)$. Then if one considers the set of points in the space R_x^m that satisfy the equation

$$y(x) = y(x_0) = y_0 ,$$

the subset of these points that lie in a sufficiently small neighborhood $|x - x_0| < r_x$ can be mapped one-to-one and differentiably onto an open region G_u^p of a p-dimensional parameter space R_u^p. There consequently exists a function

$$x = x(u) \qquad (x_0 = x(u_0))$$

defined and continuous in G_u^p such that on the one hand $y(x(u)) \equiv y_0$ and $|x(u) - x_0| < r_x$ for $u \in G_u^p$, and on the other hand, conversely, corresponding to each x in the neighborhood $|x - x_0| < r_x$ that satisfies the equation $y(x) = y_0$ there is a unique $u \in G_u^p$ such that $x(u) = x$; moreover, the operator $x'(u_0)$ is regular.

Hint. Take for the parameter space R_u^p the kernel of the operator $y'(x_0)$. A linearly independent e.g. orthogonal complement R_v^n of R_u^p in R_x^m is then of dimension $m - p = n$, so that each $x \in R_x^m$ can be represented in a unique way as a sum $x = u + v$. Then $y(x) = y\ (u+v) \equiv y(u, v)$, and here the partial derivative y_v in the direction of the subspace R_v^n is regular for $x_0 = u_0 + v_0$, so that

$$\inf_{|v|=1} |y'(x_0)\, v| = \mu > 0\,.$$

By the theorem of 4.9 the equation $y(x) = y(x_0) = y_0$ can be solved in a neighborhood of the point $x_0 = u_0 + v_0$ for v, which implies the assertion for $x = u + v(u) \equiv x(u)$.

Remark. Phrased geometrically the above theorem states that the equation $y(x) = y(x_0) = y_0$ defines in a neighborhood of the regular point x_0 a "regular surface" of dimension $m - n = p$ which is embedded in the space R_x^m. The kernel R_u^p defined by

$$y'(x_0)\, u = 0$$

is the tangent space and the orthogonal complement R_v^n the normal space to the surface at the point x_0.

Each one-to-one continuously differentiable mapping $u = u(\bar{u})$, $\bar{u} = \bar{u}(u)$ of the region G_u^p into a region $G_{\bar{u}}^p$ of the same or of another p-dimensional parameter space $R_{\bar{u}}^p$ leads to a new parametric representation $x = x(u) = x(u(\bar{u})) \equiv \bar{x}(\bar{u})$ of the required sort.

2. We consider in the following, as a supplement to the case of a regular point x_0, treated in the text and in the previous exercise, a "degenerate" function $y(x)$ which has no regular points at all in its domain of definition. For each x, therefore, the dimension of the kernel of $y'(x)$ is positive if $m \leqq n$, and $> m - n$ provided $m > m$.

Let x_0 then be a point where the dimension of the kernel of $y'(x)$ reaches its minimum p; according to hypothesis x_0 is also an irregular point of the function $y(x)$, consequently $q = m - p < n$. Prove:

In a sufficiently small neighborhood $|x - x_0| < r_x$ a function

$$z = z(x)$$

can be defined whose range lies in a space R_z^q of dimension $q = m - p$, so that in this neighborhood y becomes a function of z

$$y = y(x) \equiv \bar{y}(z(x))$$

which in the image region $z(|x - x_0| < r_x)$ is everywhere regular.

Hint. Suppose, as earlier, that R_u^p is the p-dimensional kernel of $y'(x_0)$ and that R_v^q is a linearly independent complement, so that $x = u + v$ and $y(x) = y(u + v) \equiv y(u, v)$. Then for $x_0 = u_0 + v_0$

$$\inf_{|v|=1} |y'(x_0)\, v| > 0\,,$$

and the operator $y'(x_0)$ therefore maps R_v^q one-to-one onto a q-dimensional strict subspace $R_z^q = y'(x_0)\, R_v^q$ of R_y^n. Let $P\, y$ be the projection of y on this subspace and

$$P\, y(x) = z(x) \equiv z(u, v)\,.$$

Hence $z'(x) = P\, y'(x)$, and in particular $z'(x_0) = y'(x_0)$. Consequently

$$0 < \inf_{|v|=1} |z'(x_0)\, v| = \inf_{|v|=1} |y'(x_0)\, v|\,.$$

In consequence of the continuity of $y'(x)$ for $x = x_0$, r_x can therefore be chosen so small that for $|x - x_0| < r_x$

$$0 < \inf_{|v|=1} |z'(x)\, v| \leq \inf_{|v|=1} |y'(x)\, v|\,. \tag{a}$$

Then the dimension of the kernel of $y'(x)$, which by hypothesis is $\geqq p$, is precisely $= p$ for $|x - x_0| < r_x$; for it follows from (a) that this dimension can be at most p. If $R_{u_x}^p$ stands for this kernel ($u_{x_0} = u$), then for each x in the above neighborhood the kernel $R_{u_x}^p$, which varies with x, and the fixed space R_v^q are hence linearly independent complements, i.e.,

$$R_x^m = R_{u_x}^p + R_v^q\,. \tag{b}$$

According to the inverse function theorem of 4.2, the equation $z = z(u, v)$ can be solved for v in a neighborhood of the points x_0, $z_0 = z(x_0)$, so that

$$y = y(u, v(u, z)) \equiv \bar{y}(u, z)$$

and

$$z \equiv z(u, v(u, z)) \tag{c}$$

identically. We claim that $\bar{y}$ only depends on z, so that

$$\bar{y}_u(u, z)\, du = y'(x)\, du + y'(x)\, d_u v \tag{d}$$

vanishes for $|x - x_0| < r_x$ and for each du of the kernel R_u^p.

For the proof, observe that according to (b) the differential du can be uniquely represented as a sum

$$du = du_x + dv \tag{e}$$

for each x of the sphere $|x - x_0| < r_x$, where du_x stands for a vector from the kernel of $y'(x)$, while dv is a vector from the fixed space R_v^q. By (c)

$$0 = z'(x)\, du + z'(x)\, d_u v ,$$

from which it follows, in view of (e), that

$$0 = z'(x)\, du_x + z'(x)\, dv + z'(x)\, d_u v = z'(x)\, (dv + d_u v);$$

for $z'(x)\, du_x = P\, y'(x)\, du_x = 0$, since du_x is a vector of the kernel of $y'(x)$. But then because of (a)

$$dv + d_u v = 0 , \tag{f}$$

and it therefore results from equations (d), (e), (f) that

$$\bar{y}_u(u, z)\, du = y'(x)\, du_x + y'(x)\, dv + y'(x)\, d_u v = y'(x)\, (dv + d_u v) = 0 ,$$

so that in fact $\bar{y}$ only depends on z.

The function $y = \bar{y}(z)$ has nothing but regular points. For from

$$\bar{y}'(z)\, dz = y'(x)\, d_z v = 0$$

it follows by (a) that $d_z v$ must $= 0$, which according to (c) further implies that

$$dz = z'(x)\, d_z v = 0 ,$$

which was to be proved.

3. Using the notation of exercise 1, let x_0 be a regular point of the differentiable function $y(x)$ and R_u^p the kernel ($p = m - n$) of $y'(x_0)$. Let $z(x)$ be a second function which is differentiable in a neighborhood of x_0 with a range in the space R_z^l. Prove:

For $z(x)$ to be stationary with respect to $y(x)$ at the point x_0, so that from $y'(x_0)\, du = 0$ it always follows that $z'(x_0)\, du = 0$, and the kernel of $y'(x_0)$ is thus contained in the kernel of $z'(x_0)$, it is necessary and sufficient that a linear mapping

$$z = A\, y$$

of the space R_y^n into the space R_z^l exist such that for each differential dx from R_x^m

$$\big(z'(x_0) - A\, y'(x_0)\big)\, dx = 0 .$$

Hint. As before, let R_v^n be a linearly independent complement of the kernel R_u^p. The equation $y = y(x) \equiv y(u, v)$ can in a neighborhood of the points x_0, $y_0 = y(x_0)$ be solved for v, $v = v(u, y)$, so that we have identically $y \equiv y\big(u, v(u, y)\big)$. Because $y'(x_0)\, du = 0$ it follows that $dy = y_v(x_0)\, dv$, and therefore $dv = \big(y_v(x_0)\big)^{-1}\, dy$. Now if $z'(x_0)\, du = 0$ also, then for an arbitrary $dx = du + dv$

$$z'(x_0)\, dx = z'(x_0)\, du + z'(x_0)\, dv = z'(x_0)\, dv = z'(x_0)\, \big(y_v(x_0)\big)^{-1}\, dy .$$

Thus if one sets the linear mapping of the space R_y^n into R_z^l

$$z'(x_0)\,(y_v(x_0))^{-1} = A\,,$$

then $z'(x_0)\,dx = A\,dy = A\,y'(x_0)\,dx$ and $(z'(x_0) - A\,y'(x_0))\,dx = 0$.

The condition in the theorem is thus necessary. That it is also sufficient is a result of the above identity for $dx = du$; for

$$z'(x_0)\,du = A\,y'(x_0)\,du = 0\,.$$

Remark. If for given functions $y(x)$ and $z(x)$ one seeks those points x_0 where $z(x)$ is stationary with respect to $y(x)$, then by the above it is necessary to make use of the identity

$$(z'(x_0) - A\,y'(x_0))\,dx \equiv 0$$

in the unknowns x_0 and A. For the m real unknowns x_0 and $l\,n$ real unknowns A we have from here $l\,m$ real equations. Hence if the problem generally is to have a solution, $l\,m + n$ must be $\leqq m + l\,n$, i.e.,

$$l\,p \leqq p\,,$$

which is the case only for $l = 1$, that is, for a *real* function $z(x)$. Then $A\,y$ is a real linear form of y, and the above result contains as a special case the "method of Lagrange multipliers" for determining the stationary points of a real function $z(x)$ on the surface $y(x) = 0$.

III. Integral Calculus

§ 1. The Affine Integral

1.1. Alternating operators and differentials. Let $A(x)$ be a p-linear *alternating operator* defined in the open domain G_x^m of the linear space R_x^m: for each fixed x in G_x^m

$$y = A(x)\, h_1 \ldots h_p = A(x)\, d_1x \ldots d_px \tag{1.1}$$

is a p-linear alternating function of the vectors $h_i = d_ix \in R_x^m$ with range in a linear space R_y^n. Such a function is called an *alternating differential* of pth degree.

Even for $p = 1$ we interpret this differential as "alternating." It will be shown that the concepts and theorems developed in this section which refer to alternating differentials of pth degree also remain meaningful and valid for $p = 1$. The same holds for $p = 0$ if a differential of degree zero is understood to be an ordinary vector function $A(x)$.

For $p > m$ the differentials d_ix are linearly dependent, and the alternating differential vanishes identically; therefore suppose $p \leqq m$. Let U^p be a p-dimensional subspace of R_x^m and $D\, d_1x \ldots d_px$ stand for the real alternating fundamental form of this subspace, which is uniquely determined up to a real factor. Then one has ($x \in G_x^m$, $d_ix \in U^p$)

$$A(x)\, d_1x \ldots d_px \equiv a(x)\, D\, d_1x \ldots d_px\,, \tag{1.2}$$

where $a(x) \in G_y^n$ stands for a vector function that is uniquely defined for a fixed U^p in G_x^m[1]. Since this function consequently depends on the oriented subspace U^p, $a(x; U^p)$ would be a more fitting designation. However, we wish to retain the above shorter notation, as long as there is no fear of misunderstanding.

The definitions of the continuity and the differentiability of a multilinear operator were already given in II.1.9. By these definitions the alternating operator $A(x)$ is differentiable at the point x if

$$A\,(x + h)\, h_1 \ldots h_p = A(x)\, h_1 \ldots h_p + A'(x)\, h\, h_1 \ldots h_p + |h|\, (h; x)\, h_1 \ldots h_p\,,$$

where in the present case the $(p + 1)$-linear derivative operator $A'(x)$ is alternating in the p vectors $h_1, \ldots, h_p$ and the norm $|(h; x)|$ of the likewise alternating operator $(h; x)$ vanishes for $|h| \to 0$. We recall in

---

[1] For the case of a *real* form $A(x)\, d_1x \ldots d_px$ this follows from I.5.2; only in the present case the density a will be a function $a = a(x)$ of the point x. The arguments of I.5.2 remain valid without essential modifications when the value of the given alternating p-linear form varies in a linear space R_y^n.

this connection the following fact (cf. II.1.13, exercise 6): if the derivative operator $A'(x)$ is *continuous* on a closed subregion $\bar{G}_x^m$ of G_x^m, then in $\bar{G}_x^m$ $|(h; x)| \to 0$ even *uniformly* for $|h| \to 0$.

1.2. The affine integral of an alternating differential. We now suppose that the alternating differential (1.1) is continuous on a p-dimensional closed simplex

$$s^p = s^p(x_0, \ldots, x_p)$$

of the domain G_x^m.

The edges $h_i = x_i - x_0$ emanating from the vertex x_0 generate in R_x^m a p-dimensional subspace U^p which is parallel to the plane $x_0 + U^p$ of the simplex. If $p = m$, $U^m = x_0 + U^m = R_x^m$. We orient the simplex s^p and the other simplexes of the plane $x_0 + U^p$ with the real alternating fundamental form

$$D\, h_1 \ldots h_p = \Delta(x_0, \ldots, x_p)$$

of the space U^p. We are concerned with the definition and the existence of the integral of the alternating differential (1.2) taken over the simplex s^p.

Taking as a model the Cauchy-Riemann integral concept, we consider a simpliciat decomposition

$$s^p = \sum_j s_j^p(x_0^j, \ldots, x_p^j) \tag{1.3}$$

of the simplex s^p into finitely many subsimplexes s_j^p, fix in each subsimplex an interior or boundary point x_j^* and form the sum

$$\sum_j A(x_j^*)\, h_1^j \ldots h_p^j = \sum_j a(x_j^*)\, D\, h_1^j \ldots h_p^j\,, \tag{1.4}$$

where $h_i^j = x_i^j - x_0$ and the function $a(x)$, because the vectors $h_1^j, \ldots,$ h_p^j for each j span the same subspace

$$(h_1^j, \ldots, h_p^j) = (h_1, \ldots, h_p) = U^p\,,$$

is uniquely determined. We then have this

Theorem. *If under the assumed continuity of the operator $A(x)$ on the closed simplex s^p all subsimplexes are oriented with respect to D the same as s^p, then for unrestricted refinement of the decomposition the sum* (1.4) *approaches a uniquely determined limit vector J in the space R_y^n.*

Here the "unrestricted refinement" as well as the existence of the limit vector can be understood either in the sense of the natural topologies of the linear spaces R_x^m and R_y^n or with respect to arbitrary Minkowski metrics. If δ stands for the greatest side length in the subsimplexes, then the theorem asserts the existence of a unique vector $J \in R_y^n$ such that for each $\varepsilon > 0$

$$\Big|\sum_j a(x_j^*)\, D\, h_1^j \ldots h_p^j - J\Big| < \varepsilon\, |D\, h_1 \ldots h_p|\,,$$

provided δ is sufficiently small.

The proof can either be based on the Cauchy convergence criterion, which is valid in metric spaces of finite dimension, or also, if the n real components of the sum are considered separately, on the investigation of lower and upper sums, and goes, upon taking the additivity of D into account, in the well-known fashion.[1]

The limit vector J, which according to this theorem exists, we call the *affine integral* of the alternating differential taken over the simplex s^p, and we write[2]

$$J = \int_{s^p} A(x)\, d_1x \ldots d_px = \int_{s^p} a(x)\, D\, d_1x \ldots d_px .$$

For $p = 1$ we have the line integral

$$\int_{s^1} A(x)\, dx = \int_{x_0 x_1} A(x)\, dx .$$

For $p = 0$ the differential degenerates into a vector function $A(x)$. For "the integral over the zero-dimensional simplex $s^0(x_0) = x_0$" it is convenient to understand simply the vector $A(x_0)$.

The following properties are, just as with the Cauchy-Riemann integral concept, immediate consequences of the definition of the affine integral.

First of all, the integral changes its sign when the simplex s_p is reoriented.

Further, the affine integral is additive in the following sense: If (1.3) is a decomposition of the simplex s_p into like-oriented subsimplexes s^p_j, then one has

$$\int_{s^p} A(x)\, d_1x \ldots d_px = \sum_j \int_{s^p_j} A(x)\, d_1x \ldots d_px .$$

Finally, if the inequality $|a(x)| \leqq \alpha$ holds on s^p, because of the additivity of D,

$$\left|\int_{s^p} A(x)\, d_1x \ldots d_px\right| = \left|\int_{s^p} a(x)\, D\, d_1x \ldots d_px\right| \leqq \alpha\, |D\, h_1 \ldots h_p| .$$

The following representation of the vector function $a(x)$ is a result of this inequality.

If x^* is a point of the simplex s^p and one sets

$$a(x) = a(x^*) + \varepsilon(x) ,$$

---

[1] For the proof of the uniqueness of J one considers two sequences of simplicial subdivisions of s^p. Their intersection is a subdivision of s^p, composed of convex polyhedrons. By induction one shows that the polyhedrons can be divided in simplexes which form a common simplicial subdivision of s^p and of the two given subdivisions (cf. T. Nieminen [1]). The uniqueness of J follows as a consequence of the continuity of $A(x)$.

[2] This integral agrees (except for the notation) with the integral introduced by É. Cartan in his *Calcul extérieur*.

then as a consequence of the continuity of a the length $|\varepsilon(x)|$ is smaller than an arbitrarily small $\varepsilon > 0$ as soon as the edges of the simplex are sufficiently small. In the equation

$$\int_{s^p} A(x)\, d_1x \ldots d_px = a(x^*)\, D\, h_1 \ldots h_p + \int_{s^p} \varepsilon(x)\, D\, d_1x \ldots d_px \,,$$

therefore, according to the above inequality, the norm of the integral on the right is smaller than $\varepsilon\, |D\, h_1 \ldots h_p|$, and consequently

$$a(x^*) = \lim_{s^p \to x^*} \left(\frac{1}{D\, h_1 \ldots h_p} \int_{s^p} A(x)\, d_1x \ldots d_px \right), \tag{1.5}$$

where $s^p \to x^*$ indicates that s^p shrinks in the *fixed* plane $x^* + U^p$ to the point x^*. One sees that $a(x^*) = a(x^*; U^p)$ has the character of a *density* of the operator $A(x)$, in the direction U^p.

1.3. Computation of affine integrals. If the alternating operator $A(x)$ is *independent of the point* x on $s^p(x_0, \ldots, x^p)$, $A(x) \equiv A =$ const., then it follows from (1.4), because of the additivity of the real fundamental form D, that A is also additive, and therefore

$$\int_{s^p} A\, d_1x \ldots d_px = A\, h_1 \ldots h_p \,. \tag{1.6}$$

Second, we consider the case where the operator *depends linearly on* x, $A(x) \equiv A\, x$. Then the density of A in U^p is a linear vector function $a\, x$ of x, and hence

$$A\, x\, d_1x \ldots d_px = a\, x\, D\, d_1x \ldots d_px \,.$$

Decompose s^p barycentrically r-times in succession into $N = ((p+1)!)^r$ subsimplexes $s_j^p(x_0^j, \ldots, x_p^j)$, which according to exercise 6 in I.5.9 have the same affine measure

$$|D\, h_1^j \ldots h_p^j| = \frac{1}{N}\, |D\, h_1 \ldots h_p| \qquad (j = 1, \ldots, N)\,.$$

If these subsimplexes are oriented like s^p and one takes in the sum

$$\sum_{j=1}^{N} a\, x_j^*\, D\, h_1^j \ldots h_p^j = \frac{1}{N} D\, h_1 \ldots h_p \sum_{j=1}^{N} a\, x_j^* = D\, h_1 \ldots h_p\, a \left(\frac{1}{N} \sum_{j=1}^{N} x_j^* \right)$$

x_j^* to be the center of gravity $x_j^* = \bar{x}_j = (1/(p+1)) \sum_{i=0}^{p} x_i^j$ of the subsimplex s_j^p, then (cf. I.5.9, exercise 6)

$$\frac{1}{N} \sum_{j=1}^{N} \bar{x}_j = \frac{1}{p+1} \sum_{i=0}^{p} x_i = \bar{x} \,,$$

where $\bar{x}$ stands for the center of gravity of $s^p(x_0, \ldots, x_p)$, and the above sum is for each r equal to

$$\frac{1}{p+1} D\, h_1 \ldots h_p \sum_{i=0}^{p} a\, x_i = \frac{1}{p+1} \sum_{i=0}^{p} A\, x_i\, h_1 \ldots h_p\,.$$

Consequently,

$$\int_{s^p} A\, x\, d_1x \ldots d_px = \frac{1}{p+1} \sum_{i=0}^{p} A\, x_i\, h_1 \ldots h_p = A\, \bar{x}\, h_1 \ldots h_p\,. \qquad (1.7)$$

Finally, we indicate some general formulas for the calculation of affine integrals which will later be of use to us. For the sake of brevity we content ourselves with a differential geometric argument (Fig. 3).

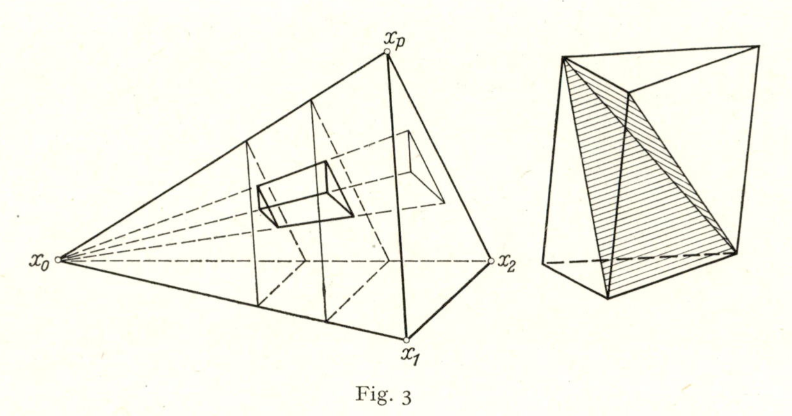

Fig. 3

We assume the side simplex

$$s_0^{p-1}(x_1, \ldots, x_p)$$

of s^p opposite to x_0 to be decomposed somehow into infinitesimal subsimplexes. Let the subsimplex of the former side simplex which contains the point x have edges $d_1x, \ldots, d_{p-1}x$, whereby we so orient the subsimplexes that $D\, h_1\, d_1x \ldots d_{p-1}x$ has the sign of $D\, h_1\, (h_2 - h_1) \ldots (h_p - h_1) = D\, h_1\, h_2 \ldots h_p$. We join the vertices of these subsimplexes of s_0^{p-1} with x_0 and cut the thus resulting pyramids with planes that are parallel to s_0^{p-1} into infinitely thin truncated pyramids which in the limit behave like prisms. We decompose each of these prisms following the classical method of Euclid into p infinitesimal p-dimensional simplexes of equal affine measure (cf. I.5.9, exercise 7).

The prisms between the planes through the points

$$x_0 + \tau\, (x - x_0) \quad \text{and} \quad x_0 + (\tau + d\tau)\, (x - x_0) \qquad (0 \leqq \tau < 1,\, d\tau > 0)\,,$$

each contribute an amount

$$\begin{aligned} &p\, a(x_0 + \tau\,(x - x_0))\, D\, d\tau\,(x - x_0)\,\tau\, d_1 x \ldots \tau\, d_{p-1} x \\ &\qquad = p\,\tau^{p-1}\, d\tau\, a(x_0 + \tau\,(x - x_0))\, D\,(x - x_0)\, d_1 x \ldots d_{p-1} x \\ &\qquad = d(\tau^p)\, A\,(x_0 + \tau\,(x - x_0))\,(x - x_0)\, d_1 x \ldots d_{p-1} x \end{aligned}$$

to the affine integral. For the affine integral taken over s^p we therefore obtain the formula

$$\begin{aligned} &\int_{s^p} A(x)\, d_1 x \ldots d_p x \\ &\quad = \int_0^1 d(\tau^p) \int_{s_0^{p-1}} A\,(x_0 + \tau\,(x - x_0))\,(x - x_0)\, d_1 x \ldots d_{p-1} x \\ &\quad = \int_{s_0^{p-1}} \left(\int_0^1 d(\tau^p)\, A\,(x_0 + \tau\,(x - x_0))\,(x - x_0) \right) d_1 x \ldots d_{p-1} x \\ &\quad = \int_{s_0^{p-1}} \varphi\, A(x)\, d_1 x \ldots d_{p-1} x\,, \end{aligned} \tag{1.8}$$

where the operator $\varphi\, A$ is defined by the operator A of degree p according to the equation

$$\varphi\, A(x) = \int_0^1 d(\tau^p)\, A(x_0 + \tau\,(x - x_0))\,(x - x_0)$$

as an alternating operator of degree $p - 1$. Thus φ is a linear functional that assigns to an alternating operator of degree p an alternating operator of degree $p - 1$.

Repeated application of this reduction formula ultimately yields a representation for the affine integral over the simplex s^p as an ordinary p-fold Cauchy integral over the unit cube in the p-dimensional number space, from which further representations can then be deduced by means of suitable variable transformations. However, since we do not need these, we shall leave them to the reader as exercises.

1.4. Exercises. 1. If $A_p(x)$ is a p-linear alternating operator which is continuous on the simplex $s^p = s^p(x_0, \ldots, x_p)$, then for $q < p$

$$\int_{s^p} A_p(x)\, d_1 x \ldots d_p x = \int_{s^{p-q}} A_{p-q}(x)\, d_1 x \ldots d_{p-q} x\,,$$

where, for $j = p, \ldots, p - q + 1$, $A_{p-q}(x)$ is defined by

$$A_{j-1}(x) = \varphi\, A_j(x) = \int_0^1 d(\tau^j)\, A_j\,(x_{p-j} + \tau\,(x - x_{p-j}))\,(x - x_{p-j})$$

as a $(p - q)$-linear alternating operator on the simplex $s^{p-q} = s^{p-q}(x_q, \ldots, x_p)$ which is provided with the orientation induced by the orientation of s^p. For $q = p$,

$$\int_{s^p} A_p(x)\, d_1x \ldots d_px = A_0$$

$$= \int_0^1 d\tau\, A_1\left(x_{p-1} + \tau\,(x_p - x_{p-1})\right)\,(x_p - x_{p-1})$$

$$= \int_{x_{p-1}x_p} A_1(x)\, dx\,.$$

2. Let the p-linear alternating operator $A(x)$ be continuous on the simplex $s^p = s^p(x_0, \ldots, x_p)$. Further let $\bar{x} = \bar{x}(x)$, $x = x(\bar{x})$ be an affine mapping which transforms s^p into $\bar{s}^p = \bar{s}^p(\bar{x}_0, \ldots, \bar{x}_p)$ so that $\bar{x}_i = \bar{x}(x_i)$ $(i = 0, \ldots, p)$. Then

$$\int_{s^p} A(x)\, d_1x \ldots d_px = \int_{\bar{s}^p} \bar{A}(\bar{x})\, d_1\bar{x} \ldots d_p\bar{x}\,,$$

where the p-linear alternating operator $\bar{A}(\bar{x})$ is obtained by the substitution $x = x(\bar{x})$:

$$\bar{A}\, d_1\bar{x} \ldots d_p\bar{x} = A\,\frac{dx}{d\bar{x}}\,d_1\bar{x} \ldots \frac{dx}{d\bar{x}}\,d_p\bar{x}\,.$$

§ 2. Theorem of Stokes

2.1. Formulation of the problem. Let

$$s^{p+1}(x_0, \ldots, x_{p+1}) \qquad (x_i - x_0 = h_i)$$

be a closed $(p + 1)$-dimensional simplex in the space R_x^m $(p \leqq m - 1)$. We determine the orientation of the simplex s^{p+1} from the sign of the real $(p + 1)$-linear alternating fundamental form

$$D\, h_1 \ldots h_{p+1} = \Delta(x_0, \ldots, x_{p+1})$$

of the subspace U^{p+1} spanned by the edges $h_1, \ldots, h_{p+1}$ of the simplex; suppose this orientation is positive, i.e., for the above ordering of the vertices $D\, h_1 \ldots h_{p+1} > 0$.

On s^{p+1} we consider a p-linear alternating form

$$A(x)\, k_1 \ldots k_p \in R_y^n \qquad (x \in s^{p+1},\ k_i \in U^{p+1})\,.$$

If the operator $A(x)$ is continuous on s^{p+1}, we can form the integral of the differential form $A(x)\, d_1x \ldots d_px$ over the boundary ∂s^{p+1} of s^{p+1} as the sum of the integrals over the $p + 2$ boundary simplexes $s_i^p(x_0, \ldots, \hat{x}_i, \ldots, x_{p+1})$ $(i = 0, \ldots, p + 1)$. The induced orientation of the boundary simplex s_i^p has the sign $(-1)^i$, so that

$$\int_{\partial s^{p+1}} A(x)\, d_1x \ldots d_px = \sum_{i=0}^{p+1} (-1)^i \int_{s_i^p} A(x)\, d_1x \ldots d_px\,. \tag{2.1}$$

The theorem of Stokes transforms this boundary integral into a $(p+1)$-fold integral over s^{p+1}. To derive this theorem we analyze the boundary integral (2.1) more carefully.

2.2. Special cases. We first consider two simple kinds of operators $A(x)$.

First, suppose $A(x) \equiv A$ is *independent of* x. Then according to (1.6), for $i = 0$,

$$\begin{aligned}\int_{s_0^p} A\, d_1x \ldots d_px &= A\,(x_2 - x_1) \ldots (x_{p+1} - x_1)\\ &= A\,(h_2 - h_1) \ldots (h_{p+1} - h_1)\\ &= \sum_{i-1}^{p+1} (-1)^{i-1} A\, h_1 \ldots \hat{h}_i \ldots h_{p+1},\end{aligned}$$

and for $i = 1, \ldots, p+1$

$$(-1)^i \int_{s_i^p} A\, d_1x \ldots d_px = (-1)^i A\, h_1 \ldots \hat{h}_i \ldots h_{p+1}.$$

Thus by (2.1)

$$\int_{\partial s^{p+1}} A\, d_1x \ldots d_px = 0\,. \tag{2.2}$$

Second, suppose $A(x) \equiv A\,x$ is *linear in* x. By (1.7), for $i = 0$, one has

$$\begin{aligned}\int_{s_0^p} A\,x\, d_1x \ldots d_px &= A\,\bar{x}_0\,(h_2 - h_1) \ldots (h_{p+1} - h_1)\\ &= \sum_{i-1}^{p+1} (-1)^{i-1} A\,\bar{x}_0\, h_1 \ldots \hat{h}_i \ldots h_{p+1},\end{aligned}$$

and for $i = 1, \ldots, p+1$

$$(-1)^i \int_{s_i^p} A\,x\, d_1x \ldots d_px = (-1)^i A\,\bar{x}_i\, h_1 \ldots \hat{h}_i \ldots h_{p+1},$$

whereby $\bar{x}_i = (1/(p+1))\,(x_0 + \cdots + \hat{x}_i + \cdots + x_{p+1})$ $(i = 0, \ldots, p+1)$ stands for the center of gravity of s_i^p. This allows (2.1) to be written

$$\int_{\partial s^{p+1}} A\,x\, d_1x \ldots d_px = \frac{1}{p+1} \sum_{i=1}^{p+1} (-1)^{i-1} A\, h_i\, h_1 \ldots \hat{h}_i \ldots h_{p+1}\,. \tag{2.3}$$

The expression on the right is an alternating form on the $p+1$ vectors $h_1, \ldots, h_{p+1}$.

2.3. The differential formula of Stokes. We now go on to the general case where $A(x)$ is a p-linear alternating operator with the following properties:

1. *$A(x)$ is continuous on the simplex s^{p+1}.*

2. *$A(x)$ is differentiable at an arbitrary interior or boundary point x^* of the simplex s^{p+1}.*

Hence, with arbitrary vectors $d_1x, \ldots, d_px$ from the space R^m_x, after the introduction of a Minkowski metric,

$$\begin{aligned} &A(x)\, d_1x \ldots d_px \\ &= A(x^*)\, d_1x \ldots d_px + A'(x^*)\,(x - x^*)\, d_1x \ldots d_px \\ &+ |x - x^*|\,(x - x^*; x^*)\, d_1x \ldots d_px\,, \end{aligned} \tag{2.4}$$

where the norm of the alternating operator $(x - x^*; x)$ converges to zero for $|x - x^*| \to 0$.

To compute the boundary integral (2.1) substitute the expression (2.4). For $i = 0, \ldots, p + 1$

$$\begin{aligned} &\int_{s_i^p} A(x)\, d_1x \ldots d_px \\ &= \int_{s_i^p} A(x^*)\, d_1x \ldots d_px + \int_{s_i^p} A'(x^*)\,(x - x^*)\, d_1x \ldots d_px + r_i\,, \end{aligned}$$

where

$$r_i = \int_{s_i^p} |x - x^*|\,(x - x^*; x^*)\, d_1x \ldots d_px\,.$$

We let δ stand for the length of the greatest diameter of s^{p+1}. Then for each $x \in s^{p+1}$, $|x - x^*| \leqq \delta$. Further, let

$$\varepsilon(s^{p+1}) = \sup_{x \in s^{p+1}} |(x - x^*; x^*)|\,. \tag{2.5}$$

To estimate the remainder term r_0, using the orienting fundamental form $D\, k_1 \ldots k_{p+1}$ of the space U^{p+1}, one sets

$$(x - x^*; x^*)\, d_1x \ldots d_px \equiv \varepsilon_0(x)\, D\, h_1\, d_1x \ldots d_px$$

on s_0^p. Then $\varepsilon_0(x) \in R^n_y$ is uniquely determined on the latter side simplex, and for $d_ix = h_{i+1} - h_1$ $(h_i = x_i - x_0)$ we have

$$\begin{aligned} (x - x^*; x^*)\,(h_2 - h_1) \ldots (h_{p+1} - h_1) &= \varepsilon_0(x)\, D\, h_1\,(h_2 - h_1) \ldots (h_{p+1} - h_1) \\ &= \varepsilon_0(x)\, D\, h_1 \ldots h_{p+1}\,. \end{aligned}$$

Thus using the definitions of δ and $\varepsilon(^{p+1})$ we see that

$$\begin{aligned} |\varepsilon_0(x)|\, D\, h_1 \ldots h_{p+1} &\leqq |(x - x^*; x^*)|\, |h_2 - h_1| \ldots |h_{p+1} - h_1| \\ &\leqq \delta^p\, \varepsilon(s^{p+1})\,. \end{aligned}$$

Because of the additivity of the absolute value of the p-linear alternating differential $D_0\, d_1x \ldots d_px \equiv D\, h_1\, d_1x \ldots d_px$ it follows from here that

$$\begin{aligned} |r_0| &\leqq \int_{s_0^p} |x - x^*|\, |\varepsilon_0(x)|\, |D\, h_1\, d_1x \ldots d_px| \\ &\leqq \frac{\delta^{p+1}}{D\, h_1 \ldots h_{p+1}}\, \varepsilon(s^{p+1}) \int_{s_0^p} |D\, h_1\, d_1x \ldots d_px| = \delta^{p+1}\, \varepsilon(s^{p+1})\,. \end{aligned}$$

In order to find an estimate for the remainder terms r_i ($i = 1, \ldots, p+1$), one writes on s_i^p

$$(x - x^*; x^*)\, d_1x \ldots d_px \equiv \varepsilon_i(x)\, D\, d_1x \ldots d_{i-1}x\, h_i\, d_ix \ldots d_px\,,$$

which uniquely determines $\varepsilon_i(x) \in R_y^n$ on the above side simplex. In particular, for the vectors $d_jx = h_j$ ($j = 1, \ldots, i-1$), $d_jx = h_{j+1}$ ($j = i, \ldots, p$) one has

$$(x - x^*; x^*)\, h_1 \ldots \hat{h}_i \ldots h_{p+1} = \varepsilon_i(x)\, D\, h_1 \ldots h_{p+1}\,.$$

Because of the additivity of the absolute value of the p-linear alternating differential $D_i\, d_1x \ldots d_px \equiv D\, d_1x \ldots d_{i-1}x\, h_i\, d_ix \ldots d_px$ one obtains in this way precisely the same estimate for r_i as we did above for r_0,

$$|r_i| \leqq \delta^{p+1}\, \varepsilon(s^{p+1})\,.$$

Summation over i now permits (2.1) to be written

$$\begin{aligned}&\int_{\partial s^{p+1}} A(x)\, d_1x \ldots d_px \\ &= \int_{\partial s^{p+1}} (A(x^*) - A'(x^*)\, x^*)\, d_1x \ldots d_px + \int_{\partial s^{p+1}} A'(x^*)\, x\, d_1x \ldots d_px + r\,,\end{aligned}$$

with

$$|r| \leqq (p+2)\, \delta^{p+1}\, \varepsilon(s^{p+1})\,, \tag{2.6}$$

where $\varepsilon(s^{p+1})$ is a number that converges to zero when the simplex s^{p+1} is allowed to converge in the fixed plane $x^* + U^{p+1}$ to the point x^*.

By (2.2) the first integral on the right vanishes. According to (2.3) the contribution made by the second integral becomes

$$\int_{\partial s^{p+1}} A'(x^*)\, x\, d_1x \ldots d_px = \frac{1}{p+1} \sum_{i=1}^{p+1} (-1)^{i-1} A'(x^*)\, h_i\, h_1 \ldots \hat{h}_i \ldots h_{p+1}\,,$$

and finally one finds

$$\begin{aligned}&\int_{\partial s^{p+1}} A(x)\, d_1x \ldots d_px \\ &= \frac{1}{p+1} \sum_{i=1}^{p+1} (-1)^{i-1} A'(x^*)\, h_i\, h_1 \ldots \hat{h}_i \ldots h_{p+1} + r\,.\end{aligned} \tag{2.7}$$

Equation (2.7) together with the estimate (2.6) of the remainder term contain the *differential formula of Stokes*, which has been established under hypotheses 1 and 2 of this section.

2.4. The exterior differential. The rotor. By means of the above analysis, associated with the alternating differential $\omega = A(x)\, h_1 \ldots h_p$ of degree p there is an alternating differential form

$$\begin{aligned}&\wedge A'(x)\, h_1 \ldots h_{p+1} \\ &= \frac{1}{p+1} \sum_{i=1}^{p+1} (-1)^{i-1} A'(x)\, h_i\, h_1 \ldots \hat{h}_i \ldots h_{p+1}\end{aligned} \tag{2.8}$$

of degree $p + 1$, the alternating part of the differential of the given form ω. This $(p + 1)$-form is (up to the factor $1/(p + 1)$) the same as the *exterior differential* of the form ω introduced by É. Cartan[1].

The operator defined by the exterior differential is the *rotor* of the operator $A(x)$:

$$\operatorname{rot} A(x) = \wedge A'(x) . \tag{2.8'}$$

2.5. Coordinate representation of the rotor. We start from the expression (2.8) for the rotor and restrict the differentials $d_i x$ to a $(p + 1)$-dimensional subspace U^{p+1} of the linear space R_x^m and consequently consider the rotor in the direction U^{p+1}.

If in a linear coordinate system $e_1, \ldots, e_{p+1}$ of this subspace

$$d_i x = \sum_{j=1}^{p+1} d\xi_i^j e_j \qquad (i = 1, \ldots, p + 1) ,$$

then

$$A(x)\, d_1 x \ldots \hat{d_i x} \ldots d_{p+1} x = \sum_{j=1}^{p+1} \Delta_i^j \alpha_j(x) ,$$

where

$$\alpha_j(x) = A(x)\, e_1 \ldots \hat{e}_j \ldots e_{p+1}$$

and Δ_i^j stands for the subdeterminant of the complete $(p + 1)$-rowed determinant

$$\Delta = \begin{vmatrix} d\xi_1^1 & \ldots & d\xi_{p+1}^1 \\ \vdots & & \vdots \\ d\xi_1^{p+1} & \ldots & d\xi_{p+1}^{p+1} \end{vmatrix}$$

associated with the differential $d\xi_i^j$.

Let $A(x)\, k_1 \ldots k_p$ $(k_i \in U^{p+1})$, and thereby also the vector functions $\alpha_j(x)$, be real. Then by the above

$$A'(x)\, d_i x\, d_1 x \cdots \hat{d_i x} \ldots d_{p+1} x = \sum_{j=1}^{p+1} \sum_{k=1}^{p+1} \frac{\partial \alpha_j}{\partial \xi^k} d\xi_i^k \Delta_i^j ,$$

and consequently

$$\operatorname{rot} A(x)\, d_1 x \ldots d_{p+1} x = \frac{1}{p+1} \sum_{j=1}^{p+1} \sum_{k=1}^{p+1} \frac{\partial \alpha_j}{\partial \xi_k} \sum_{i=1}^{p+1} (-1)^{i-1} d\xi_i^k \Delta_i^j .$$

---

[1] Our presentation of the "exterior calculus" deviates from that of Cartan in that we proceed in a coordinate-free fashion. A second formal difference is found in the differing notations for the exterior differential. Usually one writes $d\omega$ for the exterior differential of the form ω. Since we use the symbol d only for the *ordinary* differentiation operator, we prefer the more explicit notation $\wedge\, d\omega$ (the alternating part of the ordinary differential $d\omega$).

Here the sum over i on the right vanishes for $k \neq j$, and for $k = j$ it is equal to $(-1)^{j-1} \Delta$. Therefore,

$$\operatorname{rot} A(x)\, d_1x \ldots d_{p+1}x = \frac{\Delta}{p+1} \sum_{j=1}^{p+1} (-1)^{j-1} \frac{\partial \alpha_j}{\partial \xi^j},$$

which we can write

$$\operatorname{rot} A(x)\, d_1x \ldots d_{p+1}x = \varrho(x)\, D\, d_1x \ldots d_{p+1}x\,,$$

where D stands for the real alternating fundamental form of the subspace U^{p+1} with $D\, e_1 \ldots e_{p+1} = 1$, and

$$\varrho(x) = \frac{1}{p+1} \sum_{j=1}^{p+1} (-1)^{j-1} \frac{\partial \alpha_j}{\partial \xi^j}$$

is the coordinate representation of the *rotor density* in the direction of this subspace.

2.6. Extension of the definition of the rotor. Suppose the p-linear alternating operator $A(x)$ is continuous in a neighborhood of the point $x^* \in R_x^m$ and differentiable at this point. Further, let U^{p+1} be a subspace and $s^{p+1}(x_0, \ldots, x_{p+1})$ a simplex in the plane $x^* + U^{p+1}$ that lies in the neighborhood mentioned and that contains x^* as an interior or boundary point. Then by 2.3 Stokes's differential formula holds,

$$\int_{\partial s^{p+1}} A(x)\, d_1x \ldots d_px = \operatorname{rot} A(x^*)\, h_1 \ldots h_{p+1} + \delta^{p+1}(s^{p+1}; x^*)\,,$$

where δ stands for the greatest side length $|h_i| = |x_i - x_0|$ and $(s^{p+1}; x^*)$ designates a vector whose length vanishes with δ.

This formula can, conversely, be used to define $\operatorname{rot} A(x^*)$ by postulating that the formula holds. In this way we find the following, relative to (2.8), generalized

Definition. *Let the p-linear alternating operator $A(x)$ be continuous in a neighborhood of the point x^*. If a $(p+1)$-linear operator $B(x^*)$ exists such that for each simplex $s^{p+1}(x_0, \ldots, x_{p+1})$ in the space R_x^m that contains the neighborhood mentioned and the point x^* a decomposition*

$$\int_{\partial s^{p+1}} A(x)\, d_1x \ldots d_px = B(x^*)\, h_1 \ldots h_{p+1} + \delta^{p+1}(s^{p+1}; x^*) \tag{2.9}$$

holds with $|(s^{p+1}; x^*)| \to 0$ for $\delta = \max |h_i| = \max |x_i - x_0| \to 0$, then we call $B(x^*)$ the rotor of $A(x)$ at the point x^* and write $B(x^*) = \operatorname{rot} A(x^*)$.

Regarding this definition observe the following. First, it is clear that the rotor, provided it exists in the sense of the above definition at a point x^*, is *uniquely* determined by $A(x)$.

Further, it follows from the definition that the operator $B(x^*)$, for which $(p+1)$-linearity was hypothesized, is an *alternating* operator. For if two vectors h_i and h_j, and thus the vertices x_i and x_j of the simplex, are commuted in (2.9), the orientation of the simplex s^{p+1} and of the boundary ∂s^{p+1}, and with that also the left hand side of the equation changes sign. If in addition the vectors h are replaced by λh $(0 < \lambda \leq 1)$, it follows from this equation that

$$0 = B(x^*) \ldots h_i \ldots h_j \ldots + B(x^*) \ldots h_j \ldots h_i \ldots + (\lambda) ,$$

with $|(\lambda)| \to 0$ for $\lambda \to 0$, from which the assertion follows.

Therefore, at each point x where rot $A(x)$ exists in the sense of definition (2.9) one can write

$$\operatorname{rot} A(x)\, h_1 \ldots h_{p+1} = \varrho(x)\, D\, h_1 \ldots h_{p+1} , \tag{2.10}$$

where D is the real alternating fundamental form of the subspace U^{p+1} spanned by the $p + 1$ vectors h_i and $\varrho(x) = \varrho(x; A; U^{p+1})$ stands for the *rotor density* of the operator $A(x)$ in the direction U^{p+1}. The definition (2.9) yields the representation

$$\varrho(x; A; U^{p+1}) = \lim_{s^{p+1} \to x} \left(\frac{1}{D h_1 \ldots h_{p+1}} \int_{\partial s^{p+1}} A(x)\, d_1 x \ldots d_p x \right) \tag{2.11}$$

for this rotor density, where the limit $s^{p+1} \to x$ is to be taken in the fixed plane U^{p+1} in such a way that the *regularity index* of the simplex s^{p+1},

$$\operatorname{reg} s^{p+1} = \frac{\delta^{p+1}}{|D\, h_1 \ldots h_{p+1}|} = \frac{\delta^{p+1}}{V(s^{p+1})} ,$$

remains below a finite bound; here $V(s^{p+1})$ is set equal to the "volume" $|D\, h_1 \ldots h_{p+1}|$ of s^{p+1}.

Finally, the analysis carried out in 2.3 shows that for the existence of the rotor in the sense of definition (2.9) it in any case *suffices* for $A(x)$ to be *continuous* in a neighborhood of x^* and *differentiable* at this point, in which case the rotor can be represented by formula (2.8), and the coordinate representations derived in 2.5 also hold. The above definition, however, hypothesizes nothing about the differentiability of the operator $A(x)$ and therefore provides an extension of the more narrow definition (2.8).

For $p = 0$ both definitions are equivalent. Then the "0-linear alternating operator" $A(x)$ is simply a vector function $A(x)$, and by (2.8) one then has rot $A(x^*) = A'(x^*)$, while equation (2.9) degenerates into

$$A(x_1) - A(x_0) = B(x^*)\,(x_1 - x_0) + |x_1 - x_0|\,(x_1 - x_0; x^*)$$

with $|(x_1 - x_0; x^*)| \to 0$ for $x_0, x_1 \to x^*$. For $x_0 = x^*$ the equation states that $A(x)$ is differentiable at x^* with the derivative $A'(x^*) = B(x^*)$, so that also according to the second definition rot $A(x^*) = A'(x^*)$. By this latter definition the differential operator rot can therefore be thought of as a formal generalization of the derivative as one passes from a 0- to a p-linear alternating operator.

2.7. The transformation formula of Stokes. We now assume that the p-linear alternating operator $A(x)$ satisfies the following conditions on the closed simplex $s^{p+1}(x_0, \ldots, x_{p+1})$:

1. $A(x)$ *is continuous on* s^{p+1}.

2. rot $A(x)$ *exists in the sense of the extended definition* (2.9) *at each point* x *of the simplex* s^{p+1}.

3. rot $A(x)$ *is continuous on* s^{p+1}.

According to the discussion in 2.3 and 2.6 it is sufficient for this that $A(x)$ be continuously differentiable on the closed simplex s^{p+1}. In what follows we only make use of the three above assumptions, which say nothing about the differentiability of $A(x)$, and prove

Stokes's integral theorem. *Provided the p-linear alternating operator $A(x)$ satisfies the above three conditions on the closed simplex $s^{p+1}(x_0, \ldots, x_{p+1}) \subset R_x^m$, the integral transformation formula*

$$\int_{\partial s^{p+1}} A(x)\, d_1x \ldots d_px = \int_{s^{p+1}} \operatorname{rot} A(x)\, d_1x \ldots d_{p+1}x \tag{2.12}$$

holds, where the boundary ∂s^{p+1} for a given orientation of s^{p+1} is endowed with the induced orientation.

For the proof we first remark that as a consequence of the hypotheses made on the simplex s^{p+1} as well as on each $(p+1)$-dimensional subsimplex s both of the integrals in this formula make sense. Thus for each such subsimplex s the difference

$$J(s) \equiv \int_{\partial s} A(x)\, d_1x \ldots d_px - \int_{s} \operatorname{rot} A(x)\, d_1x \ldots d_{p+1}x \tag{2.13}$$

is meaningful, and in the set (s) of these simplexes it defines a well-determined set function. Stokes's integral theorem asserts that $J(s)=0$.

The following proof uses a well-known idea which was applied by Goursat to establish Cauchy's integral theorem. It rests on the following two properties of the set function $J(s)$.

In order to formulate the first one we consider a point x^* of the closed simplex $s^{p+1}(x_0, \ldots, x_{p+1})$ and a subsimplex $s = s(y_0, \ldots, y_{p+1})$ that contains this point. Let D be the real, orienting fundamental form of the plane $x^* + U^{p+1}$ of the simplex s^{p+1}. Further let

$$V(s) = |D\, k_1 \ldots k_{p+1}| \qquad (k_i = y_i - y_0)$$

be the "volume" of s and δ the greatest side length $|k_i|$. Then if s converges to x^* so that the regularity index of s,

$$\operatorname{reg} s = \frac{\delta^{p+1}}{V(s)},$$

remains below a finite bound, then

$$\lim_{s \to x^*} \frac{|J(s)|}{V(s)} = 0 \tag{2.14}$$

at each point $x^* \in s^{p+1}$.

In fact, if $(s; x^*)$ stands for a quantity that vanishes with δ, then it follows on the one hand from the existence of rot $A(x^*)$ according to definition (2.9) that

$$\int_{\partial s} A(x)\, d_1 x \ldots d_p x = \text{rot}\, A(x^*)\, k_1 \ldots k_{p+1} + \delta^{p+1}(s; x^*)_1 .$$

On the other hand, because of the continuity of rot $A(x)$ at the point x^*,

$$\int_s \text{rot}\, A(x)\, d_1 x \ldots d_{p+1} x = \text{rot}\, A(x^*)\, k_1 \ldots k_{p+1} + V(s)\,(s; x^*)_2 .$$

Thus

$$|J(s)| = |\delta^{p+1}(s; x^*)_1 - V(s)\,(s; x^*)_2| \leqq (\text{reg}\, s\, |(s; x^*)_1| + |(s; x^*)_2|)\, V(s)$$

which implies (2.14), provided reg s remains bounded as $\delta \to 0$.

The second property of the set function $J(s)$ is its additivity, in the following sense: If Z stands for a decomposition of a simplex s in the set (s) considered into a finite number of $(p + 1)$-dimensional simplexes s_Z which are oriented the same as s, then

$$J(s) = \sum_Z J(s_Z) . \tag{2.15}$$

Now the volume $V(s)$ is also additive, and it follows from (2.15) that

$$\frac{|J(s)|}{V(s)} = \frac{1}{V(s)} \left| \sum_Z J(s_Z) \right| \leqq \frac{1}{V(s)} \sum_Z |J(s_Z)| ,$$

and if one sets

$$\max_Z \frac{|J(s_Z)|}{V(s_Z)} = M_Z ,$$

$$\frac{|J(s)|}{V(s)} \leqq \frac{1}{V(s)} \left(\sum_Z V(s_Z) \right) M_Z = M_Z . \tag{2.16}$$

Now suppose $Z_1, Z_2, \ldots$ is an infinite sequence of decompositions of the simplex s with the following properties:[1]

A. *Z_{i+1} is a subdivision of Z_i.*

B. *Z_i is refined without bound for $i \to \infty$.*

C. *The regularity indices of all subsimplexes s_Z that occur are uniformly bounded.*

Then let s_1 stand for a simplex in the decomposition Z_1 for which the maximum $M_{Z_1} = M_1$ is reached, $s_2 \subset s_1$ a simplex in the decomposition Z_2 which corresponds to the maximum $M_{Z_2} = M_2$, whereby one only takes into account those subsimplexes in the decomposition Z_2 which lie in s_1, etc. Then according to inequality (2.16)

$$\frac{|J(s)|}{V(s)} \leqq M_1 \leqq M_2 \leqq \ldots \tag{2.17}$$

Since the simplexes $s, s_1, s_2, \ldots$ are nested, there exists, because of condition B, a well-determined point $x^* \in s_i$ $(i = 1, 2, \ldots)$ such that

[1] The construction of H. Whitney (I.5.5) gives an example of such a decomposition (cf. 2.8). Cf. also T. Nieminen [1].

$s_i \to x^*$ for $i \to \infty$. Now in view of condition C and equation (2.14)

$$\lim_{i\to\infty} M_i = \lim_{i\to\infty} \frac{|J(s_i)|}{V(s_i)} = 0 ,$$

and, as a consequence of the inequalities in (2.17),

$$J(s) = 0$$

for every simplex s of the set (s), in particular also for $s = s^{p+1}$. With that Stokes's integral theorem is proved.

2.8. A sequence of subdivisions. Using the method of H. Whitney, discussed in I.5.4—5, we will construct a sequence $Z_1, Z_2, \ldots$ of subdivisions satisfying the conditions A, B, C.

Let $s^{p+1}(x_0, x_1, \ldots, x_{p+1})$ be a simplex with vertices $x_0, x_1, \ldots, x_{p+1}$. Denoting the edges

$$k_1 = x_1 - x_0 \,, k_2 = x_2 - x_1, \ldots, k_{p+1} = x_{p+1} - x_p \,,$$

the affine volume of s^{p+1} is equal to

$$\begin{aligned} \Delta(s^{p+1}) &= D\,(x_1 - x_0)\,(x_2 - x_0) \ldots (x_{p+1} - x_0) \\ &= D\, k_1\, k_2 \ldots k_{p+1} \,. \end{aligned}$$

Each simplex s_σ^{p+1} of the subdivision $D = Z_1$ of s^{p+1}, constructed by the method of section I. 5.5, has as edges $p + 1$ vectors defined by

$$x_{rt} - x_{r(t-1)} = \frac{1}{2}\left((x_r + x_t) - (x_r + x_{t-1})\right) = \frac{1}{2} k_t$$

or by

$$x_{rt} - x_{(r+1)t} = -\frac{1}{2} k_r \,,$$

i.e., the vectors

$$\pm \frac{1}{2} k_1, \ldots, \pm \frac{1}{2} k_{p+1}$$

with one combination of the signs. The affine volume of s_σ^{p+1} is therefore

$$\Delta(s_\sigma^{p+1}) = \pm \frac{1}{2^{p+1}} \Delta(s^{p+1}) \,,$$

and s_σ^{p+1} is similar to one of the simplexes with the edges $\pm k_1, \ldots, \pm k_{p+1}$. The regularity indices of similar simplexes have the same value. Thus, if ϱ is the maximal value of the regularity indices of the simplexes with the edges $\pm k_1, \ldots, \pm k_{p+1}$ (with all combinations of the signs), then ϱ is equal to the maximum of the regularity indices of all simplexes of the division Z_1. The same is true for the subdivisions $Z_2, Z_3, \ldots$, where Z_{j+1} is the subdivision of Z_j constructed by the method of Whitney. Hence property C has been proved. Condition A is valid by construction, and property B is evident, because the edges of the simplexes of Z_{j+1} are obtained by dividing the corresponding edges of Z_j in two equal parts.

2.9. Remark. As a comment on the proof of Stokes's theorem given above we note that another method of proof which has often been used in the theory of integral transformation formulas would at first glance seem more obvious than the one followed above, which in principle stems from Goursat. The former proof would go briefly as follows.

Under the above hypotheses 1, 2, 3 decompose the simplex s^{p+1} into positively oriented subsimplexes s_i^{p+1} $(i = 1, \ldots, N)$ and write the boundary integral of the given alternating differential $A(x)\, d_1x \ldots d_px$

$$\int_{\partial s^{p+1}} A(x)\, d_1x \ldots d_px = \sum_{i=1}^{N} \int_{\partial s_i^{p+1}} A(x)\, d_1x \ldots d_px\,. \tag{2.18}$$

On the ith subsimplex, which we designate by $s = s_i^{p+1}$ for short, we choose some point $x_i^* = x$ and then have by the definition (2.9) of the rotor

$$\int_{\partial s} A(x)\, d_1x \ldots d_px = \operatorname{rot} A(x)\, k_1 \ldots k_{p+1} + \delta^{p+1}(s; x)\,, \tag{2.18'}$$

where $k_1, \ldots, k_{p+1}$ are the edge vectors, whose greatest length is δ, that span the simplex s; the quantity $(s; x)$ vanishes when the simplex converges to the fixed point x.

If one proceeds in a similar way for all N subsimplexes $s = s_i^{p+1}$, the sum of the first terms on the right in (2.18') yields an expression which in view of the hypothesized continuity of the operator $\operatorname{rot} A(x)$ tends to the integral of $\operatorname{rot} A(x)$ over s^{p+1} when the decomposition is (regularly) refined without limit. Because of (2.18) Stokes's theorem is thus proved provided one succeeds in proving that the sum of the remainder terms $r = \delta^{p+1}(s; x)$ vanishes in the limit.

Let us see what can be said about this last question. For the individual remainder term r one has the estimate

$$|r| = \delta^{p+1}\, |(s; x)| \leqq M\, |D\, k_1 \ldots k_{p+1}|\, |(s; x)|\,,$$

where M is a finite upper bound given a priori for the regularity indices which appear. According to the definition (2.9) of the rotor, $(s; x)$ vanishes for a regular approach of s to the *fixed* point x. If one knew in addition that the convergence $|(s; x)| \to 0$ for $\delta \to 0$ were *uniform* with respect to all points $x \in s^{p+1}$, summation of the remainder terms r would obviously yield an expression which for indefinite refinement of the decomposition vanishes, and the proof of Stokes's theorem would then be completed.

But the uniform vanishing for $\delta \to 0$ of the quantity required in this method of proof cannot in general be deduced directly from hypotheses 1, 2, 3 in 2.7. In the case $p = 0$ it follows, of course, from the mean value theorem. But an analogous generalized mean value theorem is, in general, not available for $p > 0$; rather such a theorem results only as a corollary to Stokes's theorem (cf. 2.11, exercise 2), which is to be proved. Because of this circularity, the proof sketched above fails for $p > 0$[1].

On the other hand, the proof of Stokes's theorem starting from the general postulates 1, 2, 3 following the method of 2.7, which in no way needs the *uniform* existence of the rotor of $A(x)$ in the sense described above, succeeds. Herein lies the real point of Goursat's idea.

The advantages of the above sharpened formulation of Stokes's theorem will be clear in the following applications.

2.10. The divergence. As before, let $Dh_1 \ldots h_{p+1}$ be an alternating fundamental form on the subspace U^{p+1} spanned in the space R^m_x by the $p+1$ ($\leq m$) vectors h_i. By means of this form, for a given differentiable vector field $u(x) \in U^{p+1}$ ($x \in R^m_x$), another differential operator, the *divergence* of $u(x)$, can be defined by computing the trace of the operator $u'(x)$ (cf. I. 5.9, exercise 4):

$$\operatorname{div} u(x) \equiv Tr\, u'(x) \equiv \frac{\sum_{i=1}^{p+1} D h_1 \ldots h_{i-1} (u'(x) h_i) h_{i+1} \ldots h_{p+1}}{D h_1 \ldots h_{p+1}}. \tag{2.19}$$

There exists a simple connection between the linear operators div and rot. We have

$$\sum_{i=1}^{p+1} D h_1 \ldots h_{i-1} (u'(x) h_i) h_{i+1} \ldots h_{p+1}$$
$$= \sum_{i=1}^{p+1} (-1)^{i-1} D (u'(x) h_i) h_1 \ldots \hat{h}_i \ldots h_{p+1} .$$

The last expression is equal to the exterior differential of the p-linear alternating differential form

$$A(x) h_1 \ldots h_p \equiv D u(x) h_1 \ldots h_p$$

and can thus be denoted by

$$(p+1) \operatorname{rot} (D u(x)) h_1 \ldots h_{p+1} .$$

The divergence of $u(x)$ is therefore equal to the rotor density $\varrho(x)$ of the operator $D u(x)$ multiplied by $p+1$:

$$\operatorname{div} u(x) = (p+1) \varrho(x) = (p+1) \frac{\operatorname{rot} (D u(x)) h_1 \ldots h_{p+1}}{D h_1 \ldots h_{p+1}}. \tag{2.20}$$

---

[1] If one assumes more specially that the operator $A(x)$ is *continuously differentiable,* then an application of the expression (2.8) for rot $A(x)$ yields the property essential to the above method of proof, the uniform vanishing of $(s; x)$, and the path sketched above becomes feasible (cf. II.1.13, exercise 7).

Provided the rotor on the right exists in the extended sense 2.9) of this concept, this equation gives a corresponding extended definition of the divergence, which is obviously independent of the particular normalization of the fundamental form D of the space U^{p+1}.

If a coordinate system $e_1, \ldots, e_{p+1}$ is introduced in the subspace U^{p+1} and the vector $u(x)$ has the representation

$$u(x) = \sum_{j=1}^{p+1} \omega^j(x)\, e_j\,,$$

one obtains the usual coordinate-dependent representation for div u

$$\operatorname{div} u = \sum_{j=1}^{p+1} \frac{\partial \omega^j}{\partial \xi^j}\,,$$

where the ξ^j are the coordinates of the point $x \in R_x^m$ with respect to the coordinate system $e_1, \ldots, e_{p+1}$ of the space U^{p+1} (cf. 2.12, exercise 4).

2.11. Gauss's transformation formula. Let $u(x)$ be a vector field defined on the closed simplex

$$s^{p+1} = s^{p+1}(x_0, \ldots, x_{p+1}) \qquad (x_i - x_0 = h_i)$$

with values in the subspace U^{p+1} of R_x^m $(p + 1 \leqq m)$ spanned by $h_1, \ldots, h_{p+1}$. If one applies Stokes's transformation formula (2.12) to the alternating differential[1]

$$A(x)\, d_1x \ldots d_px \equiv D\, u(x)\, d_1x \ldots d_px\,,$$

then we get

$$\int_{\partial s^{p+1}} D\, u(x)\, d_1x \ldots d_px = \int_{s^{p+1}} \operatorname{rot}\,(D\, u(x))\, d_1x \ldots d_{p+1}x$$

or, using the above definition (2.20) of the divergence,

$$\int_{\partial s^{p+1}} D\, u(x)\, d_1x \ldots d_px = \frac{1}{p+1} \int_{s^{p+1}} \operatorname{div} u(x)\, D\, d_1x \ldots d_{p+1}x\,. \tag{2.21}$$

This is the *affine form of Gauss's transformation formula.* In order to see this and to bring the formula into its usual metric formulation, we introduce a euclidean metric in U^{p+1} and let $e_1, \ldots, e_{p+1}$ stand for an orthonormal coordinate system and normalize the fundamental form D so that $D\, e_1 \ldots e_{p+1} = 1$.

Then if in the above formula s^{p+1} is positively oriented relative to D, according to I.6.9—10

$$D\, d_1x \ldots d_px = (p + 1)!\, dv_{p+1}\,,$$

[1] This, according to Stokes's theorem, is permitted provided div $u(x)$ exists on s^{p+1} in the above mentioned extended sense and is continuous there, hence, in particular, provided $u(x)$ is continuously differentiable.

where dv_{p+1} stands for the euclidean volume of the simplex spanned by the vectors $d_1x, \ldots, d_{p+1}x$. The right side of formula (2.21) can therefore be written

$$p! \int_{s^{p+1}} \operatorname{div} u(x)\, dv_{p+1} .$$

On the left hand side we determine on each of the $p+2$ side simplexes the "positive" unit normal n_i so that on the simplex $s_i^p(x_0, \ldots, \hat{x}_i, \ldots, x_{p+1})$ $(i = 0, \ldots, p+1)$ the expression

$$(-1)^{i-1} D\, n_i\, d_1x \ldots d_px$$

turns out, for example, to be positive, whereby the orientation of the side simplexes hypothesized in Stokes's transformation formula is taken into account. Then if $u(x)$ is decomposed on the side simplex s_i^p according to

$$u(x) = v_i(x)\, n_i + p_i(x) ,$$

where $p_i(x)$ stands for the orthogonal projection of $u(x)$ onto s_i^p, then $p_i(x)$ and the differentials d_ix for the side simplex are linearly dependent, and the contribution of this side simplex to the boundary integral, according to the definition of the simplex volume given in I.6.10, becomes

$$-p! \int_{s_i^p} v_i(x)\, dv_p ,$$

where dv_p stands for the euclidean element of the side simplex.

Thus, altogether, with the above definition of the "positive" normal component $v(x)$ of $u(x)$ on the boundary ∂s^{p+1},

$$\int_{\partial s^{p+1}} v(x)\, dv_p + \int_{s^{p+1}} \operatorname{div} u(x)\, dv_{p+1} = 0 . \tag{2.22}$$

This is *Gauss's transformation formula in the customary euclidean formulation.*

For other formulations of the formulas of Stokes and Gauss we refer to I. S. Louhivaara [2], C. Müller [2] and P. Hermann [5].

2.12. Exercises. 1. Let R_x^3 be a 3-dimensional linear space in which a euclidean metric is defined by means of the inner product (x_1, x_2). Let $y(x)$ be a differentiable vector function of the variable $x \in R_x^3$, with range in R_x^3, and set

$$A(x)\, h \equiv (y(x), h) \qquad (h \in R_x^3) .$$

Show that the relation

$$\operatorname{rot} A(x)\, h_1\, h_2 = \frac{1}{2} (\operatorname{rot} y(x), [h_1, h_2])$$

holds between the rotor concept defined by (2.8) and the rotor rot $y(x)$ of the ordinary vector analysis, where $[h_1, h_2]$ stands for the "exterior product" of the vectors $h_1, h_2 \in R_x^3$.

2. Prove the following theorem, which can be considered as a generalization of the mean value theorem in II.1.6.

Let $y = A(x)\, d_1 x \ldots d_p x$ be an alternating differential on the closed simplex $s^{p+1}(x_0, \ldots, x_{p+1})$ of the space R_x^m with values from the euclidean space R_y^n. Suppose this differential satisfies these assumptions:

1. $A(x)$ is continuous on s^{p+1}.
2. rot $A(x)$ exists in the sense of (2.9) and is continuous on s^{p+1}.

Then for each vector e from R_y^n there exists at least one point $x_e \in s^{p+1}$ such that $(h_i = x_i - x_0)$

$$\left(\int_{\partial s^{p+1}} A(x)\, d_1 x \ldots d_p x, e\right) = \left(\operatorname{rot} A(x_e)\, h_1 \ldots h^{p+1}, e\right).$$

Hint. Let $D\, h_1 \ldots h_{p+1}$ be the real fundamental form of the subspace parallel to s^{p+1} and $\varrho(x)$ the rotor density (cf. (2.10)) in the direction of this subspace. From Stokes's transformation formula (2.12) it then follows that

$$\left(\int_{\partial s^{p+1}} A(x)\, d_1 x \ldots d_p x, e\right) = \left(\int_{s^{p+1}} \operatorname{rot} A(x)\, d_1 x \ldots d_{p+1} x, e)\right)$$

$$= \left(\int_{s^{p+1}} \varrho(x)\, D\, d_1 x \ldots d_{p+1} x, e\right) = \int_{s^{p+1}} \left(\varrho(x), e\right) D\, d_1 x \ldots d_{p+1} x,$$

which in view of the continuity of the real function $(\varrho(x), e)$ on s^{p+1} yields the assertion.

3. Prove with the aid of the above mean value theorem the following generalization of the theorem proved in II.1.13, exercise 6:

Provided the p-linear alternating operator $A(x)$ is continuous in the region G_x^m of the space R_x^m and rot $A(x)$ exists in the sense of (2.9) at each point x^* of this region, then rot $A(x)$ is continuous in G_x^m if and only if the equation

$$\lim_{\delta \to 0} |(s^{p+1}; x^*)| = 0$$

holds *uniformly* with respect to x^* on each compact subregion of G_x^m.

4. Prove the formula

$$\operatorname{div} u = \sum_{j=1}^{p+1} \frac{\partial \omega^j}{\partial \xi^j},$$

where $x \in R_x^m$ and $u = u(x) \in U^{p+1}$ are vectors with coordinates ξ^j, ω^j $(j = 1, \ldots, p+1)$ in the coordinate system $e_1, \ldots, e_{p+1}$ of the subspace U^{p+1} of R_x^m.

5. Let $D\,h_1 \ldots h_n$ be the euclidean volume of the parallelpiped spanned by the vectors $h_i \in R_x^n$ $(i = 1, \ldots, n)$.

$$(D\,h_1 \ldots h_n)^2 = \det\left((h_i, h_j)\right),$$

and $\pi^n \subset R_x^n$ a polyhedron. Then the "index" J of the origin $x = 0$ with respect to the boundary $\partial\pi^n$ of π^n:

$$J = \int\limits_{\partial\pi^n} \frac{D\,x\,d_1x \ldots d_{n-1}\,x}{|x|^n}$$

equals zero provided $x = 0$ lies outside of π^n and equals the (oriented) volume

$$\omega_{n-1} = \int\limits_{|x|=1} D\,x\,d_1x \ldots d_{n-1}x$$

of the surface of the unit sphere provided x lies inside of π^n.

Hint. Prove that $\operatorname{rot}\left(D\,\frac{x}{|x|^n}\right) = 0$ for $x \neq 0$. The first part of the claim then follows from Stokes's formula. In the second case remove a small sphere $|x| \leqq r$ from π^n; Stokes's theorem then yields

$$J = \frac{1}{r^n} \int\limits_{|x|=r} D\,x\,d_1x \ldots d_{n-1}x\,,$$

independently of r, and the theorem follows for $r = 1$.

6. Let G be a finite region in R_x^2 bounded by a piecewise regular curve ∂G. Prove Stokes's formula

$$\int\limits_{\partial G} A(x)\,dx = \int\limits_{G} \operatorname{rot} A(x)\,d_1x\,d_2x\,,$$

where $A(x)$ is continuously differentiable on $G + \partial G$.

7. Let R_x^2 be the complex $(x = \xi + i\,\eta)$-plane and G $(\subset R_x^2)$ a polygonal region. If the complex-valued function $y = y(x) = u(x) + i\,v(x)$ is continuously differentiable on $G + \partial G$, then (the formula of Morera-Pompeiù)

$$\int\limits_{\partial G} y(x)\,dx = \iint\limits_{G} E_{yx}(x)\,d\xi\,d\eta\,,$$

where

$$E_{yx} \equiv i\left(\left(\frac{\partial u}{\partial \xi} - \frac{\partial v}{\partial \eta}\right) + i\left(\frac{\partial u}{\partial \eta} + \frac{\partial v}{\partial \xi}\right)\right).$$

Hint. The product $y(x)\,dx$ is a linear differential form, and Stokes's formula yields

$$\int\limits_{\partial G} y(x)\,dx = \frac{1}{2}\int\limits_{G} (d_1y\,d_2x - d_2y\,d_1x)\,,$$

where d_1x and d_2x stand for two arbitrary differentials of x and d_1y and d_2y are the corresponding differentials of $y(x)$. The rotor density

$$E_{yx} = \frac{1}{2}\,\frac{d_1y\,d_2x - d_2y\,d_1x}{D\,d_1x\,d_2x}\,,$$

where the numerator is an arbitrary real alternating form ($\neq 0$), is independent of the differentials d_1x, d_2x. For example, if one sets $d_1x = 1$ and $d_2x = i$ and chooses for D the oriented area of the triangle spanned by the vectors d_1x and d_2x:

$$D\, d_1x\, d_2x = \frac{1}{2} |d_1x|\, |d_2x| \sin [d_1x, d_2x] ,$$

where the expression in brackets represents the angle between d_1x and d_2x, then $D\, d_1x\, d_2x = 1/2$ and $E_{yx} = i\, d_1y - d_2y$, from which the claim can be inferred.

8. Prove under the hypotheses of the previous exercise the formula

$$2\pi i\, y(t) = \int_{\partial G} \frac{y(x)}{x - t}\, dx - \int\int_G \frac{E_{yx}(x)}{x - t}\, d\xi\, d\eta ,$$

where t is an interior point of G.

§ 3. Applications of Stokes's Theorem

3.1. Symmetry of the second derivative. For $p = 0$ a p-linear alternating operator degenerates into a vector function $y(x)$. Assuming that this vector function is continuously differentiable on the closed simplex $s^1(x_0, x_1)$, i.e., on the segment $x = x_0 + \tau\,(x_1 - x_0)$ $(0 \leqq \tau \leqq 1)$, Stoke's theorem (2.12), whose proof remains valid even for $p = 0$, states that

$$y(x_1) - y(x_0) = \int_{x_0 x_1} y'(x)\, dx . \tag{3.1}$$

This formula can also be proved, as in the elementary calculus, with the aid of the mean value theorem.

After this preliminary remark, we consider a vector function $y(x)$ in the space R_x^m that satisfies these conditions:

1. *$y(x)$ is continuously differentiable in a neighborhood of the point x_0.*
2. *The second derivative operator $y''(x)$ exists.*

We assert that even under these conditions, which assume considerably less than the ones indicated in II.1.11, the second derivative is symmetric:

For arbitrary vectors h, k of the space R_x^m

$$y''(x_0)\, h\, k = y''(x_0)\, k\, h . \tag{3.2}$$

In fact, it follows from hypothes 1 and (3.1) that

$$\int_{\partial s^2} y'(x)\, dx = 0$$

for every 2-dimensional simplex s^2 in the vicinity of x_0. Consequently, according to the definition (2.9) of the rotor, rot $y'(x_0)$ exists and has the value zero. Furthermore, since $y''(x_0)$ exists, one has, in view of expressions (2.8) and (2.8′),

$$\operatorname{rot} y'(x_0)\, h\, k = \frac{1}{2}\left(y''(x_0)\, h\, k - y''(x_0)\, k\, h\right) = 0\,,$$

which was to be proved.

3.2. The equation rot rot $A(x) = 0$. The symmetry of the second derivative can be thought of as a special case ($p = 1$) of the following general theorem:

Let $A(x)\, h_1 \ldots h_{p-1} \in R_y^n$ $(x, h_i \in R_x^m)$ *be a* $(p-1)$*-linear differential that satisfies the following conditions*:

1. $A(x)$ *is continuous in a neighborhood of* x_0.
2. rot $A(x)$ *exists and is continuous in the former neighborhood.*

Then rot rot $A(x)$ *exists and vanishes.*

Let s^{p+1} be an arbitrary simplex in the mentioned neighborhood that contains the point x_0. Then

$$\int_{\partial s^{p+1}} \operatorname{rot} A(x)\, d_1x \ldots d_px = \sum_{i=0}^{p+1} (-1)^i \int_{s_i^p} \operatorname{rot} A(x)\, d_1x \ldots d_px\,.$$

As a consequence of the hypotheses, Stokes's integral transformation formula (2.12) can be applied to each term on the right. One thus obtains for the boundary integral on the left a double sum of integrals of the differential $A(x)\, d_1x \ldots d_{p-1}x$ over the side simplexes s_{ij}^{p-1} in which each integral appears twice with opposite signs. Consequently, for each simplex s^{p+1} of the kind mentioned

$$\int_{\partial s^{p+1}} \operatorname{rot} A(x)\, d_1x \ldots d_px = 0\,.$$

But according to the definition (2.9) that means that rot rot $A(x_0)$ exists and has the value zero.

3.3. Integration of the equation $dy(x) = A(x)\,dx$. Let G_x^m be an open region in the space R_x^m that is "starlike" with respect to the point x_0: thus along with x, the entire segment $x\,x_0$ lies in G_x^m. In G_x^m let a differential $A(x)\, dx \in R_y^n$ $(dx \in R_x^m)$ be defined that satisfies the following conditions:

1. $A(x)$ *is continuous in* G_x^m.
2. rot $A(x)$ *exists and vanishes in* G_x^m.

The problem is to completely integrate the differential equation

$$dy(x) = A(x)\, dx\,. \tag{3.3}$$

As a consequence of the theorem in the previous sections condition 2 is necessary for the solvability of the problem we have posed. It will turn out that this integrability condition is also sufficient.

Our differential equation can also be written

$$\operatorname{rot} y(x)\,dx = A(x)\,dx \tag{3.3'}$$

and can therefore be conceived of as a special case ($p = 1$) of the general differential equation

$$\operatorname{rot} Y(x)\,d_1x \ldots d_px = A(x)\,d_1x \ldots d_px\,,$$

which is to be considered later. For this reason we treat the present special case so as to make the unity of the integration method used, which is based essentially on the application of Stokes's transformation formula (2.12), apparent for all dimensions $1 \leqq p \leqq m$.

For each x in the starlike region G_x^m the simplex $s^1(x_0, x_1)$, i.e., the closed segment

$$t = x_0 + \tau\,(x_1 - x_0) \qquad (0 \leqq \tau \leqq 1)$$

lies in G_x^m.

Supposing the existence of a solution of the differential equation (3.3) in G_x^m that assumes an arbitrarily given value $y_0 \in R_y^n$ at x_0, we make use of Stokes's theorem (2.12) for $s^1(x_0, x_1)$ and $y(x)$ in order to derive an expression for this hypothetical solution. In the present case, because $\operatorname{rot} y(t) = y'(t)$, this formula degenerates into

$$y(x) - y(x_0) = \int_{x_0 x} y'(t)\,dt\,,$$

and it is therefore necessary that

$$y(x) = y_0 + \int_{x_0 x} A(t)\,dt = y_0 + \int_0^1 A\bigl(x_0 + \tau\,(x - x_0)\bigr)\,(x - x_0)\,d\tau\,. \tag{3.4}$$

On the other hand, this expression defines a vector function in the region G_x^m that for x_0 assumes the value y_0. We claim that under conditions 1 and 2 this function actually satisfies the differential equation (3.3) presented and is therefore the only solution with the given initial value y_0 at x_0.

For the proof, let $x = x_1$ ($\neq x_0$) be an arbitrary point in G_x^m. For a sufficiently small differential $dx = h$, which we take to be linearly independent of $x_1 - x_0$, the simplex $s^2(x_0, x_1, x_2)$ ($x_2 = x_1 + h$) then also lies in G_x^m. As a consequence of assumptions 1 and 2 we can apply Stokes's transformation formula (2.12), according to which

$$\int_{\partial s^2} A(t)\,dt = \int_{s^2} \operatorname{rot} A(t)\,d_1t\,d_2t\,.$$

Because of assumption 2 the integral on the right vanishes, and therefore

$$\int_{\partial s^2} A(t)\,dt = \int_{x_0 x_1} A(t)\,dt + \int_{x_1 x_2} A(t)\,dt + \int_{x_2 x_0} A(t)\,dt = 0\,,$$

and consequently, according to the definition (3.4) of $y(x)$,

$$y(x_2) - y(x_1) = y(x + h) - y(x) = \int_{x_1 x_2} A(t)\, dt = \int_{x(x+h)} A(t)\, dt\,,$$

an equation that valid even if h and $x - x_0$ are linearly dependent, thus in particular also for $x = x_0$. Because of the continuity of $A(t)$ at the point $t = x$, it follows from here that

$$y(x + h) - y(x) = A(x)\, h + |h|\, (h; x)\,,$$

with $|(h; x)| \to 0$ for $|h| \to 0$, which implies the assertion $y'(x) = A(x)$.

Under conditions 1 and 2 the differential equation (3.3) presented therefore has a solution in G_x^m that assumes an arbitrarily preassigned value at x_0 and is thereby uniquely deteımined. When the region G_x^m is in particular convex, the point x_0 can be chosen arbitrarily in this region.

If the open connected region G_x^m is not starlike relative to the point x_0, then one obtains a solution of the differential equation (3.3) in G_x^m if one first constructs a solution $y_0(x)$ with $y_0(x_0) = y_0$ in a convex subregion G_0 containing x_0 and then continues it in the well-known fashion to an arbitrary point x of the region G_x^m. For this, join the points x_0 and x with a finite chain $G_0, \ldots, G_j$ of open convex regions which are chosen so that the intersections $G_i \cap G_{i+1}$ $(i = 0, \ldots, j-1)$ are not empty. Then if in each of these intersections one takes a point x_{i+1}, there exists a unique solution $y_0(x)$ in G_0 with $y_0(x_0) = y_0$, a unique solution $y_1(x)$ in G_1 with $y_1(x_1) = y_0(x_1)$, etc. Since the solutions $y_i(x)$ and $y_{i+1}(x)$ assume the same value $y_i(x_{i+1})$ at the point x_{i+1} of the convex region of intersection $G_i \cap G_{i+1}$, they are by the above identical in the entire intersection and consequently continuations of one another. $y_j(x)$ therefore is a solution element in G_j that uniquely continues $y_0(x)$ along the above chain to the point x.

The integral function thus obtained is unique locally. In general it is true that the integral experiences a zero increase on any path in G_x^m which is homologous to zero. On a closed path which is not homologous to zero the continuation of the integral can yield nonzero *periods* ω. The periods form an abelian group which is homomorphic to the homology group of the region G_x^m.

3.4. Integration of the equation rot $Y(x) = A(x)$. Now suppose, more generally, that $A(x)\, d_1x \ldots d_px \in R_y^n$ $(d_ix \in R_x^m,\ 1 \leqq p \leqq m)$ is a p-linear alternating differential defined in the region G_x^m of R_x^m which is starlike with respect to x_0 that satisfies both of the conditions in the previous section:

1. *$A(x)$ is continuous in G_x^m.*
2. rot $A(x)$ *exists and vanishes in G_x^m.*

Our task is to solve the differential equation

$$\text{rot } Y(x)\, d_1x \ldots d_px = A(x)\, d_1x \ldots d_px\,. \tag{3.5}$$

Supposing that $Y(x)$ is a $(p-1)$-linear alternating continuous solution of this equation, we wish, in generalization of the method pursued for $p = 1$, to first apply the p-dimensional Stokes transformation formula (2.12) to this solution in order to derive an expression for $Y(x)$ and to then show by means of the $(p+1)$-dimensional Stokes's theorem that the expression actually gives the general solution of the above equation in G_x^m.

For this, let $x = x_1$ $(\neq x_0)$ be an arbitrary point of the region G_x^m and $h_1, \ldots, h_{p-1}$ vectors that we first assume to be linearly independent of $x_1 - x_0$ and moreover so small that the entire p-dimensional simplex $s^p = s^p(x_0, \ldots, x_p)$, where $x_{i+1} = x_1 + h_i$ $(i = 1, \ldots, p-1)$, lies in G_x^m.

The hypothetical solution Y was assumed to be continuous. Further, since according to (3.5) rot Y exists and as a consequence of assumption 1 is continuous in G_x^m, we can apply Stokes's theroem (2.12) to Y and thus obtains, because rot $Y = A$,

$$\int_{\partial s^p} Y(t)\, d_1t \ldots d_{p-1}t = \int_{s^p} A(t)\, d_1t \ldots d_pt$$

or

$$\int_{s_0^{p-1}} Y(t)\, d_1t \ldots d_{p-1}t$$

$$= \sum_{i=1}^{p} (-1)^{i-1} \int_{s_i^{p-1}} Y(t)\, d_1t \ldots d_{p-1}t + \int_{s^p} A(t)\, d_1t \ldots d_pt\,, \tag{3.6}$$

where the $p+1$ side simplexes of s^p are denoted by $s_i^{p-1} = s_i^{p-1}(x_0, \ldots, \hat{x}_i, \ldots, x_p)$ $(i = 0, \ldots, p)$.

The left hand side can, because of the continuity of $Y(t)$ for $t = x_1 = x$, be written

$$\int_{s_0^{p-1}} Y(t)\, d_1t \ldots d_{p-1}t = Y(x)\, h_1 \ldots h_{p-1} + \delta^{p-1}(s_0^{p-1}; x)\,, \tag{3.7}$$

where $\delta = \max |h_i|$ and $(s_0^{p-1}; x)$ in general stands for a quantity that vanishes with δ.

We apply formula (1.8) to the right side of (3.6). The last integral becomes

$$\int_{s^p} A(t)\, d_1t \ldots d_pt = \int_{s_0^{p-1}} \varphi\, A(t)\, d_1t \ldots d_{p-1}t\,,$$

with the $(p-1)$-linear alternating operator

$$\varphi\, A(t) = \int_0^1 d(\tau^p)\, A\left(x_0 + \tau\,(t - x_0)\right)\,(t - x_0)\,. \tag{3.8}$$

Because of the continuity of $A(t)$ for $t = x_1 = x$

$$\int_{s^p} A(t)\, d_1 t \ldots d_p t = \varphi\, A(x)\, h_1 \ldots h_{p-1} + \delta^{p-1}(s_0^{p-1}; x) \tag{3.9}$$

follows from this.

The sum on the right in (3.6) is similarly transformed:

$$\Sigma = \sum_{i=1}^{p} (-1)^{i-1} \int_{s_i^{p-1}} Y(t)\, d_1 t \ldots d_{p-1} t$$

$$= \sum_{i=1}^{p} (-1)^{i-1} \int_{s_{i_0}^{p-2}}^{p} \varphi\, Y(t)\, d_1 t \ldots d_{p-2} t\,,$$

where $s_{i\,o}^{p-2} = s_{i\,o}^{p-2}(x_1, \ldots, \hat{x}_i, \ldots, x_p)$ $(i = 1, \ldots, p)$, and

$$\varphi\, Y(t) = \int_0^1 d(\tau^{p-1})\, Y\big(x_0 + \tau(t - x_0)\big)\, (t - x_0) \tag{3.10}$$

is a $(p-2)$-linear alternating operator. It further results from this, since the simplexes $s_{i\,0}^{p-2}$ $(i = 1, \ldots, p)$ form the boundary ∂s_0^{p-1} of s_0^{p-1}, that

$$\Sigma = \int_{\partial s_0^{p-1}} \varphi\, Y(t)\, d_1 t \ldots d_{p-2} t\,. \tag{3.11}$$

As a résumé of formulas (3.7), (3.9), (3.11) it follows from (3.6) that

$$\int_{\partial s_0^{p-1}} \varphi\, Y(t)\, d_1 t \ldots d_{p-2} t$$

$$= Y(x)\, h_1 \ldots h_{p-1} - \varphi\, A(x)\, h_1 \ldots h_{p-1} + \delta^{p-1}(s_0^{p-1}; x)\,,$$

which according to the definition (2.9) of the rotor states that rot $\varphi\, Y$ exists and is equal to $Y - \varphi\, A$. Consequently, one has

$$Y(x)\, h_1 \ldots h_{p-1} = \operatorname{rot} \varphi\, Y(x)\, h_1 \ldots h_{p-1} + \varphi\, A(x)\, h_1 \ldots h_{p-1} \tag{3.12}$$

for each $x \in G_x^m$.

If our problem can be solved at all, then there is associated with each continuous solution Y a well-determined $(p-2)$-linear alternating operator $\varphi\, Y$ for which the rotor exists so that equation (3.12) holds. Consequently rot $\varphi\, Y$ is even continuous, and according to the theorem proved in 3.2 rot rot $\varphi\, Y = 0$. Thus, because rot $Y = A$, it is necessarily true that

$$\operatorname{rot} \varphi\, A(x) = A(x)\,, \tag{3.13}$$

and $\varphi\, A$ is therefore a particular solution of equation (3.5).

In the following section it will be shown that, under integrability condition 2, $\varphi\, A$ actually satisfies equation (3.13). Thus if $B(x)$ is an arbitrary $(p-1)$-linear alternating operator defined in G_x^m for which rot B exists and vanishes, then

$$Y(x)\, h_1 \ldots h_{p-1} = B(x)\, h_1 \ldots h_{p-1} + \varphi\, A(x)\, h_1 \ldots h_{p-1} \quad (3.14)$$

is the *general solution* to the problem posed.

It follows from the above derivation that the operator B is connected with the solution Y by the formula

$$B(x) = \mathrm{rot}\, \varphi\, Y(x)\, .$$

Concerning the meaning of this arbitrary operator for the problem, let the following be remarked: From the definition (3.8) it follows at once that the differential $\varphi\, A(x)\, h_1 \ldots h_{p-1}$ vanishes provided the p vectors $x - x_0, h_1, \ldots, h_{p-1}$ are linearly dependent. But the result (3.14) remains valid even in this case, and one thus sees that B plays the role of an "initial operator" the giving of which uniquely determines the solution.

3.5. Computation of rot $\varphi\, A(x)$. In order to prove equation (3.13), we take an arbitrary point $x = x_1$ $(\neq x_0)$ in G_x^m and the vectors $k_1, \ldots, k_p$ $(1 \leqq p \leqq m-1)$ from R_x^m so that they with $x_1 - x_0$ form a linearly independent system and so that the simplex $s^{p+1} = s^{p+1}(x_0, \ldots, x_{p+1})$ $(x_{i+1} = x_1 + k_i)$ lies in G_x^m. The operator $A(x)$ is according to hypothesis 1 continuous in G_x^m. Further, since the rotor, according to 2, exists in G_x^m and because rot $A = 0$, we can apply Stokes's integral transformation (2.12) for the operator A and the simplex s^{p+1} and find

$$\int_{\partial s^{p+1}} A(x)\, d_1x \ldots d_px = 0\, . \quad (3.15)$$

Because of the additivity of the alternating differential

$$A(x)\, d_1x \ldots d_px$$

this result also holds if the vectors $x_1 - x_0, k_1, \ldots, k_p$ form a linearly dependent system and the simplex degenerates. Our argument hence remains valid in the case $p = m$, and this is true for $p \leqq m$ at the point $x_1 = x_0$ too.

Equation (3.15) yields

$$\int_{s_0^p} A(x)\, d_1x \ldots d_px = \sum_{i=1}^{p+1} (-1)^{i-1} \int_{s_i^p} A(x)\, d_1x \ldots d_px\, , \quad (3.16)$$

where $s_i^p = s_i^p(x_0, \ldots, \hat{x}_i, \ldots, x_{p+1})$ $(i = 0, \ldots, p+1)$ stands for the side simplexes of s^{p+1}. For $i = 1, \ldots, p+1$ one obtains by (1.8)

$$\int_{s_i^p} A(x)\, d_1x \ldots d_px = \int_{s_{i0}^{p-1}} \varphi\, A(x)\, d_1x \ldots d_{p-1}x\, ,$$

where $s_{i\,0}^{p-1} = s_{i\,0}^{p-1}(x_1, \ldots, \hat{x}_i, \ldots, x_{p+1})$ and φA is the operator (3.8). The simplexes $s_{i\,0}^{p-1}$ $(i = 1, \ldots, p+1)$ form the boundary ∂s_0^p of s_0^p, and summation on i in (3.16) gives, in view of the continuity of $A(x)$,

$$\int_{\partial s_0^p} \varphi\, A(x)\, d_1 x \ldots d_{p-1} x$$
$$= \int_{s_0^p} A(x)\, d_1 x \ldots d_p x = A(x_1)\, k_1 \ldots k_p + \delta^p (s_0^p; x_1)\,,$$

where $\delta = \max |k_i|$ and $(s_0^p; x_1)$ vanishes with δ. But this relation, according to definition (2.9), states that

$$\operatorname{rot} \varphi\, A(x_1)\, k_1 \ldots k_p = A(x_1)\, k_1 \ldots k_p\,.$$

Here x_1 is an arbitrary point in G_x^m. Since the equation is valid as a consequence of the linearity for arbitrary vectors $k_i \in R_x^m$, provided it holds for sufficiently small ones, assertion (3.13) is therefore proved.

3.6. Exercise. Let $A(x)$ be a p-linear alternating operator that is continuous in G_x^m and whose rotor exists in G_x^m in the sense of definition (2.9) and is continuous. If rot $A(x)$ does not vanish identically, the value of the boundary integral (3.15) is no longer zero. Transform this integral according to Stokes's theorem and then carry out transformation (1.8). By passing to the limit (analogously to 3.5) further prove the formula

$$\operatorname{rot} \varphi\, A = A - \varphi \operatorname{rot} A\,,$$

which is important for certain questions in the theory of tensor fields on differentiable manifolds.

IV. Differential Equations

In this chapter the first order differential equation

$$\frac{dy}{dx} = f(x, y) \tag{0.1}$$

is to be investigated. Here x is a vector in an m-dimensional linear space R_x^m, and $y = y(x)$ is a vector function, which is to be determined. The range of y lies in an n-dimensional space R_y^n, while $f(x, y)$ stands for a linear operator, which maps the space R_x^m into the space R_y^n. In differential form the equation is

$$dy = f(x, y)\, dx\,. \tag{0.1$'$}$$

If a basis $a_1, \ldots, a_m$ is introduced in R_x^m and $b_1, \ldots, b_n$ in R_y^n and one sets

$$x = \sum_{i=1}^{m} \xi^i a_i\,, \qquad y = \sum_{j=1}^{n} \eta^j b_j\,,$$

the equation transforms into a system of mn partial differential equations of first order for the like number of partial derivatives $\partial\eta^j/\partial\xi^i$ of the n unknown functions $\eta^j = \eta^j(\xi^1, \ldots, \xi^m)$:

$$\frac{\partial\eta^j}{\partial\xi^i} = f_i^j\,(\xi^1, \ldots, \xi^m, \eta^1, \ldots, \eta^n) \qquad (i = 1, \ldots, m;\; j = 1, \ldots, n)\,,$$

where (f_i^j) is the matrix associated with the operator f. Conversely, such a system can be summarized vectorially in the form (0.1).

§ 1. Normal Systems

1.1. Definition and problem. If the space R_x^m of the independent variable x is *one-dimensional*, $m = 1$, the differential equation

$$\frac{dy}{dx} = f(x, y) \tag{1.1}$$

is called a *normal equation*. Written in coordinates ($x = \xi\, e$, $e \in R_x^1$) it becomes

$$\frac{d\eta^j}{d\xi} = f^j(\xi, \eta^1, \ldots, \eta^n) \qquad (j = 1, \ldots, n)$$

and is therefore equivalent to a *normal system* of n equations for the equally many functions $\eta^j = \eta^j(\xi)$.

Keeping the vector notation y, equation (1.1) can also be written in the form

$$\frac{dy}{d\xi} = f_1(\xi, y), \tag{1.1'}$$

where $f_1(\xi, y) = f(x, y)\, dx/d\xi = f(x, y)\, e$. For the sake of unity we prefer to use the general vector form (1.1) even in the present special case of a one-dimensional x-space.

In what follows the uniqueness and existence of the solution to a normal system is to be investigated under the following assumption:

The linear operator $f(x, y)$ is continuous for $|x - x_0| \leqq r_x < \infty$, $|y - y_0| \leqq r_y < \infty$.

We exclude the trivial case where $f(x, y)$ only depends on x and the integration of the differential equation is reduced to a quadrature.

1.2. The uniqueness problem. We consider in the following solutions of the normal system, i. e., functions $y = y(x)$ that are differentiable for $|x - x_0| \leqq r_x$ and whose values fall in the sphere $|y - y_0| \leqq r_y$, so that

$$y'(x) = \frac{dy(x)}{dx} = f(x, y(x))$$

and

$$y(x_0) = y_0. \tag{1.2}$$

The problem is to determine under which assumptions there exists precisely *one* solution.

For this purpose we associate with the operator $f(x, y)$ the upper variation of f with respect to y, which is defined for each $\varrho \geqq 0$ by means of

$$\varphi(\varrho) \equiv \sup |f(x, y_1) - f(x, y_2)| \tag{1.3}$$

for $|x - x_0| \leqq r_x$, $|y_j - y_0| \leqq r_y$ $(j = 1, 2)$, $|y_1 - y_2| \leqq \varrho$. This function satisfies these conditions:

1. $\varphi(\varrho)$ $(\geqq 0)$ increases monotonically with ϱ, $\varphi(0) = 0$ and $\varphi(\varrho) \equiv \varphi(2\, r_y)$ for $\varrho \geqq 2\, r_y$.
2. $\varphi(\varrho)$ is subadditive: $\varphi(\varrho_1 + \varrho_2) \leqq \varphi(\varrho_1) + \varphi(\varrho_2)$.
3. $\varphi(\varrho)$ is continuous.

Property 1 is evident from the definition of φ. To prove 2, let ϱ_1 and ϱ_2 be two arbitrary numbers $\geqq 0$. Then choose two points y_1, y_2

in the ball $|y - y_0| \leqq r_y$ so that $|y_1 - y_2| \leqq \varrho_1 + \varrho_2$. Moreover, let $y = y_3$ be that point on the segment that joins y_1 and y_2 for which

$$|y_1 - y_3| = \frac{\varrho_1}{\varrho_1 + \varrho_2} |y_1 - y_2| \qquad (\leqq \varrho_1) ,$$

$$|y_3 - y_2| = \frac{\varrho_2}{\varrho_1 + \varrho_2} |y_1 - y_2| \qquad (\leqq \varrho_2).$$

Then

$$|f(x, y_1) - f(x, y_2)| \leqq |f(x, y_1) - f(x, y_3)| + |f(x, y_3) - f(x, y_2)|$$
$$\leqq \varphi(\varrho_1) + \varphi(\varrho_2) .$$

This relation is valid for any pair of values y_1, y_2 with $|y_1 - y_2| \leqq \varrho_1 + \varrho_2$. Hence $\varphi(\varrho_1 + \varrho_2) \leqq \varphi(\varrho_1) + \varphi(\varrho_2)$.

Properties 1 and 2 hold independently of the assumption of the continuity of f. If the continuity requirement is satisfied, then $f(x, y)$ is uniformly continuous on the closed point set $|x - x_0| \leqq r_x$, $|y - y_0| \leqq r_y$, and $\varphi(\varrho) \to 0$ for $\varrho \to 0$; i.e., φ is continuous for $\varrho = 0$. The subadditivity then yields the continuity of $\varphi(\varrho)$ for every $\varrho \geqq 0$.

Observe that $\varphi(\varrho) > 0$ for $\varrho > 0$, because $f(x, y)$ would otherwise depend only on x, which was excluded above.

1.3. Osgood's uniqueness theorem. Now let $y_1(x)$ and $y_2(x)$ be two solutions of the normal equation (1.1) that satisfy the initial condition (1.2). We define $y(x) \equiv y_1(x) - y_2(x)$ and for $r \leqq r_x$

$$m(r) \equiv \sup_{|x - x_0| \leqq r} |y(x)| .$$

Then fix two numbers r, Δr so that $0 \leqq r < r + \Delta r \leqq r_x$, and choose x and Δx in such a way that $|x - x_0| = r \leqq |x + \Delta x - x_0| \leqq r + \Delta r$. Then

$$y'(x) = \frac{dy(x)}{dx} = f(x, y_1(x)) - f(x, y_2(x)) ,$$

and

$$|y(x + \Delta x)| = |y(x) + \int_{x(x+\Delta x)} (f(t, y_1(t)) - f(t, y_2(t)))\, dt|$$
$$\leqq |y(x)| + \int_{x(x+\Delta x)} |f(t, y_1(t)) - f(t, y_2(t))|\, |dt|$$
$$\leqq m(r) + \int_{x(x+\Delta x)} \varphi(|y(t)|)\, |dt|$$
$$\leqq m(r) + \varphi(m(r + \Delta r)) \int_{x(x+\Delta x)} |dt|$$
$$\leqq m(r) + \varphi(m(r + \Delta r))\, \Delta r .$$

If the maximum $m(r + \Delta r)$ of $|y(t)|$ ($|t - x_0| \leqq r + \Delta r$) is reached at a point x' of the interval $r < |t - x_0| \leqq r + \Delta r$, then for $x + \Delta x = x'$

$$m(r + \Delta r) \leqq m(r) + \varphi(m(r + \Delta r))\, \Delta r\,.$$

In the other case $m(r + \Delta r) = m(r)$. The above inequality is thus valid in general, and consequently, setting $m(r) = \varrho(r) = \varrho$,

$$\Delta\varrho = \varrho(r + \Delta r) - \varrho(r) \leqq \varphi(\varrho + \Delta\varrho)\, \Delta r\,.$$

We let r_0 ($\leqq r_x$) stand for the least upper bound of those numbers $r < r_x$ for which $m(r) = 0$ and assume $r_0 < r_x$. Then $\varrho(r) \to 0$ as $r \to r_0 + 0$, and $\varrho(r) > 0$ for $r_0 < r \leqq r_x$. Consequently $\varphi(\varrho + \Delta\varrho) > 0$, so that the above inequality can be written

$$\frac{\Delta\varrho}{\varphi(\varrho + \Delta\varrho)} \leqq \Delta r$$

for these values of r. From this it follows that for $r_0 < r \leqq r_x$

$$\int_{\varrho(r)}^{\varrho(r_x)} \frac{d\varrho}{\varphi(\varrho)} \leqq r_x - r < r_x\,.$$

Since $\varrho(r) \to 0$ as $r \to r_0 + 0$, this implies that

$$\int_0^{m(r_x)} \frac{d\varrho}{\varphi(\varrho)} < r_x\,.$$

Thus if one assumes that the integral

$$\int_0 \frac{d\varrho}{\varphi(\varrho)} = \infty\,,$$

then necessarily $r_0 = r_x$. Consequently $\varrho(r) = m(r) = 0$ for $r < r_x$, and therefore

$$y(x) \equiv y_1(x) - y_2(x) = 0$$

on the entire interval $|x - x_0| < r_x$.

This result implies the

Uniqueness theorem of W. F. Osgood. *Provided the integral*

$$\int_0 \frac{d\varrho}{\varphi(\varrho)} \tag{1.4}$$

diverges, the normal system (1.1) *has at most one solution* $y = y(x)$ *with the given initial value* $y(x_0) = y_0$.

The theorem also includes the case where $f(x, y)$ only depends on x, because then $\varphi(\varrho) \equiv 0$.

Osgood's condition is certainly fulfilled when $f(x, y)$ satisfies a Lipschitz condition ($K = \text{const.}$)

$$|f(x, y_1) - f(x, y_2)| \leqq K\, |y_1 - y_2|$$

with respect to y. For then $\varphi(\varrho) \leqq K\varrho$, and Osgood's integral diverges. More generally, uniqueness holds provided for sufficiently small $|y_1 - y_2|$

$$|f(x, y_1) - f(x, y_2)| \leqq K\,|y_1 - y_2| \log \frac{1}{|y_1 - y_2|} \cdots \log_p \frac{1}{|y_1 - y_2|},$$

where $\log_p$ is the p-fold iterated logarithm.

1.4. Inversion of Osgood's theorem. The question arises whether Osgood's sufficient condition is also necessary for the uniqueness of the solution. This is actually the case, in the following sense.

Let $\varphi(\varrho)$ be a continuous function with properties 1 and 2 of 1.2. We consider the set $\{f\}$ of all operators $f(x, y)$ continuous for $|x - x_0| \leqq r_x$, $|y - y_0| \leqq r_y$ which satisfy the inequality

$$\varphi_f(\varrho) \leqq \varphi(\varrho)\,,$$

where $\varphi_f(\varrho)$ stands for the upper variation of $f(x, y)$ defined by (1.3). This theorem is then valid:

If the Osgood integral (1.4) *converges, there is an operator* $f \in \{f\}$ *such that the differential equation* (1.1) *has at least two solutions with the initial condition* $y(x_0) = y_0$.

For the proof we restrict the discussion to those special differential equations in the class $\{f\}$ for which the linear operator $f(x, y)$ carries the differential $dx \in R_x^1$ over into a one-dimensional subspace R_y^1 of the space R_y^n. We then are concerned with the one dimensional case ($m = n = 1$). After introduction of arbitrary unit vectors in R_x^1 and R_y^1 the differential equation can be written in the coordinate form

$$dy = f(x, y)\,dx\,,$$

where x, y and f are now real.

This subclass of $\{f\}$ contains, in particular, the (real) operator

$$f(x, y) \equiv \varphi(|y|)\,,$$

where φ is the function defined above for all $|y|$. In fact, for this function

$$|f(x, y_1) - f(x, y_2)| = \varphi(|y_1|) - \varphi(|y_2|)\,,$$

provided $|y_1| \geqq |y_2|$, and in view of inequality 2 in 1.2

$$\varphi(|y_1|) = \varphi\big(|y_2| + (|y_1| - |y_2|)\big) \leqq \varphi(|y_2|) + \varphi(|y_1| - |y_2|)\,.$$

But here $|y_1| - |y_2| \leqq |y_1 - y_2|$, and therefore $\varphi(|y_1| - |y_2|) \leqq \varphi(|y_1 - y_2|)$. Hence

$$|f(x, y_1) - f(x, y_2)| \leqq \varphi(|y_1 - y_2|)\,,$$

and f thus belongs to the class $\{f\}$ considered.

But the real differential equation

$$dy = \varphi(|y|)\,dx$$

has two solutions that vanish for $x = 0$, namely the trivial solution $y = 0$ and the function $y = y(x)$ which for $y \geqq 0$, $x \geqq 0$ is defined as the inverse function of the integral

$$x = \int_0^y \frac{dy}{\varphi(y)}$$

and for $x < 0$ by $y(x) = -y(-x)$.

1.5. Existence proof by the polygonal method. We now consider the question of the *existence* of a solution of the normal system (1.1). This is to be established using the classical *polygonal method of Cauchy*. We assume for the time being only the continuity condition of 1.1. Let $|f(x, y)| \leqq M < \infty$ in the region $|x - x_0| \leqq r_x$, $|y - y_0| \leqq r_y$; we set $r_0 = \min[r_x, r_y/M]$.

Let the function associated with $f(x, y)$ in 1.2 be $\varphi(\varrho)$. Then on the domain just mentioned

$$|f(x, y_1) - f(x, y_2)| \leqq \varphi(|y_1 - y_2|) .$$

Let $x = a$ be one end point of the interval $|x - x_0| \leqq r_0$. On the closed segment $x_0\, a$ insert a monotonic sequence of points D: $x_0, \ldots, x_N = a$, $|x_1 - x_0| < \cdots < |x_N - x_0|$. We define, beginning with the initial value y_0, the sequence of points

$$y_j = y_{j-1} + \int_{x_{j-1}x_j} f(t, y_{j-1})\, dt \qquad (j = 1, \ldots, N) .$$

That this procedure is meaningful follows from the relation

$$|y_j - y_0| \leqq M\, |x_j - x_0| \leqq M\, r_0 \leqq r_y ,$$

which can be established by induction. For if $j = 0$ it is trivial, and if it holds for $y_0, \ldots, y_{j-1}$, then

$$\begin{aligned} |y_j - y_0| &\leqq |y_j - y_{j-1}| + |y_{j-1} - y_0| \\ &\leqq \int_{x_{j-1}x_j} |f(t, y_{j-1})|\, |dt| + M\, |x_{j-1} - x_0| \\ &\leqq M\, (|x_j - x_{j-1}| + |x_{j-1} - x_0|) = M\, |x_j - x_0| . \end{aligned}$$

For $|x_{j-1} - x_0| \leqq |x - x_0| \leqq |x_j - x_0|$ we now set

$$y_D(x) = y_{j-1} + \int_{x_{j-1}x} f(t, y_{j-1})\, dt . \tag{1.5}$$

By this equation $y_D(x)$ is defined on the closed segment $x_0\, a$ as a continuous function such that $y_D(x_j) = y_j$ $(j = 0, \ldots, N)$. For $|x_{j-1} - x_0| < |x - x_0| < |x_j - x_0|$ it has the derivative

$$y_D'(x) = f(x, y_{j-1}) , \tag{1.6}$$

and for $x = x_j$ the one-sided derivatives $f(x_j, y_{j-1})$ and $f(x_j, y_j)$. The derivative is continuous (except at the points x_j)[1] and bounded ($|y_D'(x)| \leqq M$). One further deduces from (1.5) that

$$\begin{aligned}|y_D(x) - y_0| &\leqq |y_D(x) - y_{j-1}| + |y_{j-1} - y_0| \\ &\leqq \int_{x_{j-1} x} |f(t, y_{j-1})|\, |dt| + M\, |x_{j-1} - x_0| \\ &\leqq M\,(|x - x_{j-1}| + |x_{j-1} - x_0|) \leqq M\, |x - x_0| \\ &\leqq M\, r_0 \leqq r_y\,.\end{aligned}$$

Now we write

$$y_D'(x) = f(x, y_D(x)) + r_D(x)\,,$$

where according to (1.6)

$$|r_D(x)| = |f(x, y_D(x)) - f(x, y_{j-1})| \leqq \varphi\,(|y_D(x) - y_{j-1}|)$$

for $|x_{j-1} - x_0| < |x - x_0| < |x_j - x_0|$. Here, by (1.5),

$$\begin{aligned}|y_D(x) - y_{j-1}| &\leqq \int_{x_{j-1} x} |f(t, y_{j-1})|\, |dt| \\ &\leqq \int_{x_{j-1} x_j} |f(t, y_{j-1})|\, |dt| \leqq M\, |x_j - x_{j-1}|\,.\end{aligned}$$

Suppose δ_D is the largest of the numbers $|x_j - x_{j-1}|$ $(j = 1, \ldots, N)$. Then

$$|r_D(x)| \leqq \varphi(M\,\delta_D)\,.$$

Therefore at the points of continuity $x \neq x_j$ $(j = 1, \ldots, N)$ of the derivative[2]

$$y_D'(x) = f(x, y_D(x)) + \langle\varphi(M\,\delta_D)\rangle\,,$$

and the same inequality also holds for the two one-sided derivatives at the points $x = x_j$. The integration of the relation between x and $x + \Delta x$ further yields

$$y_D(x + \Delta x) = y_D(x) + \int_{x(x+\Delta x)} f(t, y_D(t))\, dt + |\Delta x|\, \langle\varphi\,(M\,\delta_D)\rangle\,. \quad (1.7)$$

Now let δ be an arbitrary positive number. We consider two divisions D_1 and D_2 of the segment $x_0\, a$, so fine that $\delta_{D_1}, \delta_{D_2} \leqq \delta$. If one then sets

$$\bar{y}(x) \equiv y_{D_1}(x) - y_{D_2}(x) \qquad (|\bar{y}(x)| \leqq 2\, M\, r_0)$$

and applies the relation (1.7) for $D = D_1$ and $D = D_2$, one obtains by subtraction

$$\begin{aligned}&\bar{y}(x) + \Delta x) \\ &= \bar{y}(x) + \int_{x(x+\Delta x)} (f(t, y_{D_1}(t)) - f(t, y_{D_2}(t)))\, dt + 2\, |\Delta x|\, \langle\varphi(M\,\delta)\rangle\,. \quad (1.8)\end{aligned}$$

---

[1] At the point $x = x_0$, $y_D'(x)$ is continuous and $y_D'(x_0) = f(x_0, y_0)$.

[2] We use $\langle\varrho\rangle$ to designate any quantity whose norm $|\langle\varrho\rangle| < \varrho$.

We now complete the argument, more or less as in 1.3, in the following way. This time let

$$m(r) \equiv \sup_{|x-x_0| \leqq r} |\bar{y}(x)| \qquad (r \leqq r_0) .$$

If one again takes $0 \leqq r < r + \Delta r \leqq r_0$ and $|x - x_0| = r \leqq |x + \Delta x - x_0| \leqq r + \Delta r$, then in view of (1.8)

$$|\bar{y}(x + \Delta x)| \leqq |\bar{y}(x)| + \int_{x(x+\Delta x)} |f(t, y_{D_1}(t)) - f(t, y_{D_2}(t))| \, |dt| + 2 |\Delta x| \, \varphi(M \delta)$$

$$\leqq m(r) + \int_{x(x+\Delta x)} \varphi(|\bar{y}(t)|) \, |dt| + 2 \, |\Delta x| \, \varphi(M \delta) .$$

Here $|\bar{y}(t)| \leqq m(r + \Delta r)$, and one concludes just as in 1.3 that

$$m(r + \Delta r) \leqq m(r) + \big(\varphi(m(r + \Delta r)) + 2 \varphi(M \delta)\big) \Delta r .$$

Taking into account the monotonicity of φ it follows from here that

$$\Delta m = m(r + \Delta r) - m(r) \leqq \big(\varphi(m(r + \Delta r)) + 2 \varphi(M \delta)\big) \Delta r$$

$$\leqq 3 \varphi\big(m(r + \Delta r) + M \delta\big) \Delta r$$

and

$$\frac{\Delta m}{\varphi(m(r + \Delta r) + M\delta)} \leqq 3 \Delta r .$$

If one sets

$$\varrho = m(r) + M \delta ,$$

one finds, as in 1.3,

$$\int_{M\delta}^{M\delta + m(r_0)} \frac{d\varrho}{\varphi(\varrho)} \leqq 3 r_0 . \tag{1.9}$$

1.6. We consider now for a given $\alpha > 0$ the integral

$$\int_{\alpha}^{\alpha+\beta} \frac{d\varrho}{\varphi(\varrho)}$$

as a function of $\beta \geqq 0$ (cf. Fig. 4, where ϱ is the abscissa and $1/\varphi(\varrho)$ is the ordinate). Since it vanishes for $\beta = 0$ and increases continuously, monotonically and without bound as β increases, there is a well-determined value

$$\beta = \beta(\alpha)$$

such that

$$\int_{\alpha}^{\alpha+\beta(\alpha)} \frac{d\varrho}{\varphi(\varrho)} = 3 r_0 .$$

$\beta(\alpha)$ is defined for $\alpha > 0$ as a positive, monotonically increasing and continuous function of α. Hence the limit

$$\lim_{\alpha \to 0} \beta(\alpha) = \beta_0 \geqq 0$$

exists, and $\beta_0 > 0$ or $\beta_0 = 0$ according as the Osgood integral (1.4) converges or diverges.

1.7. We return to inequality (1.9). If one sets $M\,\delta = \alpha$, then

$$\int\limits_{\alpha}^{\alpha+m(r_0)} \frac{d\varrho}{\varphi(\varrho)} \leqq 3\,r_0 = \int\limits_{\alpha}^{\alpha+\beta(\alpha)} \frac{d\varrho}{\varphi(\varrho)},$$

from which follows:

$$m(r_0) \leqq \beta(\alpha) = \beta(M\,\delta)\,.$$

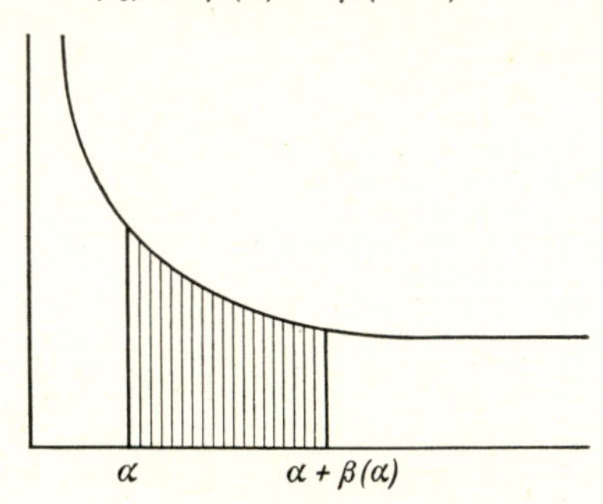

Fig. 4

Now if the *integral* (1.4) *is divergent*, $\beta(M\,\delta)$ tends to zero as $\delta \to 0$, and since then for $|x - x_0| \leqq r_0$

$$|\bar{y}(x)| = |y_{D_1}(x) - y_{D_2}(x)| \leqq m(r_0) \leqq \beta(M\,\delta)\,, \tag{1.10}$$

it follows from the Cauchy convergence criterion that for unlimited refinement of the partition D the approximating function tends to a well-defined limit function, uniformly on the interval $|x - x_0| \leqq r_0$.

The function thus constructed is a solution to our problem. First $y(x_0) = y_0$ and $|y(x) - y_0| \leqq r_y$ for $|x - x_0| \leqq r_0$, because each approximating function possesses this property. Further, according to (1.7) $(x = x_0,\ \Delta x = x - x_0)$,

$$y_D(x) = y_0 + \int\limits_{x_0 x} f(t, y_D(t))\,dt + r_0\langle\varphi(M\,\delta_D)\rangle\,,$$

and refining D, one finds

$$y(x) = y_0 + \int\limits_{x_0 x} f(t, y(t))\,dt\,.$$

Thus

$$\frac{dy(x)}{dx} = f(x, y(x))$$

for $|x - x_0| \leqq r_0 = \min\,[r_x, r_y/M]$. This completes the proof.

1.8. Summary. The preceding investigation has led to this result: *Let $f(x, y)$ be an operator defined for*

$$|x - x_0| \leqq r_x < \infty\,, \quad |y - y_0| \leqq r_y < \infty\,,$$

which maps the space R_x^1 linearly into the space R_y^n, with the following properties:

A. *$f(x, y)$ is continuous and bounded* ($|f(x, y)| \leqq M < \infty$).

B. *The Osgood integral*

$$\int_0 \frac{d\varrho}{\varphi(\varrho)}$$

diverges, where $\varphi(\varrho)$ is the function associated with $f(x, y)$ by means of equation (1.3).

Under these assumptions the differential equation

$$\frac{dy}{dx} = f(x, y)$$

has one and only one solution on the interval $|x - x_0| \leqq r_0 = \min [r_x, r_y/M]$ *such that* $y(x_0) = y_0$ *and* $|y(x) - y_0| \leqq r_y$.

1.9. The above analysis of the Cauchy polygonal method also provides information about the rapidity of the limit process (inequality (1.10)). If, for example, one assumes the Lipschitz condition, $\varphi(\varrho) = K\varrho$, then

$$\int_\alpha^{\alpha+\beta} \frac{d\varrho}{\varphi(\varrho)} = \frac{1}{K} \log\left(1 + \frac{\beta}{\alpha}\right),$$

and $\beta(\alpha) = \beta(M\,\delta) = \alpha\,(e^{3Kr_0} - 1) \sim 3\,K\,r_0\alpha = 3\,K\,r_0\,M\,\delta$. For $\varphi(\varrho) = K\,\varrho\,\log\,(1/\varrho)$ one finds

$$\beta(M\,\delta) = (M\,\delta)^\mu - M\,\delta \qquad \text{with} \qquad \mu = e^{-3Kr_0}.$$

The slower the Osgood integral diverges, the slower the convergence of the polygonal method.

Finally, observe that the above existence proof also includes the Osgood uniqueness theorem. In fact, by the above derivation, inequality (1.10) also holds if $\bar{y}(x)$ is understood to be the difference between a given solution and an approximating function $y_D(x)$ constructed according to the polygonal method. If δ is the length of the greatest subsegment in the partition D, then

$$|y(x) - y_D(x)| \leqq \beta(M\,\delta)\,.$$

If the Osgood integral (1.4) now diverges, $|y(x) - y_D(x)|$ vanishes as $\delta \to 0$, from which the uniqueness of the solution follows.

1.10. Exercises. 1. In the differential equation

$$dy = f(x, y)\,dx \qquad (x \in R^1_x,\ y \in R^n_y)$$

let $f(x, y)$ be a linear operator that is continuous for $|x - x_0| \leqq r_x$, $|y - y_0| \leqq r_y$ ($|f(x, y)| \leqq M$) and satisfies a Lipschitz condition

$$|f(x, y_1) - f(x, y_2)| \leqq K\,|y_1 - y_2|$$

($K = $ const.).

Following *Picard's method of successive approximations* this normal equation is solved with the initial condition $y(x_0) = y_0$ by taking as the first approximation to the solution $y = y(x)$ the constant $y_0(x) \equiv y_0$ and determining the sequence of approximations $y_i(x)$ $(i = 1, 2, \ldots)$ by means of the recursion formula

$$y_i(x) = y_0 + \int_{x_0 x} f(t, y_{i-1}(t))\, dt\,.$$

Show that this sequence converges uniformly for $|x - x_0| \leqq r_0 = \min\,[r_x, r_y/M]$ and that the limit function $y(x) = \lim y_i(x)$ solves the differential equation, where $y(x_0) = y_0$ and $|y(x) - y_0| \leqq r_y$.

Hint. By means of induction, show for each i, first, that

$$|y_i(x) - y_0| \leqq r_y$$

and further that

$$|y_i(x) - y_{i-1}(x)| \leqq M\, K^{i-1} \frac{|x - x_0|^i}{i!}\,.$$

2. For the linear differential equation

$$dy = A(x)\, dx\, y \qquad (x \in R^1_x, \quad y \in R^n_y)\,,$$

where $A(x)$ is continuous for $|x - x_0| \leqq r_0$, the Picard method yields the solution

$$y(x) = \sum_{i=0}^{\infty} A_i(x)\, y_0\,,$$

where the operator $A_i(x)$ is calculated recursively from

$$A_0(x) \equiv I\,, \quad A_i(x) = \int_{x_0 x} A(t)\, dt\, A_{i-1}(t) \qquad (i = 1, 2, \ldots)\,.$$

Hint. The series $\sum_{i=0}^{\infty} A_i(x)$ has the exponential series

$$\sum_{i=0}^{\infty} \frac{(|A|\, |x - x_0|)^i}{i!}\,,$$

where $|A| = \max |A(x)|$ for $|x - x_0| \leqq r_0$, as a majorant.

3. The method given above for the existence and the uniqueness of the integral of the equation $dy = f(x, y)\, dx$ gives also a bound for the difference $|y_1[x] - y_2(x)|$ of two solutions $y_1(x)$ and $y_2(x)$, corresponding to two different initial values $y_1(x_0) = y_1$, $y_2(x_0) = y_2$.

§ 2. The General Differential Equation of First Order

2.1. Uniqueness of the solution. In the following we treat the differential equation

$$dy = f(x, y)\, dx\,, \tag{2.1}$$

investigated above for $m = 1$, in the general case where the dimension of the space R_x^m of the independent variable is arbitrary and $m > 1$. We make the following hypothesis concerning the linear operator $f(x, y)$:

1. *The operator* $f(x, y)$ *is continuously differentiable for* $|x - x_0| \leqq r_x$, $|y - y_0| \leqq r_y$.

That means: for each pair of points x, y in the balls mentioned

$$f(x + h, y + k) = f(x, y) + f_x(x, y)\, h + f_y(x, y)\, k + \delta(\delta)$$

holds with the continuous operators f_x and f_y, where $|\delta|^2 = |h|^2 + |k|^2$ and the norm of the operator (δ) converges to zero for $\delta \to 0$.

Then if K stands for the least upper bound of the norm of the continuous operator $f_y(x, y)$ in the above closed balls, it follows from the mean value theorem that $f(x, y)$ satisfies the Lipschitz condition

$$|f(x, y_1) - f(x, y_2)| \leqq K\, |y_1 - y_2| \tag{2.2}$$

for $|x - x_0| \leqq r_x$, $|y_j - y_0| \leqq r_y$ $(j = 1, 2)$.

In what follows we are concerned with the solvability of equation (2.1) under the additional conditions $y(x_0) = y_0$,

$$|y(x) - y_0| \leqq r_y \text{ for } |x - x_0| \leqq r_0 = \min(r_x, r_y/M)\,,$$

$$M = \max |f(x, y)| \text{ for } |x - x_0| \leqq r_x, \quad |y - y_0| \leqq r_y\,.$$

Assuming that such a solution $y(x)$ exists, then on the connecting segment $t = x_0 + \tau\,(x - x_0)$ $(0 \leqq \tau \leqq 1)$, $y(t)$ is a solution of the normal equation $dy = f(t, y)\, dt$. In view of (2.2) it follows from exercise 1 in 1.10 that this normal equation has one and, as a consequence of Osgood's uniqueness theorem, only one solution $y(t)$ such that $y(x_0) = y_0$ and $|y(t) - y_0| \leqq r_y$ for $|t - x_0| \leqq |x - x_0| \leqq r_0$. For $\tau = 1$ this solution has the well-determined value $y(x)$, and one concludes that this radial integral, constructed for $|x - x_0| \leqq r_0$, is the only solution of (2.1) with the initial value $y(x_0) = y_0$ and the supplementary property $|y(x) - y_0| \leqq r_y$, if such a solution exists at all.

2.2. Integrability condition. Now the question arises whether the above constructed radial integral $y(x)$ actually solves the differential equation (2.1). For dimensions $m > 1$ of R_x^m this is not the case unless besides 1 an additional integrability condition is satisfied. For if $y(x)$ is to solve equation (2.1) the operator $y'(x)$ exists, and for an arbitrary differential $dx = h \in R_x^m$ we have

$$y'(x)\, h = f(x, y(x))\, h\,.$$

In view of hypothesis 1 it follows from this that $y(x)$ is continuously differentiable even twice, so that for an arbitrary second differential $k \in R_x^m$

$$y''(x)\,k\,h = f_x(x, y(x))\,k\,h + f_y(x, y(x))\,(y'(x)\,k)\,h$$
$$= f_x(x, y(x))\,k\,h + f_y(x, y(x))\,(f(x, y(x))\,k)\,h\,.$$

But then, because of the symmetry of the continuous operator $y''(x)$, we have

$$\wedge\,(f_x(x, y(x))\,h\,k + f_y(x, y(x))\,f(x, y(x)\,h)\,k) = 0\,.$$

We see: in order for the radial integral $y(x)$ constructed in 2.1 to satisfy the differential equation (2.1) it is *necessary* that the operator $f(x, y)$ fulfills, besides 1., the following integrability condition:

2. *For* $y = y(x)$ *the bilinear function alternating* (*in* h *and* k)

$$R(x, y)\,h\,k = \wedge\,(f_x(x, y)\,h\,k + f_y(x, y)\,f(x, y)\,h\,k) \equiv 0\,.$$

2.3. Conditions 1 and 2 are sufficient. In this and the following section we show that integrability condition 2 if taken with the general continuity assumption 1 is not only necessary but also sufficient for the radial integral $y(x)$ to satisfy the differential equation (2.1).

As already mentioned, because of the Lipschitz condition (2.2) we can construct the function $y(x)$, applying Picard's method of successive approximations (exercise 1 of 1.10).

We thus join each point x of the ball $|x - x_0| \leqq r_0$ with x_0 by the segment $t = x_0 + \tau\,(x - x_0)$ $(0 \leqq \tau \leqq 1)$ and for each $n \geqq 0$ set, starting with $y_0(x) = y_0$, $y'_{n+1}(t)\,dt = f(t, y_n(t))\,dt$, $y_{n+1}(x_0) = y_0$, so that

$$y_{n+1}(x) = y_0 + \int_{\tau=0}^{1} f(t, y_n(t))\,dt = y_0 + \int_{\tau=0}^{1} F_n(t)\,dt\,, \tag{2.3}$$

whereby $dt = d\tau\,(x - x_0)$. According to the exercise mentioned $y_n(x)$ converges as $n \to \infty$ to the function $y(x)$ and in fact *uniformly* in the ball $|x - x_0| \leqq r_0$; consequently, $y(x)$ is continuous in this ball. It further follows from $y_n(x_0) = y_0$ and $|y_n(x) - y_0| \leqq r_y$ that these last relationships are also valid for the limit function $y(x)$.

It follows from hypotheses 1 by means of induction that the approximating functions $y_n(x)$ are continuously differentiable for $|x - x_0| \leqq r_0$. For a differential $dx = d\lambda\,(x - x_0)$, which depends linearly on $x - x_0$, we have by construction

$$y'_{n+1}(x)\,dx = f(x, y_n(x))\,dx = F_n(x)\,dx\,, \tag{2.4}$$

and for an arbitrary differential dx, because $t'(x)\,dx = \tau\,dx$,

$$y'_{n+1}(x)\,dx = \int_0^1 \left(\tau\,F'_n(t)\,dx\,dt + F_n(t)\,dx\,d\tau\right) .$$

This we write, after addition and subtraction of $\tau\,F'_n(t)\,dt\,dx$,

$$y'_{n+1}\,dx = \int_{\tau=0}^1 \left(\tau\,F'_n(t)\,dt + d\tau\,F_n(t)\right) dx + r_n(x)\,dx ,$$

where

$$r_n(x)\,dx = 2 \int_{\tau=0}^1 \tau \wedge F'_n(t)\,dx\,dt . \tag{2.5}$$

In the expression for $y'_{n+1}(x)\,dx$ the integrand in the first term is

$$\left(\tau\,d_\tau\,F_n(t) + d\tau\,F_n(t)\right) dx = d_\tau\left(\tau\,F_n(t)\right) dx .$$

Consequently,

$$y'_{n+1}(x)\,dx = F_n(x)\,dx + r_n(x)\,dx = f\left(x, y_n(x)\right) dx + r_n(x)\,dx , \tag{2.6}$$

which implies (2.4), because according to (2.5),

$$r_n(x)\,dx = 0 \text{ for } dx = d\lambda\,(x - x_0),\ dt = d\tau\,(x - x_0).$$

In the following section we show that as a consequence of integrability condition 2

$$\lim r_n(x) = 0 \tag{2.7}$$

uniformly for $|x - x_0| \leqq r_0$. Since $y_n(x)$ also converges uniformly in this ball to the limit function, which is continuous there, it follows from (2.6) and (2.7), because of the continuity of the operator $f(x, y)$, that $y'_{n+1}(x)$ converges for $n \to \infty$ *uniformly* in the ball $|x - x_0| \leqq r_0$ to the continuous limit operator $f\left(x, y(x)\right)$. But then according to exercise 7 in II.1.13, the operator $y'(x)$ exists. Thus for each $dx \in R^m_x$

$$y'(x)\,dx = f\left(x, y(x)\right) dx ,$$

which, up to the proof of (2.7), brings us to our goal.

2.4. Proof of $r_n(x) \to 0$. We must still settle the key point, the uniformity of the convergence (2.7) for $|x - x_0| \leqq r_0$.

Completely written, formula (2.5) for $r_n(x)$ is

$$r_n(x)\,dx = 2 \int_{\tau=0}^1 \tau \wedge \{f_t\left(t, y_n(t)\right) dx\,dt + f_y\left(t, y_n(t)\right) \left(y'_n(t)\,dx\right) dt\} .$$

For $n > 0$, by (2.6), $y'_n(t)\,dx = f\left(t, y_{n-1}(t)\right) dx + r_{n-1}(t)\,dx$, and by (2.4), for $dt = d\tau\,(x - x_0)$, $y'_n(t)\,dt = f\left(t, y_{n-1}(t)\right) dt$. We introduce this in the above formula and set for arbitrary differentials h, k set

$$\wedge \{f_t\left(t, y_n(t)\right) h\,k + f_y\left(t, y_n(t)\right) \left(f\left(t, y_{n-1}(t)\right) h\right) k\} = R_n(t)\,h\,k . \tag{2.8}$$

Then for $n > 0$

$$r_n(x)\,dx = 2\int_{\tau=0}^{1} \tau\, R_n(t)\,dx\,dt + \int_{\tau=0}^{1} \tau\, f_y\big(t, y_n(t)\big)\,\big(r_{n-1}(t)\,dx\big)\,dt\,, \tag{2.9}$$

while for $n = 0$

$$r_0(x)\,dx = 2\int_{\tau=0}^{1} \tau\, R_0(t)\,dx\,dt\,, \tag{2.9'}$$

with

$$R_0(t)\,h\,k = \wedge f_t(t, y_0)\,h\,k\,. \tag{2.8'}$$

From the uniform convergence $y_n(t) \to y(t)$ for $|t - x_0| \leqq r_0$ and the continuity hypothesis 1 it follows, according to (2.8), that as $n \to \infty$ the norm $|R_n(t)|$ tends uniformly in the ball mentioned to the norm $|R(t, y(t))|$. Since the latter *according to integrability condition 2 vanishes identically,* the maximum R_n of $|R_n(t)|$ in the ball $|t - x_0| \leqq r_0$ also converges to zero:

$$\lim R_n = 0\,. \tag{2.10}$$

From here it furthermore follows that R_n is bounded:

$$R_n \leqq R\ (< \infty)\,. \tag{2.11}$$

After this preparation we go on to estimate $r_n(x)$. For $n = 0$ it follows from (2.9′), because $|R_0(t)| \leqq R_0$ and $dt = d\tau\,(x - x_0)$, that

$$|r_0(x)\,dx| \leqq R_0\,|x - x_0|\,|dx| \int_0^1 2\tau\,d\tau = R_0\,|x - x_0|\,|dx|\,.$$

Therefore $|r_0(x)| \leqq R_0|x - x_0|$, which, with regard for what follows, we write

$$\frac{1}{2}\,|r_0(x)| \leqq \frac{R_0}{2!}\,|x - x_0|\,.$$

If one takes into account that $|f_y(t, y_n(t))| \leqq K$ for each n and that according to the above $(1/2)\,|r_0(t)| = (R_0/2!)\,|t - x_0| = (R_0/2!)\,|x-x_0|\tau$, then

$$\frac{1}{2}\,|r_1(x)| \leqq \frac{R_1}{2!}\,|x - x_0| + \frac{R_0}{3!}\,K\,|x - x_0|^2$$

follows from (2.9) for $n = 1$. Hence for $n = 0$ and $n = 1$

$$\frac{1}{2}\,|r_n(x)| \leqq \sum_{i=0}^{n} \frac{R_{n-i}}{(i+2)!}\,K^i\,|x - x_0|^{i+1}\,. \tag{2.12}$$

If this is true for an arbitrary n, then it follows from (2.9) that

$$\frac{1}{2}\,|r_{n+1}(x)\,dx| \leqq \frac{R_{n+1}}{2!}\,|x - x_0|\,|dx| \int_0^1 2\tau\,d\tau\, +$$

$$\frac{1}{2}\int_0^1 K \sum_{i=1}^{n+1} \frac{R_{n+1-i}}{(i+1)!}\,K^{i-1}\,|x - x_0|^{i+1}\,|dx|\,\tau^{i+1}\,d\tau\,,$$

and therefore

$$\frac{1}{2}|r_{n+1}(x)| \leqq \sum_{i=0}^{n+1} \frac{R_{n+1-i}}{(i+2)!} K^i |x - x_0|^{i+1} .$$

Hence, the estimate in (2.12) holds for every $n \geqq 0$.

Now let $\varepsilon > 0$ be given. As a consequence of (2.10) there exists an n_0 such that $R_{n-i} < \varepsilon$ for $n - i > n_0$, while according to (2.11) $R_{n-i} \leqq R$ for $n - n_0 \leqq i \leqq n$. Accordingly we decompose the sum (2.12) into two parts: $\sum_1$ from $i = 0$ to $n - n_0 - 1$ and $\sum_2$ for $n - n_0 \leqq i \leqq n$. We then have

$$\sum\nolimits_1 \leqq \varepsilon\, |x - x_0| \sum_{i=0}^{n-n_0-1} \frac{K^i|x-x_0|^i}{(i+2)!} < \varepsilon\, |x - x_0|\, e^{K|x-x_0|} ,$$

and consequently for each x in the ball $|x - x_0| \leqq r_0$

$$\sum\nolimits_1 < \varepsilon\, r_0\, e^{K r_0} .$$

For $\sum_2$ one obtains by (2.11)

$$\sum\nolimits_2 \leqq R\, |x - x_0| \sum_{n-n_0}^{n} \frac{K^i|x-x_0|^i}{(i+2)!} < \frac{R|x-x_0|}{(n-n_0+2)!} (K\,|x - x_0|)^{n-n_0} e^{K|x-x_0|} ,$$

and for $|x - x_0| \leqq r_0$

$$\sum\nolimits_2 < \frac{R(K\, r_0)^{n-n_0}}{(n-n_0+2)!} r_0\, e^{K r_0} .$$

In summary, in the ball $|x - x_0| \leqq r_0$

$$\frac{1}{2}|r_n(x)| \leqq \sum\nolimits_1 + \sum\nolimits_2 < \left(\varepsilon + \frac{R(K\, r_0)^{n-n_0}}{(n-n_0+2)!}\right) r_0\, e^{K r_0} ,$$

and

$$\limsup_{n\to\infty} |r_n(x)| \leqq 2\, r_0\, e^{K r_0}\, \varepsilon .$$

Since this holds in the ball $|x - x_0| \leqq r_0$ for each $\varepsilon > 0$ no matter how small, the uniformity of (2.7) is established.

§ 3. The Linear Differential Equation of Order One

We are going to investigate more carefully the special case of a linear homogeneous equation

$$dy = A(x)\, dx\, y , \tag{3.1}$$

where $A(x)$ is a bilinear operator that maps the space $R^n_y \times R^m_x$ into R^n_y.

The functions denoted in § 1 by f^j_i are now linear homogeneous functions of the coordinates $\eta^1, \ldots, \eta^n$ of y, and equation (3.1) becomes

$$\frac{\partial \eta^j}{\partial \xi^i} = \sum_{k=1}^{n} A^j_{k i}(\xi^1, \ldots, \xi^m)\, \eta^k \qquad (i = 1, \ldots, m;\ j = 1, \ldots, n) .$$

3.1. Uniqueness of the solution. In the present case hypothesis 1 of 2.1 reduces to

1. *The operator $A(x)$ is continuously differentiable in a ball $|x - x_0| \leqq r$ of the space R_x^m.*

Instead of the ball one could also take an arbitrary finite compact region G_x^m, starlike with respect to x_0.

The validity of the Lipschitz condition follows from the continuity of $A(x)$, the constant K being the maximum of the norm $|A(x)|$ for $x \in G_x^m$. The radial integral $y(x)$ can thus be constructed with an arbitrary initial value $y(x_0) = y_0 \in R_y^n$ by means of the Picard method. According to exercise 2 of 1.10 we have

$$y_\nu(x) = \sum_{i=0}^{\nu} A_i(x)\, y_0\,,$$

where, setting $t = x_0 + \tau\,(x - x_0)$,

$$A_0(x) = I \qquad \text{and} \qquad A_{i+1} = \int_{\tau=0}^{1} A(t)\, dt\, A_i(t)\,.$$

As $\nu \to \infty$, $y_\nu(x)$ converges in G_x^m uniformly to the radial integral $y(x)$, which by 2.1 is the only possible solution of equation (3.1), with the initial value $y(x_0) = y_0$.

Because (3.1) has the trivial solution $y(x) \equiv 0$, it follows from the uniqueness theorem that a nontrivial solution $y(x)$ nowhere vanishes.

We now consider the set of all solutions of (3.1). From the linearity of this equation it follows that any linear combination of particular solutions $y_1(x), \ldots, y_l(x)$ is likewise a solution. If the particular solutions are linearly dependent for one value $x = x_0$, so that

$$y(x_0) = \sum_{i=1}^{l} \lambda^i\, y_i(x_0) = 0 \qquad (\lambda^i \neq 0 \text{ for some } i = 1, \ldots, l)\,,$$

then $y(x) \equiv 0$, and the solutions $y_i(x)$ are linearly dependent for all $x \in G_x^m$. This implies that the values $y = y(x)$ of the solution space form a subspace $L_y^q(x)$ of R_y^n whose dimension q $(0 \leqq q \leqq n)$ is independent of x.

3.2. Integrability condition. The necessary and sufficient condition 2 found in § 2 for a radial integral $y(x)$ $(y(x_0) = y_0)$ to satisfy the differential equation is in the present linear homogeneous case

$$\bigwedge_{hk} \left(A'(x)\, h\, k + A(x)\, k\, A(x)\, h\right) y(x) = 0\,,$$

which also can be written

$$R(x)\, h\, k\, y(x) = \bigwedge_{hk} (A'\, h\, k - A\, h\, A\, k)\, y(x) = 0\,. \tag{3.2}$$

This condition must hold for each $x \in G_x^m$ and arbitrary differentials $dx = h, k$.

The expression $R(x)\,h\,k\,y$ is a trilinear form in h, k and y; it is alternating in the first two. For a given pair h, k

$$R(x)\,h\,k = \bigwedge_{h\,k} \big(A'(x)\,h\,k - A(x)\,h\,A(x)\,k\big)$$

is a linear transformation of the space R_y^n into itself. For fixed values of x, h, k, the solution space of the equation

$$R(x)\,h\,k\,y = 0 \tag{3.3}$$

is a linear subspace of R_y^n, namely the kernel $K_y(x; h, k)$ of the linear transformation $R(x)\,h\,k$. For fixed x let the differentials h and k run through the space R_x^m. The intersection of all the kernels $K_y(x; h, k)$ thus obtained is then a subspace $K_y^p(x)$ of R_y^n of dimension p $(0 \leqq p \leqq n)$. The set of all of the points in this subspace yield the complete solution of equation (3.3) for a given $x \in R_x^m$ and for freely varying differentials h, k.

On the other hand we have seen in 3.1 that for a given $x \in G_x^m$ all the values $y = y(x)$ of the solution set of equation (3.1) form a linear subspace $L_y^q(x)$. From integrability condition (3.2), which holds for $x \in G_x^m$ and for arbitrary differentials $dx = h, k$, it follows that $y(x)$ belongs to the subspace $K_y^p(x)$. Thus for each $x \in G_x^m$, the inclusions

$$L_y^q(x) \subset K_y^p(x) \subset R_y^n \tag{3.4}$$

hold.

3.3. Complete integrability. We say the differential equation (3.1) is *completely* integrable if it has a solution with the initial value $y(x_0) = y_0$ for any $y_0 \in R_y^n$. This is the case precisely when the radial integral represents a solution of the differential equation (3.1) however the initial value y_0 is chosen. It follows from the definition of the subspace $L_y^q(x)$ that for complete integrability the dimension q must be equal to n for each x. Then, by the inclusions (3.4), for $x \in G_x^m$ the dimension $p = n$, and thus $L_y^n(x) = K_y^n(x) = R_y^n$. Therefore, in order for (3.1) to be completely integrable, it is necessary that $p = n$ and that

$$R(x) = 0$$

for $x \in G_x^m$.

This condition is also sufficient. For if the equation holds for $h, k \in R_x^m$ and $y \in R_y^n$, integrability condition (3.2) is satisfied.

Later (in 3.12) we shall see that this result is valid under less restrictive hypotheses relative to the operator $A(x)$ than postulated in 3.1. For example, it suffices to assume the simple (not necessarily continuous) differentiability of A.

3.4. Differential equations with constant coefficients. We now consider the case where the operator $A(x) = A$ is *constant*, i.e., independent of x. The integrability condition for the equation $dy = A\,dx\,y$ is now

$$R\,h\,k\,y \equiv \frac{1}{2}(A\,k\,A\,h\,y - A\,h\,A\,k\,y) = 0\,,$$

or more briefly,

$$A\,h\,A\,k = A\,k\,A\,h\,. \tag{3.5}$$

The operator $A\,x$ defines for each $x \in R_x^m$ a linear self-mapping of the space R_y^n. The integrability condition thus states that *the product of two such transformations $A\,h$ and $A\,k$ is commutative.*

In order to integrate the differential equation $dy = A\,dx\,y$ under this assumption, we fix two points x_0 and x in the space R_x^m and carry out the integration along the segment $x_0\,x$, assigning an arbitrary initial value $y_0 \in R_y^n$ to the value $x = x_0$. Using the *method of successive approximations* (cf. 1.10, exercises 1—2) one arrives at the point x with the value

$$y(x) = \sum_{j=0}^{\infty} A_j(x)\,y_0\,,$$

where $A_0(x) \equiv I$ is the identity transformation and

$$A_j(x) = \int_{x_0 x} A\,dt\,A_{j-1}(t) \qquad (j = 1, 2, \ldots)\,.$$

These integrals can be computed successively for $j = 1, 2, \ldots$ in an elementary way. One finds

$$A_j(x) = \frac{1}{j!} A\,(x - x_0) \ldots A\,(x - x_0) = \frac{1}{j!}\,(A\,(x - x_0))^j\,;$$

$(A\,x)^j$ is the jth power of the operator $A\,x$, and thus

$$y(x) = f\,(x - x_0)\,y_0\,,$$

where $f(x)$ stands for the linear self-mapping

$$f(x) = \sum_{j=0}^{\infty} \frac{(A\,x)^j}{j!} \quad ((A\,x)^0 \equiv I) \tag{3.6}$$

of the space R_y^n.

It follows from the general theory of linear homogeneous differential equations developed above that the function $y(x) = f\,(x - x_0)\,y_0$, which is defined in the space R_x^m, is actually the solution; it is uniquely determined by the initial value $y(x_0) = y_0$. In the simple case at hand this can also be seen directly (cf. II.1.13, exercise 8). The operator series $f\,(x - x_0)$ and the series obtained from it by termwise differentiation are uniformly convergent for $|x - x_0| \leqq r < \infty$. In view of the commutativity of the operators $A\,x$,

$$df\,(x - x_0) = A\,dx\,f\,(x - x_0)\,, \quad dy(x) = d\,(f\,(x - x_0)\,y_0) = A\,dx\,y(x)\,.$$

The operator $f(x)$ can be differentiated infinitely often, and for $dx = x$ one has

$$d^j f(0) = (A\,dx)^j = (A\,x)^j ,$$

and the differential quotient $d^j f(0)/dx^j$ is defined by means of the equation

$$\frac{d^j f(0)}{dx^j} x^j = (A\,x)^j .$$

The expansion (3.6) is the Maclaurin expansion of the operator $f(x)$.

3.5. Functional equation. Let a, b, c be three points of the space R^m_x. If y_1, y_2, y_3 are the three corresponding values of a solution $y = y(x)$ of the differential equation $dy = A\,dx\,y$ (A constant), then one has

$$y_2 = f(b-a)\,y_1 , \qquad y_3 = f(c-b)\,y_2 = f(c-a)\,y_1 ,$$

hence $y_3 = f(c-b)\,f(b-a)\,y_1 = f(c-a)\,y_1$. Since this relation holds for each choice of the initial value y_1, $f(c-a) = f(c-b)\,f(b-a)$, or, if we set $x_1 = b - a$, $x_2 = c - b$,

$$f(x_1 + x_2) = f(x_2)\,f(x_1) = f(x_1)\,f(x_2) .$$

Since $f(0) = I$, we have the special result

$$f(x)\,f(-x) = I .$$

The linear self-transformations $f(x)$ are therefore *regular*, and we have

$$\big(f(x)\big)^{-1} = f(-x) .$$

The automorphisms $f(x)$ ($x \in R^m_x$) consequently form an abelian topological group[1].

3.6. The case $m = n$. It is of interest to ask under which conditions the solution $y = y(x)$ of the differential equation $dy = A\,dx\,y$ (A constant) provides a local, i.e., in the neighborhood of each point x, one-to-one mapping $x \to y$. It follows from the theory of implicit functions (cf. II.4.2) that this is the case precisely when the derivative $y'(x)$ is regular and provides a one-to-one mapping between the spaces R^m_x and R^n_y. This implies that the dimensions m and n are *equal.*

A further analysis of the problem shows that, in the case $m = n$, $y'(x)$ can be regular only in the two lowest dimension cases $m = n = 1$ and $m = n = 2$. In the former case one is essentially dealing with the elementary real exponential function. In the second case, $y(x)$ is simply periodic, and the mapping $x \to y$ is then isomorphic to the one provided by the complex exponential function; by an appropriate choice of the coordinate systems in the x- and y-planes one can bring the equa-

---

[1] In this regard, see the article of G. Pólya [1].

tion $y = y(x)$ into the form $w = a^z$, where a is real and z and w are complex variables.

The proofs of these assertions can be carried out based on exercise 5 in 3.14.

3.7. Generalizations of the existence theorem. We turn back to the general linear differential equation

$$dy = A(x)\,dx\,y\,.$$

It was shown in 3.1—3, *assuming the continuous differentiability of the bilinear operator* $A(x)$, that in order for the differential equation to have a solution that assumes an *arbitrarily* preassigned value $y_0 \in R_y^n$ at an arbitrary point $x_0 \in G_x^m$ it is sufficient for the operator

$$R(x)\,h\,k \equiv \wedge\,\big(A'(x)\,h\,k - A(x)\,h\,A(x)\,k\big)$$

to vanish identically.

In what follows this condition for complete integrability of the differential equation will be derived in a new way and at the same time sharpened[1].

3.8. The transformations T and U. Our method is based essentially on certain general structural properties of the integrals of a linear normal system. Let us first summarize these rules.

We restrict ourselves in what follows to a convex region G_x^m, where for the present we only require the *continuity* of the operator function $A(x)$. The differential equation can then be integrated as a normal system along any oriented piecewise regular curve l. The initial value $y_1 \in R_y^n$ at the initial point $x = x_1$ of l can be prescribed arbitrarily, and one arrives at the end point $x = x_2$ of l with a well-determined value $y_2 \in R_y^n$. From the linear structure of the differential equation it follows, in view of the uniqueness theorem, that the final value y_2 for a given l depends *linearly* on the initial value y_1; for if the final value of the integral corresponding to the initial value $\bar{y}_1$ is $\bar{y}_2$, then $\lambda\,y(x) + \bar{\lambda}\,\bar{y}(x)$ (λ and $\bar{\lambda}$ real) is the normal integral on l uniquely determined by the initial value $\lambda\,y_1 + \bar{\lambda}\,\bar{y}_1$, and consequently the final value corresponding to this initial value is $\lambda\,y(x_2) + \bar{\lambda}\,\bar{y}(x_2) = \lambda\,y_2 + \bar{\lambda}\,\bar{y}_2$.

Hence there is associated with every oriented piecewise regular curve l in the region G_x^m a well-determined *linear transformation* T_l of the space R_y^n: the integral $y(x)$ of the normal system $dy = A(x)\,dx\,y$ along l, uniquely determined by the initial value $y(x_1) = y_1$, has at the end point x_2 of l the final value

$$y_2 = y_1 + \int_l dy(x) = T_l\,y_1\,.$$

[1] R. Nevanlinna [6, 7].

If I stands for the identity transformation of the space R_y^n, then the increase of the integral $y(x)$ along l

$$\int_l dy(x) = \int_l A\, dx\, y(x) = (T_l - I)\, y_1 = U_l\, y_1$$

is likewise a linear transformation of the initial value y_1.

From the uniqueness theorem it follows that the transformations T_l have the following properties:

1. If $l_1 = x_1 x_2$, $l_2 = x_2 x_3$ are two paths and $l_2 l_1 = x_1 x_2 x_3$ is the "product path" of l_1 and l_2, then

$$T_{l_2 l_1} = T_{l_2} T_{l_1} .$$

2. If $l^{-1} = x_2 x_1$ stands for the path reciprocal to $l = x_1 x_2$, which results from l by reorientation, then

$$T_l T_{l^{-1}} = T_{l^{-1}} T_l = I .$$

From the second rule it follows in particular that the transformations T_l are regular: the inverse transformation $T_l^{-1} = T_{l^{-1}}$ exists.

Besides these group theoretical properties, the transformations T_l in addition satisfy the following metric condition, which has the character of a Lipschitz inequality:

3. If the paths l are restricted to those in a compact subregion $\overline{G}$ of G_x^m with arc lengths $|l| \leq \varrho_0 < \infty$, then for any metrics on the spaces R_x^m and R_y^n there exists a constant $M = M(G, \varrho_0) > 0$ such that the norm of the transformation U_l satisfies the inequality

$$|U_l| = |T_l - I| \leq M\, |l| .$$

This condition means that the inequality

$$\left|\int_l dy(x)\right| = |y_2 - y_1| \leq M\, |l|\, |y_1|$$

holds for the increase in the integral $y(x)$ constructed along l with the initial value y_1, for any path in $\overline{G}$ and every y_1 in R_y^n, provided $|l| \leq \varrho_0$ (cf. 3.14, exercises 3—4).

Let γ_a be a closed path, beginning and ending at the point a. The corresponding transformation $U_{\gamma_a} = U_a$ then yields for each $y_1 \in R_y^n$ the increase

$$U_a\, y_1 = \int_{\gamma_a} dy\, (x)$$

in the integral $y(x)$, integrated around the curve γ_a with the initial value y_1 at a. If instead of a one chooses another beginning and end point b, which is associated with the transformation $U_{\gamma_b} = U_b$, and if we denote the portion of the curve from a to b (oriented in the same way as γ) by l, then by rules 1 and 2 the relations

$$U_b = T_l\, U_a\, T_{l-1} , \qquad U_a = T_{l^{-1}}\, U_b\, T_l , \tag{3.7}$$

hold. In particular this implies: If for one a on γ U_a represents the null transformation of the space R^n_y, then this is true for every choice of the initial point a. Therefore

$$\int_\gamma dy(x) = \int_\gamma A(x)\, dx\, y(x) = 0\,,$$

however the initial value of the integral $y(x)$ is chosen at an arbitrary point on the closed curve γ.

Besides (3.7) we shall use another transformation formula, which likewise is an immediate consequence of rules 1 and 2.

On an oriented closed curve γ we take in the order determined by the orientation p points $x_1, \ldots, x_p$. We connect these points with a point x_0 outside of or on γ by means of curves $l_i = x_0\, x_i$. The curves l_i, l_{i+1} (i modulo p) and the arc $x_i\, x_{i+1}$ of γ then form a closed curve γ_i, which we describe in the sequence $x_0\, x_i\, x_{i+1}\, x_0$ from x_0 to x_0. Then (setting $T\gamma_i = T_i$, $U\gamma_i = U_i$ and $T_{l_1} = T_l$)

$$T_\gamma = T_l\, T_p \ldots T_1\, T_{l^{-1}}\,.$$

If $T_i = I + U_i$ is introduced, then for $U_\gamma = T_\gamma - I$ one obtains the expansion

$$U_\gamma = T_l \left(\sum_{i=1}^{p} S_i \right) T_{l^{-1}}\,, \tag{3.8}$$

where

$$S_i = \sum U_{j_1} \ldots U_{j_i}\,,$$

taken over all combinations $p \geqq j_1 > \cdots > j_i \geqq 1$. Because the products of the transformations U are in general not commutative, the factors in the individual terms of this sum must be taken in the order indicated.

3.9. The integrability condition $U_\gamma = 0$. After this preparation we prove

Lemma 1. *If the operator $A(x)$ is continuous in the convex region G^m_x, then for the complete integrability of the differential equation*

$$dy = A(x)\, dx\, y$$

it is necessary and sufficient that the transformation

$$U_\gamma = 0$$

for every closed, piecewise regular curve γ in G^m_x.

It follows from (3.7), as already remarked, that it makes no difference at which point of γ one starts along the curve.

The condition is obviously necessary. For if one constructs that normal integral of the differential equation that assumes a preassigned value y_0 at a point x_0, then the increase of this integral for one revolution is

$$\int_\gamma dy(x) = U_\gamma \, y_0 .$$

On the other hand, for complete integrability of the differential equation this normal integral, as a consequence of the uniqueness theorem, must agree on γ with that solution which assumes the value y_0 at x_0; and since this solution is unique in G_x^m, the increase calculated above is in fact equal to zero for each $y_0 \in R_y^n$.

The condition is also sufficient. In order to see this, we integrate the differential equation as a normal system along the segment $x_0 \, x$, beginning with an arbitrary initial value y_0 at a point $x_0 \in G_x^m$. One thus obtains a function $y(x)$ which is well-determined in G_x^m as the only possible solution of the differential equation with the property $y(x_0) = y_0$.

If the integrability condition in the lemma is assumed, this function in fact satisfies our differential equation. For the proof let x and $x + h$ be two arbitrary points of the region G_x^m and $\gamma = x \, x_0 \, (x + h) \, x$ the boundary of the simplex determined by the points x_0, x and $x + h$, described in the direction indicated. On this boundary let $\bar{y}(t)$ be the integral with the initial value $\bar{y}(x) = y(x)$ for $t = x$. It follows from the uniqueness theorem that $\bar{y}(t) = y(t)$ on the entire polygonal path $x \, x_0 \, (x + h)$; consequently the corresponding increase is

$$\Delta y = y\,(x + h) - y(x) = \bar{y}\,(x + h) - \bar{y}(x) = \int_{x\,x_0(x+h)} dy(t) .$$

But according to the integrability condition

$$U_\gamma \, \bar{y}(x) = \int_\gamma d\bar{y}(t) = 0;$$

therefore

$$\Delta y = \int_{x\,x_0(x+h)} d\bar{y}(t) = \int_{x(x+h)} d\bar{y}(t) = \int_{x(x+h)} A(t)\, dt \, \bar{y}(t) ,$$

so that, because of the continuity of $A(t)$ for $t = x$,

$$\Delta y = A(x)\, h\, \bar{y}(x) + |h|\,(h; x) = A(x)\, h\, y(x) + |h|\,(h; x) ,$$

with $|(h; x)| \to 0$ as $|h| \to 0$. Hence for $x \in G_x^m$ and $h = dx$

$$dy(x) = A(x)\, dx\, y(x) ,$$

which was to be proved.

Remark. As a result of the second part of the above proof, the integrability condition $U_\gamma = 0$ can be restricted to the boundaries of *triangles* in the region G_x^m; the condition then holds for every other closed, piecewise regular curve in the region.

3.10. Reduction of the integrability condition $U_\gamma = 0$. The above integrability condition relates to the behavior of the operator $A(x)$ "in the large". The central point of our method of solution consists in being able to replace this condition with the help of the "Goursat idea", already used in II.1.8 and III.2.7—2.9, by an equivalent local differential condition which refers to the behavior of $A(x)$ at every point of the region G_x^m.

Let x_0 be a point of G_x^m and γ the boundary of a triangle $s = s(x_1, x_2, x_3)$ that contains this point. We set $x_2 - x_1 = h$, $x_3 - x_1 = k$ and denote by $\Delta = D\,h\,k$ the real fundamental form of the plane E of s, so that Δ represents the oriented (affine) area of the triangle. If the integrability condition of Lemma 1 holds, then $U_\gamma/\Delta = 0$, and the limit transformation

$$\lim \frac{U_\gamma}{\Delta} = 0$$

exists trivially when the triangle in the arbitrary but fixed plane E converges to x_0 so that the greatest side length $\delta \to 0$.

This obvious state can now be turned around, and one can even restrict oneself to *regular* convergence $\delta \to 0$, for which δ^2/Δ remains under a finite bound. That follows from

Lemma 2. *Let the operator $A(x)$ be continuous in G_x^m, so that for each point x_0 of the region and in each fixed plane E through x_0*

$$\lim \frac{U_\gamma}{\Delta} = 0 \tag{3.9}$$

when the triangle $s = s(x_1, x_2, x_3)$ with boundary γ in E converges regularly to $x_0 \in s$ ($\Delta = D\,h\,k$, $h = x_2 - x_1$, $k = x_3 - x_1$).[1]

Then the integrability condition of lemma 1 is valid for every closed, piecewise regular curve in the region G_x^m.

Proof. According to the final remark in 3.9 it suffices to show that the integrability condition $U_\gamma = 0$ holds for every *triangle* in G_x^m.

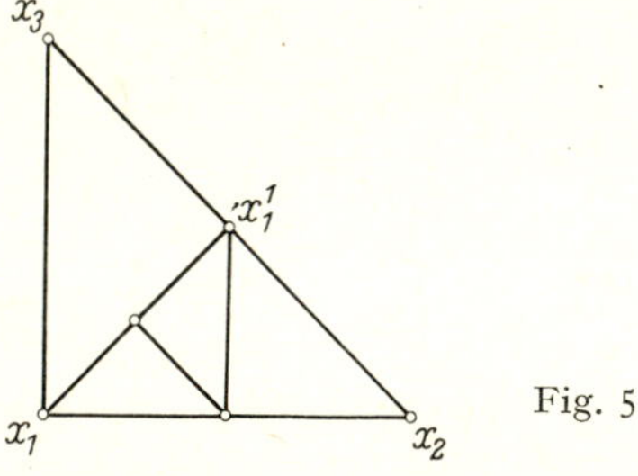

Fig. 5

Hence, let $s_0 = s_0(x_1, x_2, x_3)$ be such a triangle. In order to demonstrate the assertion $U_0 \equiv U_\gamma = 0$ for the boundary $\gamma = x_1\,x_2\,x_3\,x_1$ we

[1] Such regular sequences exist (cf. construction of Whithney, I.5.5). Cf. also T. Nieminen [1].

introduce a euclidean metric on the space R_x^m, in such a way that s_0 becomes a right isoceles triangle with the right angle at x_1. Denote by M the constant $M(s_0, \varrho_0)$ of postulate 3 in 3.8, whereby $\varrho_0 = 2\,\delta_0 = 2\,|x_3 - x_2|$ (cf. Fig. 5).

We decompose s_0 by means of the perpendicular $x_1'\,x_1$ to the hypotenuse $x_3\,x_2$ into two congruent triangles s_1 and s_1' with boundaries $\gamma_1 = x_1'\,x_1\,x_2\,x_1'$ and $\gamma_1' = x_1'\,x_3\,x_1\,x_1'$. If one sets $U_{\gamma_1} = U_1$, $U_{\gamma_1'} = U_1'$, then by (3.8)

$$U_0 = T_l\,U\,T_{l^{-1}}\,,$$

where

$$U = U_1 + U_1' + U_1'\,U_1\,.$$

According to condition 3

$$|T_l| \leqq 1 + |U_l| \leqq e^{|U_l|} \leqq e^{M|l|} = e^{M\delta_0/2}\,, \qquad |T_{l^{-1}}| \leqq e^{M\delta_0/2}\,.$$

Furthermore, provided $|U_1|$ is the larger of the norms $|U_1|$, $|U_1'|$,

$$|U| \leqq |U_1| + |U_1'| + |U_1'|\,|U_1| \leqq 2\,|U_1|\left(1 + \frac{|U_1|}{2}\right) \leqq 2\,|U_1|\,e^{|U_1|/2}$$
$$\leqq 2\,|U_1|\,e^{M|\gamma_1|/2} \leqq 2\,|U_1|\,e^{M\delta_0}\,,$$

and therefore

$$|U_0| \leqq |T_l|\,|U|\,|T_{l^{-1}}| \leqq 2\,|U_1|\,e^{2M\delta_0}\,.$$

If Δ_0 and $\Delta_1 = \Delta_0/2$ are now the areas of the triangles s_0 and s_1, it follows that

$$\frac{|U_0|}{\Delta_0} \leqq \frac{|U_1|}{\Delta_1}\,e^{2M\delta_0}.$$

This procedure can be repeated ad infinitum. One obtains a sequence of triangles $s_0 \supset s_1 \supset \cdots \supset s_j \supset \cdots$, and there exists a well-determined limit point $x_0 \in s_j$ $(j = 0, 1, \ldots)$. If $\delta_j = \delta_0/2^{j/2}$ stands for the length of the hypotenuse and Δ_j the area of s_j, the above inequality, after j iterations, yields

$$\frac{|U_0|}{\Delta_0} \leqq \frac{|U_j|}{\Delta_j}\,e^{2M(\delta_0 + \cdots + \delta_{j-1})} \leqq \frac{|U_j|}{\Delta_j}\,e^{7M\delta_0}\,.$$

The triangles s_j converge (regularly) to the limit point $x_0 \in s_0$, and since the quotient $|U_j|/\Delta_j$ converges to zero by (3.9), $U_0 = U_\gamma = 0$, which was to be proved.

3.11. Extension of the definition of the operator $\boldsymbol{R}(\boldsymbol{x})$. Assuming only the continuity of the operator $A(x)$, Lemmas 1 and 2 have supplied the necessary and sufficient condition (3.9) for the complete integrability of our differential equation. This condition is to be related to the sufficient condition $R(x) \equiv 0$ cited earlier in 3.7.

This connection is apparent from the following

Lemma 3. *Let the operator $A(x)$ be continuous in a neighborhood of the point x_0 and differentiable at this point.*

Then if $s = s(x_1, x_2, x_3)$ is a triangle ($\partial s = \gamma$) in the neighborhood mentioned containing the point x_0 and δ is the greatest side length, then ($x_2 - x_1 = h$, $x_3 - x_1 = k$)

$$\begin{aligned} U_\gamma &= R(x_0)\, h\, k + \delta^2(\delta) \\ &= \wedge \left(A'(x_0)\, h\, k - A(x_0)\, h\, A(x_0)\, k\right) + \delta^2(\delta)\,, \end{aligned} \qquad (3.10)$$

where (δ) stands for a linear transformation of the space R_y^n whose norm converges to zero as $\delta \to 0$.

The orientation of the boundary $\partial s = \gamma$ is determined by the above ordering of the vertices, while the beginning point can be chosen arbitrarily. Therefore, if the differential equation $dy = A(x)\, dx\, y$, with the initial value y_0 at an arbitrary point, is integrated as a normal system around the boundary ∂s oriented in the way described, then for every $y_0 \in R_y^n$ the increase in the integral $y(x)$ is

$$U_\gamma\, y_0 = \int_{\partial s} dy(x) = \int_{\partial s} A(x)\, dx\, y(x) = R(x_0)\, h\, k\, y_0 + \delta^2(\delta)\, y_0\,. \qquad (3.11)$$

In order to not interrupt the course of our existence proof, we shall prove Lemma 3 later (in 3.13).

Formula (3.11) leads to a generalization of the definition of the trilinear operator $R(x)$ analogous to the extension of the definition of the rotor given in III.2.6. In fact, the left hand side of this formula makes sense for every triangle s in the space R_x^m containing x_0 and each $y_0 \in R_y^n$, assuming only the continuity of the operator $A(x)$. Consequently $R(x_0)\, h\, k\, y_0$ can be uniquely defined by requiring trilinearity and the validity of relation (3.11) for each triangle s.

This definition agrees with the earlier one

$$R(x_0)\, h\, k\, y_0 = \underset{h\,k}{\wedge} \left(A'(x_0)\, h\, k - A(x_0)\, h\, A(x_0)\, k\right) y_0$$

if the operator $A(x)$ is continuous in a neighborhood of the point x_0 and in addition *differentiable* at x_0. On the other hand, the extended definition supposes only that the differential equation $dy = A(x)\, dx\, y$ can be integrated normally along piecewise regular curves. It can therefore, in certain cases, make sense assuming only the continuity of $A(x)$.

3.12. Existence theorem. Summarizing the results of Lemmas 1, 2 and 3 we obtain the following existence theorem, in which the operator $R(x)$ is to be taken in the sense of the extended definition given in 3.11.

Theorem. *Let the bilinear operator $A(x)$ be continuous in the convex region G_x^m of the space R_x^m.*

In order that the differential equation

$$dy = A(x)\, dx\, y$$

may possess a solution for every x_0 *in* G_x^m *and an arbitrary* y_0 *in* R_y^n *which for* $x = x_0$ *assumes the value* y_0, *it is necessary and sufficient that the operator* $R(x)$ *exists in* G_x^m *and vanishes identically, so that for arbitrary vectors* h *and* k *in the space* R_x^m *and every* y *in* R_y^n

$$R(x)\, h\, k\, y = 0\,.$$

If in particular $A(x)$ *is differentiable at every point of the region* G_x^m, *then*

$$R(x)\, h\, k\, y = \bigwedge_{h\,k} \big(A'(x)\, h\, k - A(x)\, h\, A(x)\, k\big)\, y\,,$$

and the identical vanishing of this expression is necessary and sufficient for the complete integrability of the differential equation.

The second part of this theorem gives in a sharpened formulation the existence theorem mentioned in 3.7 and already proved by another method in 3.1—3.3, where *continuous* differentiability of the operator was assumed. Our second method of proof, based on the Goursat idea, shows that requiring the continuity of the derivative $A'(x)$ is superfluous. As known, this sharpening is also typical in other connections, for example, in the proof of the Cauchy integral theorem in function theory, using the Goursat idea.

3.13. Proof of Lemma 3. Let $A(x)$ be continuous on $\overline{G}$: $|x - x_0| \leqq r$ and suppose $|A|$ is the least upper bound of the norm $|A(x)|$ on $\overline{G}$. Since (for a euclidean metric) any triangle $s \subset \overline{G}$ has a boundary length $\leqq 3\sqrt{3}\, r = \varrho_0$, by exercise 4 of 3.14 the Lipschitz condition $|U_l| \leqq M\, |l|$ holds, with $M = |A|\, e^{|A|\varrho_0}$, for any subsegment l of the boundary ∂s. If δ stands for the length of the longest side of the triangle $s = s(x_1, x_2, x_3)$, then because $|l| \leqq 3\,\delta$,

$$U_l = (\delta)\,, \tag{3.12}$$

where (δ) tends to zero with δ. Now integrate $dy(t) = A(t)\, dt\, y(t)$ along the oriented boundary $\gamma = \partial s$, beginning with an arbitrary boundary point $x = a$ and an initial value y_0. For a boundary point x and the corresponding subpath $ax = l$

$$y(x) = y_0 + \int_l A(t)\, dt\, y(t) = y_0 + U_l\, y_0\,,$$

which because of (3.12) implies that

$$y(x) = y_0 + (\delta)\, y_0\,. \tag{3.12'}$$

In view of the continuity of the operator $A(x)$ at the point x_0 this implies the more precise relation

$$y(x) = y_0 + A(x_0)\,(x - a)\, y_0 + \delta(\delta)\, y_0\,. \tag{3.12''}$$

By hypothesis $A(x)$ is even differentiable at the point $x = x_0$. Thus at each boundary point

$$A(x) = A(x_0) + A'(x_0)(x - x_0) + |x - x_0|(x - x_0; x_0),$$

where the norm $|(x - x_0; x_0)|$ tends to zero uniformly on ∂s when $\delta \to 0$. Therefore

$$\begin{aligned} U_{\gamma_a} y_0 &= \int_{\gamma_a} A(x)\, dx\, y(x) \\ &= \int_{\gamma_a} A(x_0)\, dx\, y(x) + \int_{\gamma_a} A'(x_0)(x - x_0)\, dx\, y(x) + \delta^2(\delta)\, y_0, \end{aligned}$$

where γ_a denotes the closed path of integration γ having a as initial and final point.

If $y(x)$ is replaced in the first integral by the expression (3.12″) and in the second by (3.12′), then

$$\begin{aligned} U_{\gamma_a} y_0 &= \int_{\gamma_a} \{A(x_0)\, dx - A(x_0)\, dx\, A(x_0)\, a - A'(x_0)\, x_0\, dx\}\, y_0 \\ &+ \int_{\gamma_a} \{A(x_0)\, dx\, A(x_0)\, x + A'(x_0)\, x\, dx\}\, y_0 + \delta^2(\delta)\, y_0. \end{aligned}$$

The first integral vanishes, and a simple direct evaluation yields

$$\{- \wedge A(x_0)\, h\, A(x_0)\, k + \wedge A'(x_0)\, h\, k\}\, y_0$$

for the second.

Thus, independently of the choice of the initial point a and of the vector y_0,

$$U_{\gamma_a} y_0 = R(x_0)\, h\, k\, y_0 + \delta^2(\delta)\, y_0,$$

which proves Lemma 3.

Remark. The method developed in 3.8—3.13 and based on Goursat's idea of successiv subdivisions of a simplex can be generalized to the *non-linear* case (cf. G. Bächli [1] and the dissertation of S. Heikkilä [1]).

3.14. Exercises. 1. If the differential equation

$$dy = A(x)\, dx\, y \qquad (x \in R_x^m,\ y \in R_y^n),$$

where the bilinear operator $A(x)$ is continuously differentiable in the region G_x^m ($\subset R_x^m$), which is starlike with respect to x_0, is integrated along a straight line from x_0 to x with an initial value $y_0 = y(x_0)$, then (cf. 1.10, exercise 2) the function thus obtained

$$y(x) = \sum_{i=0}^{\infty} A_i(x)\, y_0,$$

is continuously differentiable in G_x^m, and

$$dy(x) = y'(x)\, dx = \sum_{i=1}^{\infty} A_i'(x)\, dx\, y_0.$$

2. Under the hypotheses of exercise 1 the function $y = y(x)$ has a second differential $y''(x)\, h\, k$ for every $k \in R_x^m$ and for an h that depends linearly on $x - x_0$.

Hint. By construction of $y(x)$, $dy(x) = y'(x)\, h = A(x)\, h\, y(x)$ for each h which depends linearly on $x - x_0$. If one sets $h = x - x_0$, the identity

$$y'(x)\,(x - x_0) = A(x)\,(x - x_0)\, y(x)$$

thus holds. Therefore, since $y(x)$ is differentiable, for an arbitrary $dx = k$,

$$d\big(y'(x)\,(x - x_0)\big)$$
$$= A'(x)\, k\,(x - x_0)\, y(x) + A(x)\, k\, y(x) + A(x)\,(x - x_0)\, y'(x)\, k.$$

Applying the definition of the differential, one concludes from this that $y''(x)\, k\,(x - x_0)$ exists and is equal to $d\big(y'(x)\,(x - x_0)\big) - y'(x)\, k$. Since y'' is a bilinear operator, the differential $y''(x)\, k\, h$ hence exists for an arbitrary k and every h that depends linearly on $x - x_0$. Further, it follows from the expression for $y''(x)\, k\, h$ that it is a continuous function of x. Finally, as a result of Schwarz's theorem (cf. II.3.4), for the vectors h and k mentioned, $y''(x)\, h\, k$ also exists and equals $y''(x)\, k\, h$.

3. Let $A(x)\, dx\, y \in R_y^n$ ($x \in R_x^m$, $y \in R_y^n$) be a bilinear function such that the operator $A(x)$ is continuous in the region G_x^m ($\subset R_x^m$). Let the number $|A|$ denote the maximum of the norm $|A(x)|$ on a compact subregion $\overline{G}$ of G_x^m. Integrate the differential equation

$$dy = A(x)\, dx\, y$$

along a piecewise regular path $l \subset \overline{G}$ that connects the point $x = x_1$ with $x = x_2$ so that $y(x_1) = y_1$. If the final value of the integral $y = y(x)$ is equal to $y_2 = y(x_2)$, then

$$|y_2| \leqq |y_1|\, e^{|A||l|} \qquad (|l| = \text{length of } l)\,.$$

Hint. The claim is trivial for $y_1 = 0$. For $y_1 \neq 0$ the integral $y(x) \neq 0$ at every point x of the curve l. From

$$d|y| \leqq |dy| = |A(x)\, dx\, y| \leqq |A|\, |dx|\, |y|$$

it follows by division with $|y| \neq 0$ that

$$\frac{d|y|}{|y|} \leqq |A||\, dx|\,,$$

from which the assertion results by means of integration along l from x_1 to x_2.

4. Under the assumptions of the previous exercise,

$$|y_2 - y_1| \leqq |A|\, e^{|A||l|} \cdot |l|\, |y_1|\,,$$

and hence for $|l| \leqq \varrho_0$,

$$|y_2 - y_1| \leqq M\, |l|\, |y_1|\,,$$

where

$$M = |A|\, e^{|A|\varrho_0}\, \varrho_0\,.$$

5. Let the constant bilinear operator A be so constituted that the solutions $y = y(x)$ of the differential equation $dy = A\,dx\,y$ $(x, y \in R^n)$ which do not identically vanish provide a locally one-to-one mapping $x \to y$. Then the dimension is $n = 1$ or $n = 2$, and $y(x)$ is isomorphic to the real or complex exponential function.

Hint. Every solution $y(x)$ of the problem is $\neq 0$. For if $y(x_0) = 0$, the representation $y(x) = f(x)\,y(0)$ shows, in view of the functional equation for the transformation $f(x)$, that $y(x) = f(x)\,y(0) = f\big((x - x_0) + x_0\big)\,y(0) = f(x - x_0)\,f(x_0)\,y(0) = f(x - x_0)\,y(x_0) = 0$ (cf. 3.5), and $y(x)$ would therefore vanish identically (also cf. 3.1).

On the other hand, $y(x)$ assumes all values different from zero. We first prove:

Each locally defined element of the inverse function $x = x(y)$ of $y = y(x)$ can be continued in the region $0 < |y| < \infty$.

To see this, observe that the minimum $\alpha = \min |A\,h\,k|$ for $|h| = |k| = 1$ is *positive*. Namely, let k be an arbitrary unit vector in R^n. There is then a well-determined solution $y = \bar{y}(x) = f(x)\,k$ of our problem such that $\bar{y}(0) = k$. We have $\bar{y}'(0)\,h = A\,h\,k$, and since the mapping $x \to \bar{y}$ is by hypothesis locally one-to-one, the derivative $\bar{y}'(0)$ is *regular*, and $A\,h\,k \neq 0$ for every $|h| = 1$. It follows from this that the minimum α is positive.

Now let $x = x_1$ be chosen arbitrarily and $y_1 = y(x_1)$, where $y = y(x)$ $(\neq 0)$ is an arbitrary solution. Join y_1 $(\neq 0)$ with an arbitrary second point $y_2 \neq 0$ by means of a polygonal path l that avoids the point $y = 0$. In a neighborhood of $y = y_1$ the inverse function $x = x(y)$ $\big(x_1 = x(y_1)\big)$ is single-valued. Now continue this element $x = x(y)$ as inverse function of $y = y(x)$ along l. We claim that the continuation arrives at the end point $y = y_2$ with a well-determined element $x = x(y)$ of the inverse function of $y = y(x)$.

Otherwise, the continuation succeds only on a subarc $y_1\,c_0 \subset l$, without reaching the end point $y = c_0$. This subarc borders on c_0 with a certain line segment. Let $y = c_1$, c_2 be two arbitrary interior points of this segment, and $a_i = x(c_i)$ $(i = 1, 2)$. On the interval $c_1\,c_2$, $x = x(y)$ satisfies the equation $dy = A\,dx\,y$, from which we have

$$|dy| = |A\,dx\,y| \geqq \alpha\,|dx|\,|y| \geqq \alpha\,\varrho\,|dx|\,,$$

where $\varrho > 0$ denotes the shortest distance from $y = 0$ to l. Therefore,

$$|a_1 - a_2| \leqq \int_{c_1 c_2} |dx| \leqq \frac{1}{\alpha \varrho} \int_{c_1 c_2} |dy| = \frac{|c_1 - c_2|}{\alpha \varrho}.$$

According to the Cauchy criterion this implies the existence of a limit point $a_0 = \lim x(y)$ for $y \to c_0$. The function $x = x(y)$ as the inverse of $y = y(x)$ would thus exist in a neighborhood of the point $y = c_0$, so that $y(a_0) = c_0$, which contradicts the assumption regarding c_0. The truth of the assertion follows from here: in the continuation of $x(y)$ along l one reaches the end point $y = y_2$ with a value $x(y_2) = x_2$, and the inversion of the final element coincides with $y = y(x)$. Consequently $y(x_2) = y_2$. Since $y_2 \neq 0$ was arbitrary, it follows in particular from what was proved above that $y(x)$ assumes *all* values $y \in R^n$, with the single exception $y = 0$.

We now consider the unit sphere $|y| = 1$. According to the above, the inverse function $x = x(y)$, provided the dimension $n > 1$, can be continued without bound on this surface. We now consider the case $n = 3$. Since the sphere is simply connected (it has the null homology group), the continued function $x = x(y)$ must be single-valued not only locally, but on the entire sphere. The mapping $y \to x$ is hence one-to-one on $|y| = 1$ and has as image in the x-space a compact surface F_x.

But this leads to a contradiction. Suppose $y_0 = y(0)$. Since the form $A\, h\, y_0$ is different from zero, $A\, h\, y_0$ as a linear transformation of $h \in R^n$ is *regular*, and there is a well-determined value $h = a \neq 0$ such that $A\, a\, y_0 = y_0$. On the line $x = \xi\, a$ $(-\infty < \xi < +\infty)$ $y = e^{\xi}\, y_0$ is then a solution of the differential equation. For

$$dy = e^{\xi}\, d\xi\, y_0 = e^{\xi}\, d\xi\, A\, a\, y_0 = A (d\xi\, a)\, (e^{\xi}\, y_0) = A\, dx\, y\,.$$

Since this solution agrees with the given solution $y(x)$ for $\xi = 0$, we thus have $y(\xi\, a) = e^{\xi}\, y_0$.

Now let $x = b + \xi\, a$. Clearly

$$y\,(b + \xi\, a) = f\,(b + \xi\, a)\, y_0 = f(b)\, f(\xi\, a)\, y_0 = f(b)\, y(\xi\, a) = e^{\xi}\, f(b)\, y_0$$
$$= e^{\xi}\, y(b)\,.$$

From here one sees that the function $y = y(x)$ maps the line $x = b + \xi\, a$ one-to-one onto a half ray $(y = e^{\xi}\, y(b))$ that emanates from $y = 0$. Since for variable $b \in R^3$, $y(b)$ assumes all values $\neq 0$, these half rays cover the entire punctured space $y \neq 0$. Because the rays cut the sphere $|y| = 1$, the compact surface F_x must be intersected by every line in the parallel family $x = b + \xi\, a$ (b variable, a fixed). But for sufficiently large $|b|$ this is impossible, and the thus deduced contradiction shows that the case $n = 3$ is excluded.

It is similarly proved that the assumption $n > 3$ is also contradictory.

Therefore only the dimensions $n = 1, 2$ remain. In the case $n = 1$, $y(x) = y(\xi a)$ is equal to the exponential function $e^{\xi} y_0$. In the case $n = 2$, the following can be deduced.

Since the inverse function $x = x(y)$ can be continued without bound on the unit sphere $|y| = 1$, corresponding to one full revolution, starting at $y = c$, is a continuous arc with end points $x = b_1, b_2$ such that $y(b_1) = y(b_2) = c$. In a way similar to the above one sees that this arc cannot be closed. Hence $b_1 \neq b_2$. If one sets $\omega = b_1 - b_2$, then $y(b_1) = y(b_2)$, $f(b_1) = f(b_2)$, $f(\omega) = f(b_1 - b_2) = f(b_1) f(-b_2) = (fb_1) f((b_2))^{-1} = I$. This further yields

$$y(x + \omega) = f(x + \omega) y_0 = f(\omega) f(x) y_0 = f(x) y_0 = y(x),$$

and the function $y(x)$ is thus *periodic* with period ω. From the properties of $y(x)$ on the lines parallel to $x = \xi a$, where $y(x)$ is nonperiodic, it follows that ω is linearly independent of a. This function then has a primitive period b parallel to ω, and the vectors a, b span the space R^2.

Taking into account the integrability condition $A h A k = A k A h$ it now follows by means of a simple calculation, in which one uses a and b as a coordinate system, that $y(x)$ is isomorphic to the complex exponential function.

V. Theory of Curves and Surfaces

As an application of the preceding chapters, the basic features of the classical curve theory and of the Gaussian surface theory will be presented in this chapter. We shall consider curves and surfaces which lie in an n-dimensional linear space and investigate the properties of these structures relative to a euclidean metric introduced in the embedding space.

§ 1. Regular Curves and Surfaces

1.1. Regular arcs and surfaces. In the following we consider two linear spaces of dimensions m and n. In the former so-called parameter space R_x^m let an open m-dimensional domain G_x^m be given. Further, let

$$x \to y = y(x) \tag{1.1}$$

be a single-valued mapping from G_x^m into the space R_y^n, with the following properties:

1. *The vector function $y(x)$ is continuously differentiable.*
2. *The derivative operator $y'(x)$ is regular.*

Such a mapping defines a local "m-dimensional regular surface" F^m in the space R_y^n.

It follows from our assumptions that $m \leqq n$. The theory of implicit functions (cf. II.4.2) shows that a sufficiently small neighborhood H_x^m of a point $x \in G_x^m$ is mapped one-to-one onto a set H_y^m determined by equation (1.1)[1]. The set H_y^m is then, and only then, an n-dimensional neighborhood of the point $y = y(x)$ when $m = n$. In the case $m < n$ the surface F^m is in the proper sense of the word "embedded" in R_y^n.

For $m = 1$, (1.1) is a regular arc in R_y^n. When $m = 2$, $n = 3$, one is dealing with the main problem in the Gaussian surface theory. We shall in what follows choose the finite dimension n of the embedding space arbitrarily and the dimension of the surface in general to be $m = n - 1$. However, we do not impose this restriction at first; thus m is an arbitrary number in the interval $1 \leqq m \leqq n - 1$.

[1] If this one-to-one-ness does not hold for the entire parameter region G_x^m, then the latter region can be replaced by a subregion for which this is true, which is possible under hypotheses 1 and 2.

1.2. Transformation of parameters. Tangent space. In the above definition $y = y(x)$ is the equation of an arc or surface F^m with respect to the parameter x, which is called "admissible" in order to briefly indicate that the equation of the surface has the properties 1 and 2 mentioned above relative to this parameter.

Now let

$$\bar{x} = \bar{x}(x)\,, \qquad x = x(\bar{x})$$

be a one-to-one mapping of the open region G_x^m onto an open region $\bar{G}_{\bar{x}}^m$ of the same or of another m-dimensional linear space $\bar{R}_{\bar{x}}^m$. If this function $\bar{x}(x)$ is continuously differentiable in G_x^m and the derivative $\bar{x}'(x)$ is regular, then $\bar{x}$ is also an admissible parameter, and the equation of the given surface F^m with respect to this parameter is

$$y = y(x(\bar{x})) = \bar{y}(\bar{x}) = \bar{y}\,.$$

Because the surface F^m in the space R_y^n is given independently of the admissible parametric representations, the representing function $y(x)$, going from one admissible parameter x to another $\bar{x}$, is according *invariant*: one obtains $\bar{y}(\bar{x})$ from $y(x)$ simply by substituting $x = x(\bar{x})$ and conversely $y(x) = \bar{y}(\bar{x}(x))$.

If one sets $x \sim \bar{x}$ whenever these two points from different parametric representations correspond to the same surface point $y(x) = \bar{y}(\bar{x})$, then this relation is an equivalence whose equivalence classes

$$(x, \bar{x}, \bar{\bar{x}}, \ldots)$$

correspond one-to-one to the surface points

$$y = y(x) = \bar{y}(\bar{x}) = \bar{\bar{y}}(\bar{\bar{x}}) = \ldots$$

By the definition of the differential, $d\bar{x} = \bar{x}'(x)\,dx$, $dx = x'(\bar{x})\,d\bar{x}$. The parameter differentials $dx = h$ and $d\bar{x} = \bar{h}$ are transformed according to the formulas

$$\bar{h} = \frac{d\bar{x}}{dx}\,h\,, \qquad h = \frac{dx}{d\bar{x}}\,\bar{h}\,.$$

This is the transformation law of *contravariance*.

On the surface, there is corresponding to the parameter differential $dx = h$ the *tangent vector*

$$k = dy = y'(x)\,dx = y'(x)\,h\,.$$

By the chain rule

$$y'(x) = \bar{y}'(\bar{x})\,\bar{x}'(x), \quad \bar{y}'(\bar{x}) = y'(x)\,x'(\bar{x})\,,$$

and consequently

$$k = dy = y'(x)\,h = \bar{y}'(\bar{x})\,\bar{x}'(x)\,h = \bar{y}'(\bar{x})\,\bar{h} = d\bar{y} = \bar{k}\,.$$

This states that the derivative operator $y'(x) = A(x)$, which can also be interpreted as a vector from the linear mn-dimensional operator space, is transformed under an admissible transformation of the parameter according to the law of *covariance*

$$\bar{A} = A\frac{dx}{d\bar{x}}, \qquad A = \bar{A}\frac{d\bar{x}}{dx}.$$

If a covariant operator operates upon a contravariant vector, there results an invariant, as can be seen from the invariance of the surface tangent $k = dy = d\bar{y} = \bar{k}$. From this invariance it follows that the set of all tangent vectors at the point $y = y(x) = \bar{y}(\bar{x})$, the *tangent space*

$$y'(x)\, R^m_x = \bar{y}'(\bar{x})\, \bar{R}^m_{\bar{x}},$$

is also invariant at the above surface point. It is therefore an m-dimensional subspace of the embedding space R^n_y, independent of the choice of the parameter. The m-dimensional hyperplane

$$E^m_y(x) = y(x) + y'(x)\, R^m_x$$

is the *tangent plane* of the surface at the point $y = y(x)$.

Later we shall return to the notions of invariance, contravariance and covariance in more detail and from a more general point of view.

1.3. The surface as m-dimensional manifold. According to the definitions of 1.1 and 1.2 the surface F^m is homeomorphic (topologically equivalent) to the parameter region G^m_x (and to $\bar{G}^m_{\bar{x}}$). If "open point sets" or "neighborhoods" H^m on F^m are declared to be images of the open point sets H^m_x in G^m_x, then F^m satisfies locally the axioms for a *Hausdorff space*[1].

But more than that, F^m is an *m-dimensional manifold*, for (in addition to A and B) the essential axiom holds:

C. There is a covering set (H^m) of F^m, each neighborhood H^m, $\bar{H}^m, \ldots$ of which is homeomorphic to an open region H^m_x, $\bar{H}^m_{\bar{x}}, \ldots$ of the parameter spaces R^m_x, $\bar{R}^m_{\bar{x}}, \ldots$ Furthermore, for a nonempty intersection $H^m \cap \bar{H}^m$ (etc.) the equivalence relation $x \sim \bar{x}$, in which the parameters have the same image point p on $H^m \cap \bar{H}^m$, is topological.

If in particular the equivalence relations $x = x(\bar{x})$ (etc.) are continuously differentiable and the derivatives $x'(\bar{x})$ regular (which was

[1] F^m can be covered with a set of neighborhoods (H^m) such that:

A. (Axiom for a topological space). The union of arbitrarily many and the intersection of finitely many neighborhoods is again a neighborhood H^m.

B. (Hausdorff's separation axiom.) Two different points p of the surface have two disjoint neighborhoods H^m.

hypothesized above), then F^m is a "continuously differentiable" or "regular" manifold. This property allows the notion of a continuously differentiable vector function to be introduced on F^m. One special such invariant is the function $y = y(x)$ by means of which the manifold F^m was defined as a surface *embedded* in the vector space R^n_y.

1.4. General definition of a surface. The above considerations are based upon the especially simple assumption that the surface F^m in its full extent permits a single-valued parametric mapping $G^m_x \leftrightarrow F^m$. With such a special assumption one does not come far in the theory of surfaces. In surface theory one is not interested only in the local properties of a surface; however, if one does go on to investigate such a structure in the large, the situation becomes more complicated. Thus, for example, a two-dimensional surface in R^3_y generally cannot be mapped over its entire extent homeomorphically onto a region $G^2_x \subset R^2_x$. In order to go on, one can, following the classical procedure of Gauss, construct the surface out of adjacent surface elements that possess the above homeomorphy property C (triangulation of the surface), or, more generally: one covers the surface with a system of neighborhoods H^m so that each H^m has property C.

In order to formulate this last point of view into a precise definition of an m-dimensional surface F^m in the space R^n_y, it is convenient to first define the surface independently (of the embedding), specifically as a Hausdorff space (Axioms A, B) with the parametrization property C. After F^m is thus defined as an m-dimensional manifold, its embedding is brought about by means of a continuous (or differentiable) mapping of the surface points $p = (x, \bar{x}, \ldots)$ into the vector space R^n_y.

In what follows we shall at first stick mainly to the local properties of a curve or a surface, and for this reason it suffices to undertake our investigation relative to a fixed parameter x. Not until § 4, in connection with the above general definition of a surface, shall we return to a more careful examination of the question of parameter transformations mentioned briefly in 1.2.

1.5. Metrization of the embedding space. The above definitions have assumed no metrization of the embedding space or of the parameter spaces. We now introduce a *metric* into the embedding space R^n_y by means of a real, bilinear, symmetric, positive definite *fundamental form* (y_1, y_2), which also induces a euclidean metric in each tangent space of the surface. The goal of the differential geometric embedding theory established by Gauss is to investigate the properties of curves and surfaces relative to a euclidean embedding space.

Regarding the dimension m $(1 \leq m \leq n - 1)$, we shall restrict ourselves to the extreme cases $m = 1$ and $m = n - 1$. The case

$m = 1$ yields the theory of curves governed by the Frenet equations, while the case $m = n - 1$ is a generalization of the classical Gaussian surface theory associated with the special dimensions $m = 2$, $n = 3$.

§ 2. Curve Theory

2.1. Arc length. We consider a regular arc $y = y(x)$ of the euclidean space R^n_y. In this 1-dimensional case we can use the real number line as the parameter space, and the arc is therefore defined by a mapping

$$y = y(\xi)$$

of the interval $\alpha < \xi < \beta$ into the space R^n_y, so that $y(\xi)$ is continuously differentiable in the latter interval and the mapping is regular.

In the case at hand, the derivative can be represented in the usual way as the limit

$$y'(\xi) = \lim_{\Delta\xi \to 0} \frac{y(\xi + \Delta\xi) - y(\xi)}{\Delta\xi},$$

which for $\alpha < \xi < \beta$ is therefore continuous and different from zero.

Corresponding to the subinterval $\xi_0\,\xi$ is a subarc with *arc length*

$$\int_{\xi_0}^{\xi} |y'(\xi)|\, d\xi = \sigma(\xi)\,.$$

From $\sigma'(\xi) = |y'(\xi)| > 0$ it follows that $\sigma(\xi)$ is a monotonically increasing function which is differentiable with respect to ξ as often as the function $y(\xi)$ is continuously differentiable. The transformation

$$\sigma = \sigma(\xi)\,, \qquad \xi = \xi(\sigma)$$

is consequently admissible, and one can use the arc length σ as the parameter. This offers certain advantages, because

$$\frac{dy}{d\sigma} = \frac{dy}{d\xi}\frac{d\xi}{d\sigma} = \frac{y'(\xi)}{|y'(\xi)|},$$

and as a consequence

$$\left|\frac{dy}{d\sigma}\right| = 1\,, \qquad |dy| = d\sigma$$

hold identically.

2.2. The associated *n*-frame. For our further investigation it is to be assumed that the parametric representation $y = y(\xi)$ satisfies the following more special conditions:

1. *The first $n + 1$ derivatives of the function $y(\xi)$ exist in the interval $\alpha < \xi < \beta$.*

2. *The first n derivatives are for each ξ of this interval linearly independent.*

If

$$\bar{\xi} = \bar{\xi}(\xi) , \qquad \xi = \xi(\bar{\xi})$$

is an admissible parameter transformation which is moreover $(n + 1)$-times differentiable (as is the case, for example, for $\bar{\xi} = \sigma(\xi)$), then the above hypotheses are also fulfilled with respect to the parameter $\bar{\xi}$. The second hypothesis is a consequence of $\bar{\xi}'(\xi) \neq 0$ and of the pth derivative ($p = 1, \ldots, n + 1$) of y with respect to the one parameter being a linear combination of the first p derivatives with respect to the other parameter.

From these statements it further results that the first p derivatives at each point $y = y(\xi)$ of the arc span a p-dimensional subspace $S_y^p(\xi)$ which is independent of the choice of the parameter. This is the *p-dimensional osculating space* and $y(\xi) + S_y^p(\xi)$ the *p-dimensional osculating plane* of the curve at the point $y(\xi)$; for $p = 1$ one has the tangent to the curve at the point in question.

Now orthogonalize the linearly independent derivatives

$$y'(\xi), \ldots, y^{(n)}(\xi) ,$$

in this order, by means of the Schmidt orthogonalization process, with respect to the fundamental euclidean form (y_1, y_2) laid down in 1.5. Using the notation from I.4.4, we find

$$\begin{aligned} y^{(p)}(\xi) &= \lambda_{p1}(\xi)\, e_1(\xi) + \cdots + \lambda_{pp}(\xi)\, e_p(\xi) , \\ e_p(\xi) &= \mu_{p1}(\xi)\, y'(\xi) + \cdots + \mu_{pp}(\xi)\, y^{(p)}(\xi) . \end{aligned} \tag{2.1}$$

Here $\lambda_{pp}(\xi)\,\mu_{pp}(\xi) \equiv 1$. All coefficients λ_{ij} and μ_{ij} are, for a given parameter, uniquely determined if the sign of λ_{pp} is fixed; for example, we take $\lambda_{pp}(\xi) > 0$.

The first p orthonormal vectors $e_1(\xi), \ldots, e_p(\xi)$ generate the osculating space $S_y^p(\xi)$ at the point $y(\xi)$. Because this space is invariant, it follows that the orthogonal coordinate system

$$e_1(\xi), \ldots, e_n(\xi)$$

is uniquely determined, independently of the choice of parameter, at each point of the arc. This is the *accompanying n-frame* of the curve; e_1 is the unit tangent, e_2 the first or principal normal, e_3 the second or binormal, and so on.

If one takes the arc length σ as parameter, which will be the case in what follows, where we do not write out the argument σ, then it follows from the identities

$$(y', y') = 1 , \qquad (y', y'') = 0 ,$$

that $e_1 = y'$, $e_2 = y''/|y''|$. Therefore, in the Schmidt orthogonalization scheme, for the parameter σ,

$$\lambda_{11} = 1 , \quad \lambda_{21} = 0 , \quad \lambda_{22} = |y''| .$$

2.3. Frenet's formulas. Since the existence of the derivative $y^{(n+1)}$ was assumed, by the Schmidt orthogonalization scheme not only do the derivatives $e_1', \ldots, e_{n-1}'$ exist, but so also does e_n'. For each σ of the interval $\alpha < \sigma < \beta$, therefore, there exists a uniquely determined linear transformation $A(\sigma) = A$ of the space R_y^n such that for $p = 1, \ldots, n$

$$e_p' = A\, e_p \,. \tag{2.2}$$

A is the *Frenet operator* and the corresponding matrix (α_p^q)

$$(\alpha_p^q) = (e_p', e_q) = (A\, e_p, e_q)\,,$$

with respect to the n-frame at the point σ, the *Frenet matrix* of the arc at this point.

In order to investigate the properties of this transformation, we consider the orthogonal transformation $T(\sigma) = T$ determined by the n equations

$$e_p(\sigma) = T(\sigma)\, e_p^0 \quad (e_p^0 = e_p(\sigma_0)\,,\ \ \alpha < \sigma_0 < \beta)\,, \tag{2.3}$$

where $T(\sigma_0) = I$ is the identity transformation. By means of this transformation one obtains from

$$e_p' = T'\, e_p^0 = T'\, T^{-1}\, e_p$$

the representation

$$A = T'\, T^{-1} = T'\, T^*$$

for the Frenet operator, where $T' = dT/d\sigma$ and T^* stands for the transformation adjoint to T.

From this it is seen that A is *skew symmetric*. In fact, it follows from the identity $T\, T^{-1} = T\, T^* = I$, by means of differentiation with respect to σ, that

$$0 = T'\, T^* + T(T^*)' = T'\, T^* + T(T')^* = A + A^*\,,$$

which expresses the skew symmetry of A. Consequently, the matrix of the operator A with respect to any orthonormal coordinate system, in particular also with respect to the n-frame at the point σ, is skew symmetric:

$$\alpha_p^q + \alpha_q^p = 0\,. \tag{2.4}$$

If one further considers that the unit vector e_p, according to the Schmidt orthogonalization scheme, is a linear combination of the first p derivatives of y, and the derivative $y^{(p)}$, conversely, is a linear combination of $e_1, \ldots, e_p$, then it follows that for $q > p + 1$, $\alpha_p^q = (e_p', e_q) = (A\, e_p, e_q) = 0$. Because of (2.4), the same is also true for $q < p - 1$. Thus, if one sets $\alpha_p^{p+1} = \varkappa_p$ $(p = 1, \ldots, n-1)$, then

$$\begin{aligned} \alpha_p^{p+1} &= \varkappa_p\,, \quad \alpha_{p+1}^p = -\varkappa_p\,, \\ \alpha_p^q &= 0 \quad (q \neq p+1,\ p-1)\,. \end{aligned} \tag{2.5}$$

The Frenet matrix (α_p^q) is therefore a "skew symmetric Jacobi matrix", and the Frenet equations (2.2), written out with respect to the n-frame $e_1, \ldots, e_n$ at the point σ, are

$$\begin{aligned} e_1' &= \varkappa_1 e_2 , \\ e_p' &= -\varkappa_{p-1} e_{p-1} + \varkappa_p e_{p+1} \quad (p = 2, \ldots, n-1) , \\ e_n' &= -\varkappa_{n-1} e_{n-1} . \end{aligned} \tag{2.2'}$$

The $n - 1$ quantities $\varkappa_p = \varkappa_p(\sigma)$ are called the *curvatures* of the curve $y = y(\sigma)$ at the point σ; $\varkappa_1$ is the first or principal curvature, $\varkappa_2$ is the second or torsion, etc. As functions of the arc length σ they are uniquely determined at each point of the arc up to the sign, which depends on the orientation of the unit vectors e_p. From the Schmidt orthogonalization scheme (2.1) and the Frenet formulas (2.2′) one obtains the expression

$$\varkappa_p = \frac{\lambda_{(p+1)\,(p+1)}}{\lambda_{pp}} \tag{2.6}$$

for these curvatures, so that $\varkappa_p > 0$ provided λ_{pp} is assumed to be > 0 in equation (2.1) (cf. 2.6, exercise 1). With these formulas one also easily computes the curvatures directly from the derivatives of the function $y(\sigma)$ (cf. 2.6, exercise 1).

2.4. Integration of the Frenet equations. From (2.1) and (2.6) it can be seen that $\varkappa_p(\sigma)$ is differentiable $(n - p)$ times in the given interval, provided the curve $y = y(\sigma)$ satisfies conditions 1 and 2 of 2.2. We now show, conversely:

If the $\varkappa_p(\sigma)$ $(p = 1, \ldots, n-1)$ are defined as (positive) and each $(n-p)$-times-differentiable functions in the interval $\alpha < \sigma < \beta$, then in the euclidean space R_y^n there exists an arc $y = y(\sigma)$ satisfying conditions 1 and 2 of 2.2 with the arc element $d\sigma$ and the prescribed curvatures $\varkappa_p$.

This curve is uniquely determined up to a translation and orthogonal transformation of the space.

Assuming there exists an arc in R_y^n with the asserted properties, then its n-frame satisfies the Frenet equations (2.2′). The operator A has with respect to the n-frame at the point σ the prescribed Frenet matrix (2.5) and is in this way uniquely determined for each σ of the interval $\alpha < \sigma < \beta$ as a skew symmetric transformation of the space R_y^n. If the orthogonal transformation T is introduced by means of (2.3), then $A = T' \, T^{-1}$, and T satisfies the differential equation

$$T' = A\,T \tag{2.7}$$

in the n^2-dimensional operator space, with the initial value $T(\sigma_0) = I$.

Here the Frenet operator A depends on the accompanying n-frame of the curve $y = y(x)$. But in our converse problem, determining this curve by means of the given curvatures $\varkappa_p(\sigma)$ ($p = 1, \ldots, n$), only the alternating Frenet matrix $M(\sigma)$ is known. In order to fix the operator A, we observe that $M(\sigma)$ is invariant under the transformation of $A(\sigma)$ by the orthogonal transformation $T(\sigma)$, so that $A(\sigma)$ has the same Frenet matrix in the frame $(e_1(\sigma), \ldots, e_n(\sigma))$ as the alternating operator $B(\sigma) = T^{-1}(\sigma)\, A(\sigma)\, T(\sigma)$ in the frame $(e_1(\sigma_0), \ldots, e_n(\sigma_0))$. The equation (2.7) can therefore be written

$$T'(\sigma) = A\, T = T(\sigma)\, B(\sigma)\,. \tag{2.7'}$$

In order to integrate this equation for the unknown operator $T(\sigma)$, with the initial value I in an arbitrarily given orthonormal basis $(e_1(\sigma_0), \ldots, e_n(\sigma_0))$, we use the main theorem of the theory of linear differential equations. The solution $T(\sigma)$ is uniquely determined, and $T(\sigma_0) = I$.

This solution is orthogonal. In fact,

$$\begin{aligned}(T\, T^*)' &= T'\, T^* + T(T')^* = T\, B\, T^* + T(T\, B)^* \\ &= T\, B\, T^* + T\, B^*\, T^* = 0\,,\end{aligned}$$

because $B + B^* = 0$.

Thus $T\, T^*$ is independent of σ and equal to the constant I. That implies the orthogonality of $T(\sigma)$. Thus if the initial vectors $e_p(\sigma_0) = e_p^0$ ($p = 1, \ldots, n$) in (2.3) are orthonormalized, the equation (2.7') determines an orthonormal n-frame $(e_p(\sigma))$ ($p = 1, \ldots, n$). In particular, by $y'(\sigma) = e_1(\sigma)$, we conclude that

$$y(\sigma) = y_0 + \int_{\sigma_0}^{\sigma} e_1(\sigma)\, d\sigma \qquad (y(\sigma_0) = y_0) \tag{2.8}$$

is the only possible curve of the sort required that goes through the point y_0, with the frame $(e_1^0, \ldots, e_n^0)$.

This arc, which is uniquely determined by the prescribed curvatures $\varkappa_p$, the point y_0 and the n-frame $(e_1^0, \ldots, e_n^0)$, has, in fact, all of the required properties.

First, it follows from (2.8) that $|y'(\sigma)| = |e_1(\sigma)| = 1$. The parameter σ is thus the arc length of the constructed curve.

Further, because of the $(n - p)$-differentiability of the functions $\varkappa_p$, the derivatives $y' = e_1$, $y'' = e_1' = A\, e_1 = \varkappa_1\, e_2$, $y''' = \varkappa_1'\, e_2 + \varkappa_1\, e_2' = \varkappa_1'\, e_2 + \varkappa_1\, A\, e_2 = -\varkappa_1^2\, e_1 + \varkappa_1'\, e_2 + \varkappa_1\, \varkappa_2\, e_3$ exist, and in general for $p = 2, \ldots, n$

$$y^{(p)} = \lambda_{p1}\, e_1 + \cdots + \lambda_{pp}\, e_p\,,$$

with

$$\lambda_{pp} = \varkappa_1 \ldots \varkappa_{p-1} \qquad (> 0)\,,$$

exists; even the derivative $y^{(n+1)}$ exists. From the expression for $y^{(p)}$ it can be seen that the derivatives $y', \ldots, y^{(n)}$ are linearly independent and that the above constructed orthonormal system $e_1, \ldots, e_n$ results from these derivatives by means of the Schmidt orthogonalization process. The vectors e_p thus form the n-frame of the constructed curve at the point σ.

Further, in view of formula (2.6) it follows from the above expression for λ_{pp} that the curve constructed has the prescribed functions $\varkappa_1, \ldots, \varkappa_{n-1}$ as curvatures at each point of the interval $\alpha < \sigma < \beta$.

The claim relative to the uniqueness of the curve constructed is also a result of the construction. For if $y = \bar{y}(\sigma)$ is a second solution that goes through the point $\bar{y}(\sigma_0) = \bar{y}_0$ with the n-frame $\bar{e}_1^0, \ldots, \bar{e}_n^0$, there exists a unique orthogonal transformation T_0 such that $T_0\, \bar{e}_p^0 = e_p^0$. Then the curve

$$y = y^*(\sigma) = y_0 + T_0(\bar{y}(\sigma) - \bar{y}_0)$$

goes through the point $y^*(\sigma_0) = y_0$ with the n-frame $e_1^0, \ldots, e_n^0$. At each point σ of the given interval it has the same curvatures as $y = \bar{y}(\sigma)$ and thus also as the constructed curve $y = y(\sigma)$. But then we have identically $y^*(\sigma) \equiv y(\sigma)$, which completes the proof.

2.5. Degeneration of the curve. In the above discussion it was assumed that the first n derivatives of the function $y(\xi)$ are linearly independent at each point of a certain parameter interval.

Now assume that only the first l ($< n$) derivatives turn out to be linearly independent, while for $y^{(l+1)}(\xi)$ the linear relation

$$y^{(l+1)}(\xi) = \sum_{i=1}^{l} \lambda_i(\xi)\, y^{(i)}(\xi)$$

holds in this interval, with continuous coefficients λ_i.

For a fixed ξ_0 of the parameter interval the linearly independent derivatives $y'(\xi_0), \ldots, y^{(l)}(\xi_0)$ generate an l-dimensional subspace U_y^l of the space R_y^n. In the latter, U_y^l has an orthogonal complement of dimension $n - l$; let a be an arbitrary vector of this complement.

With the constant vector a we form the functions

$$\zeta_i(\xi) = (y^{(i)}(\xi), a) \qquad (i = 1, \ldots, l)\,.$$

These l functions now satisfy the like number of equations in the normal system of linear differential equations

$$z_i'(\xi) = z_{i+1}(\xi) \quad (i = 1, \ldots, l-1)\,, \quad z_l'(\xi) = \sum_{i=1}^{l} \lambda_i(\xi)\, z_i(\xi)\,,$$

and this is also satisfied by the identically vanishing functions $z_i(\xi) = 0$. But for $\xi = \xi_0$,

$$\zeta_i(\xi_0) = (y^{(i)}(\xi_0), a) = 0 \qquad (i = 1, \ldots, l)\,,$$

and therefore, according to the already often used uniqueness theorem, these two systems of solutions are identical.

By this, $\zeta_i(\xi) = 0$ in the entire parameter interval, and in particular

$$\zeta_1(\xi) = (y'(\xi), a) = 0 ,$$

and

$$\int_{\xi_0}^{\xi} \zeta_1(\xi)\, d\xi = (y(\xi) - y(\xi_0), a) = 0 .$$

Since this equation holds for each a from the orthogonal complement of U_y^l and each ξ in the parameter interval, the entire arc is orthogonal to this complement and therefore lies in the l-dimensional hyperplane $y(\xi_0) + U_y^l$. If the curve is shifted parallel to itself by the vector $-y(\xi_0)$, it lies in the subspace U_y^l, where the curve theory developed above can be applied to it.

2.6. Exercises. 1. Prove formula (2.6),

$$\varkappa_p = \frac{\lambda_{(p+1)(p+1)}}{\lambda_{pp}} .$$

2. Show that the curvature $\varkappa_p = \varkappa_p(\sigma)$ of the curve $y = y(\sigma)$ can be computed from

$$\varkappa_p^2 = \frac{\Delta_{(p+1)(p+1)}\, \Delta_{(p-1)(p-1)}}{\Delta_{pp}^2} \qquad (p = 1, \ldots, n-1)$$

where $\Delta_{00} = 1$, and for $p \geqq 1$

$$\Delta_{pp} = \begin{vmatrix} (y', y') & \ldots & (y', y^{(p)}) \\ \vdots & & \vdots \\ (y^{(p)}, y') & \ldots & (y^{(p)}, y^{(p)}) \end{vmatrix} .$$

Hint. According to exercise 16 in I.6.11, where one is to set $z_p = y^{(p)}$, one has for $p = 1, \ldots, n$

$$\lambda_{11}^2 \ldots \lambda_{pp}^2 = \Delta_{pp} .$$

3. Determine the curve in R_y^n (uniquely determined up to a euclidean displacement), whose curvatures $\varkappa_1, \ldots, \varkappa_{n-1}$ are (positive) constants.

Hint. Let T be the orthogonal operator defined by equation (2.3) and $\sigma_0 = 0$. Since the matrix (2.5) of the Frenet operator A with respect to the n-frame $e_1, \ldots, e_n$ is by hypothesis independent of the arc length σ, integration of the differential equation (2.7) by means of Picard's method of successive approximations (cf. IV.1.10, exercises 1—2) yields

$$T = T(\sigma) = \sum_{i=0}^{\infty} \frac{\sigma^i}{i!} A^i \qquad (A^0 = T(0) = I) ,$$

so that

$$y'(\sigma) = e_1(\sigma) = T(\sigma)\, e_1(0) = \sum_{i=0}^{\infty} \frac{\sigma^i}{i!} A^i\, e_1(0)\,. \tag{a}$$

Because of the skew symmetry of A, by exercise 13 in I.6.13, an orthogonal transformation T_0 and a fixed orthonormal system $a_1, \ldots, a_p$ exist such that for $p = 1, \ldots, n$

$$e_p(0) = T_0\, a_p\,, \qquad a_p = T_0^*\, e_p(0)$$

and for $q = 1, \ldots, m = [n/2]$

$$A\, a_{2q-1} = \varrho_q\, a_{2q}\,, \quad A\, a_{2q} = -\,\varrho_q\, a_{2q-1}\,, \tag{b}$$

where for an odd $n = 2m + 1$ the axis a_n with

$$A\, a_n = 0 \tag{b'}$$

must be added. From the special Jacobi structure of the Frenet matrix (2.5) one concludes that the numbers ϱ_q are $\neq 0$. For if one of these numbers should $= 0$, then it would follow from (b) that the kernel of the operator A would have at least dimension 2. But if one refers this operator to the coordinate system $e_1(0), \ldots, e_n(0)$, then one finds that the equation

$$A\, x = \sum_{i=1}^{n} \xi^i\, A\, e_i(0) = 0$$

holds for an even n only for $x = 0$ and for an odd $n = 2m + 1$ only for

$$x = \xi^1 \sum_{j=1}^{m} \frac{\varkappa_1 \varkappa_3 \ldots \varkappa_{2j-1}}{\varkappa_2 \varkappa_4 \ldots \varkappa_{2j}}\, e_{2j+1}(0)\,. \tag{c}$$

The kernel of A is therefore of dimension 0 in the first case, and of dimension 1 in the second.

Because equations (b) are invariant with respect to an orthogonal transformation of the plane spanned by a_{2q-1} and a_{2q}, the transformation T_0 and the orthogonal system $a_1, \ldots, a_n$ can be normalized in a unique way so that, for example,

$$e_1(0) = T_0\, a_1 = \sum_{q=1}^{m} \lambda_q\, a_{2q} + \lambda\, a_n\,,$$

where $\lambda = 0$ for an even $n = 2m$. If this is substituted in (a) a short calculation yields, because of (b),

$$y'(\sigma) = -\sum_{q=1}^{m} \lambda_q \left(\sin(\varrho_q\, \sigma)\, a_{2q-1} - \cos(\varrho_q\, \sigma)\, a_{2q}\right) + \lambda\, a_n\,,$$

from which

$$y = \sum_{q=1}^{m} \frac{\lambda_q}{\varrho_q} \left(\cos(\varrho_q\, \sigma)\, a_{2q-1} + \sin(\varrho_q\, \sigma)\, a_{2q}\right) + \lambda\, \sigma\, a_n \tag{d}$$

is obtained, if the integration constant $y(0)$ is chosen suitably.

It can be seen from the above that the constants $\lambda_1, \ldots, \lambda_m$ and (for an odd n) λ satisfy the relation

$$|y'(\sigma)|^2 = \sum_{q=1}^{m} \lambda_q^2 + \lambda^2 = 1 . \tag{e}$$

On the other hand, the constants λ_q and, for an odd n, λ too are $\neq 0$. Otherwise, as follows from (d), the curve would degenerate, which is not possible if all curvatures are different from zero. Further, it follows from equations (d) and (e) that equation (d) contains precisely $n - 1$ independent parameters, corresponding to the $n - 1$ curvatures. Conversely, if these parameters, that is, the numbers $\lambda_q \neq 0$, $\varrho_q \neq 0$ and (for an odd n) $\lambda \neq 0$, are arbitrarily prescribed according to relation (e), then σ is the arc length of the curve, and by the formulas in exercise 2 one obtains for the curvatures of the curve *constant* nonzero values that up to their sign are uniquely determined.

Remark. Equation (d) shows that for an even $n = 2\,m$ the curve lies on the sphere

$$|y| = \sum_{q=1}^{m} \frac{\lambda_q^2}{\varrho_q^2} ,$$

while for an odd $n = 2\,m + 1$ it winds itself indefinitely around the axis a_n determined by equation (c). For $n = 2$ the curve is a circle, for $n = 3$ a helix.

4. Compute the curvatures of the curve (d) in the preceding exercise for $n = 2, 3, 4$.

Hint. For the computation of the determinant Δ_{pp} in exercise 2, observe that as a consequence of equations (d) and (e) $(y', y') = 1$ and for $i + j > 2$

$$(y^{(i)}, y^{(j)}) = \sum_{q=1}^{m} \lambda_q^2 \, \varrho_q^{i+j-2} \cos \frac{(i-j)\,\pi}{2} .$$

§ 3. Surface Theory

3.1. The first fundamental form. We refer to the definition given in § 1 and consider an m-dimensional regular surface F^m embedded in the n-dimensional euclidean space R_y^n, where for the time being we assume $1 \leqq m \leqq n - 1$. Such a surface is defined by an "admissible" parametric representation (cf. 1.1 and 1.2).

The parameter space R_x^m is mapped by means of the regular linear operator $y'(x)$ one-to-one onto the m-dimensional tangent space of the surface at the surface point $y(x)$ determined by the parameter x. Corresponding to the vectors h and k of the parameter space are the tangents $y'(x)\,h$ and $y'(x)\,k$ with the inner product

$$G(x)\,h\,k \equiv \big(y'(x)\,h,\; y'(x)\,k\big) . \tag{3.1}$$

This is the *first fundamental form* of the surface theory, first introduced by Gauss. It determines the measurement of volumes and angles on the surface. At each point of the parameter region G_x^m it is a real, bilinear, symmetric, positive definite function of the parameters h and k.

The tangent $dy = y'(x)\,dx = y'(x)\,h$ has length

$$|dy| = |y'(x)\,h\,| = \sqrt{G(x)\,h\,h}\,,$$

and the angle ϑ enclosed by the tangent vectors $y'(x)\,h$ and $y'(x)\,k$ is determined up to its sign by

$$\cos\vartheta = \frac{(y'(x)\,h,\, y'(x)\,k)}{|y'(x)\,h||y'(x)\,k|} = \frac{G(x)\,h\,k}{\sqrt{G(x)\,h\,h\,G(x)\,k\,k}}\,.$$

More generally, the following is true: Corresponding to a d-dimensional simplex spanned at the point x by the linearly independent vectors $h_1, \ldots, h_d$ $(1 \leqq d \leqq m \leqq n-1)$ by means of the regular operator $y'(x)$, there is in the tangent space $y'(x)\,R_x^m$ a simplex at the surface point $y(x)$, spanned by the linearly independent tangents $y'(x)\,h_1, \ldots, y'(x)\,h_d$, which according to I.6.10 has the volume

$$\frac{1}{d!}\sqrt{\det\left(G(x)\,h_i\,h_j\right)}\,,$$

where

$$\det\left(G(x)\,h_i\,h_j\right) = \begin{vmatrix} G(x)\,h_1\,h_1 & \ldots & G(x)\,h_1\,h_d \\ \vdots & & \vdots \\ G(x)\,h_d\,h_1 & \ldots & G(x)\,h_d\,h_d \end{vmatrix}.$$

In the theory of curves it is convenient to use the arc length of the curve as the parameter, so that the parameter interval and the arc are related isometrically. Corresponding to this, in the theory of surfaces one uses at each point x of the region G_x^m the first fundamental form as the basic metric form, by means of which the linear parameter space, for a fixed x, becomes euclidean with the inner product

$$(h, k)_x = G(x)\,h\,k\,.$$

In this metric $y'(x)$ consequently provides an orthogonal mapping of the parameter space R_x^m onto the tangent space $y'(x)\,R_x^m$.

If one introduces an affine coordinate system $a_1, \ldots, a_m$ in the parameter space R_x^m and writes the differentials

$$h = d_1 x = \sum_{i=1}^{m} d_1\xi^i\,a_i\,, \qquad k = d_2 x = \sum_{j=1}^{m} d_2\xi^j\,a_j\,,$$

the result is the usual coordinate form of the first fundamental form:

$$G(x)\, d_1x\, d_2x = \sum_{i,j=1}^{m} g_{ij}(x)\, d_1\xi^i\, d_2\xi^j\,, \tag{3.1'}$$

where

$$g_{ij}(x) = G(x)\, a_i\, a_j \qquad (g_{ij} = g_{ji})\,.$$

3.2. The unit normal. From now on we restrict ourselves to *the case* $m = n - 1$ $(m \geqq 2)$.

Under this assumption the m-dimensional tangent plane of the surface at the point $y = y(x)$ defined by the equation

$$E_y^m(x) = y(x) + y'(x)\, R_x^m \tag{3.2}$$

possesses a well-determined one-dimensional orthogonal complement; this is the *normal* to the surface at the point $y = y(x)$. With the latter point as the initial point lay off two opposite unit normals (normals of length 1). For what follows it is important to fix the orientation of the normal. This can be accomplished by means of the Schmidt orthogonalization process in the following way.

Start at a point $y = y(x)$, determine here the tangent plane (3.2) and fix an arbitrary vector $y = a \neq 0$ that intersects this plane. If this vector, which in the following is to be held *constant*, is projected onto the tangent plane $E_y^m(x)$ and if the projection is $p = p(x)$, then the unit normal $e = e(x)$ can be fixed at the point x by

$$e(x) = \frac{a - p(x)}{|a - p(x)|} \qquad (|a - p(x)| > 0)\,;$$

the other is $-e(x)$. Because of the continuity of the derivative $y'(x)$ the tangent plane for a continuous displacement of the point x moves continuously, and hence the projection $p = p(x)$ is also a continuous function of the location. The condition $|a - p(x)| > 0$ is thus satisfied in the vicinity of the initial point x, and the above formula hence determines $e(x)$ uniquely as a continuous function of x.

A careful analytic proof of this plausible conclusion is easy to give with the aid of the Schmidt orthogonalization process (cf. 3.8, exercise 1). One further sees that if $y(x)$ is differentiable several times, say q times, then the projection $p(x)$ and with it also the normal $e(x)$ is differentiable $(q - 1)$ times.

In the following we fix at an arbitrary point $y = y(x)$ of the surface a definite direction of the unit normal as the "positive" one. The positive direction of the normal is then defined at another point x_1 by continuation on an arc $x_0\, x_1$.

3.3. The second fundamental form. Henceforth we are going to assume that the function $y(x)$ is *twice continuously differentiable* in the region G_x^m $(m = n - 1)$. By the above the normal $e(x)$ is consequently continuously differentiable once.

At each point $y = y(x)$ of the surface the tangent space and the positive unit normal span the entire embedding space R_y^n. Consequently, for each pair of parameter differentials h, k the vector $y''(x)\,hk$ can be decomposed in a unique way into two orthogonal components, the tangential and the normal.

We shall treat the first component later. The second component is the orthogonal projection of the vector $y''(x)\,h\,k$ onto the positive unit normal, and is therefore equal to

$$\big(y''(x)\,h\,k,\ e(x)\big)\,e(x) = L(x)\,h\,k\,e(x)\,.$$

Here

$$L(x)\,h\,k \equiv \big(y''(x)\,h\,k,\ e(x)\big) \tag{3.3}$$

is the *second fundamental form* of the surface, first introduced by Gauss. It is, like the first fundamental form, a real, bilinear, and, because of the symmetry of the second derivative, symmetric function of the differentials h and k.

3.4. The operator $\Gamma(x)$ and the derivative formula of Gauss. We now investigate the tangential component of the vector $y''(x)\,h\,k$, that is, the orthogonal projection of this vector onto the tangent space $y'(x)\,R_x^m$.

According to the above this projection, on the one hand, is equal to the difference

$$y''(x)\,h\,k - L(x)\,h\,k\,e(x)\,,$$

from which it can be seen that it depends linearly and symmetrically on the parameter differentials h and k.

On the other hand, as a vector of the tangent space, it has a unique preimage in the parameter space which at the given point x likewise will be a bilinear and symmetric function of h and of k and hence can be denoted by

$$\Gamma(x)\,h\,k = \Gamma(x)\,k\,h\,.$$

By this the tangential component of $y''(x)\,h\,k$ considered is equal to

$$y'(x)\,\Gamma(x)\,h\,k\,.$$

Comparison of both expressions yields the *derivative formula of Gauss*,

$$y''(x)\,h\,k = y'(x)\,\Gamma(x)\,h\,k + L(x)\,h\,k\,e(x)\,, \tag{3.4}$$

by means of which the decomposition of the second derivative of $y(x)$ into a tangential and a normal component is effected.

3.5. Dependence of the operator $\Gamma(x)$ on $G(x)$. The bilinear symmetric operator $\Gamma(x)$ can be computed from the first fundamental form $G(x)\, h\, k$.

To see this observe that the unit normal $e(x)$ stands perpendicular to each tangent $y'(x)\, l$, from which it follows, by means of the above derivative formula, that

$$\left(y''(x)\, h\, k,\ y'(x)\, l\right) = \left(y'(x)\, \Gamma(x)\, h\, k,\ y'(x)\, l\right),$$

and consequently

$$G(x)\, l\, \Gamma(x)\, h\, k = \left(y''(x)\, h\, k\,,\ y'(x)\, l\right).$$

On the other hand, if for brevity we do not indicate the fixed point x and observe the symmetry of the second differential,

$$\begin{aligned}(y''\, h\, k, y'\, l) &= G'\, h\, k\, l - (y'\, k, y''\, h\, l)\\ &= G'\, h\, k\, l - (y''\, l\, h, y'\, k)\\ &= G'\, h\, k\, l - G'\, l\, h\, k + (y''\, k\, l, y'\, h)\\ &= G'\, h\, k\, l - G'\, l\, h\, k + G'\, k\, l\, h - (y''\, h\, k, y' l)\,.\end{aligned}$$

Consequently

$$2\, (y''\, h\, k, y'\, l) = G'\, h\, k\, l + G'\, k\, l\, h - G'\, l\, h\, k$$

and

$$G(x)\, l\, \Gamma(x)\, h\, k = \frac{1}{2}\,(G'\, h\, k\, l + G'\, k\, l\, h - G'\, l\, h\, k)\,. \tag{3.5}$$

Since the operator $G(x)$ is not degenerate and equation (3.5) holds for an arbitrarily fixed pair of vectors h, k for each parameter vector l, it determines the vector $\Gamma(x)\, h\, k$ uniquely. For the explicit computation of this vector one can, for example, metrize the parameter space at the point x in a euclidean fashion, by using $G(x)\, h\, k$ as the fundamental metric form, and construct a coordinate system $a_1(x), \ldots, a_{n-1}(x)$ which with respect to this metric is orthonormal. Then if $\Gamma^i(x)\, h\, k$ are the components of $\Gamma(x)\, h\, k$ in this coordinate system, we have for $i = 1, \ldots, n-1$

$$\Gamma^i\, h\, k = G\, a_i\, \Gamma\, h\, k = \frac{1}{2}\,(G'\, h\, k\, a_i + G'\, k\, a_i\, h - G'\, a_i\, h\, k)\,.$$

If h and k are also expressed in coordinates

$$h = \sum_{s=1}^{n-1} \xi_1^s\, a_s\,, \qquad k = \sum_{t=1}^{n-1} \xi_2^t\, a_t\,,$$

and the coordinate representation (3.1′) for G is substituted, one finds

$$\Gamma^i\, h\, k = \sum_{s,t=1}^{n-1} \xi_1^s\, \xi_2^t\, \Gamma^i\, a_s\, a_t\,,$$

and

$$\Gamma^i a_s a_t = \frac{1}{2}\left(\frac{\partial g_{ti}}{\partial \xi^s} + \frac{\partial g_{is}}{\partial \xi^t} - \frac{\partial g_{st}}{\partial \xi^i}\right) = \begin{bmatrix} s\,t \\ i \end{bmatrix}. \tag{3.6}$$

These are the *Christoffel symbols of the first kind.*

3.6. The operator $\Lambda(x)$ and the differentiation formula of Weingarten. It was remarked above that the existence and continuity of the derivative $y''(x)$ implies the continuous differentiability of the unit normal $e(x)$. To determine the derivative $e'(x)$ we start from the identity

$$(e(x),\ e(x)) = 1$$

and from this, by means of differentiation with the differential $dx = h$, obtain the equation

$$(e(x),\ e'(x)\ h) = 0\ .$$

This identity states that for every h in the parameter space, $e'(x)\ h$ stands perpendicular to $e(x)$ and therefore lies in the tangent space $y'(x)\ R^m_x$. Consequently, this vector has a unique preimage in the parameter space that will likewise depend linearly on h and can thus be denoted by

$$-\ \Lambda(x)\ h\ .$$

Accordingly, we have

$$e'(x)\ h = -\ y'(x)\ \Lambda(x)\ h\ . \tag{3.7}$$

This is *Weingarten's differentiation formula.* In connection with the Gaussian differentiation formula it plays a role in the surface theory analogous to that of the Frenet formulas in the theory of curves.

The operator $\Lambda(x)$ defines at each point x a linear self-mapping of the parameter space. In order to find the relation of this operator to the fundamental forms G and L, we start from the equation

$$(y'(x)\ k,\ e(x)) = 0\ ,$$

which holds for $x \in G^m_x$ and for each parameter vector k. If we differentiate this identity with respect to x with the differential $dx = h$, then

$$(y''(x)\ h\ k,\ e(x)) + (y'(x)\ k,\ e'(x)\ h) = 0\ .$$

Here the first term on the left, according to the definition of the second fundamental form, is equal to $L(x)\ h\ k$, while Weingarten's formula yields

$$-\ (y'(x)\ k,\ y'(x)\ \Lambda(x)\ h) = -\ G(x)\ k\ \Lambda(x)\ h$$

for the second term. Consequently,

$$G(x)\ k\ \Lambda(x)\ h = L(x)\ h\ k\ . \tag{3.8}$$

This equation holds for an arbitrary fixed h and every k in the parameter space and therefore uniquely determines the vector $\Lambda(x)\ h$.

The coordinate system $a_1(x), \ldots, a_{n-1}(x)$, introduced in the preceding section, which is orthonormal with respect to $G(x)\,h\,k$, is most naturally chosen to be the principal axis system of the form $L(x)\,h\,k$ with respect to $G(x)\,h\,k$. Then, according to what was said in I.6.8,

$$\Lambda(x)\,a_i(x) = \varkappa_i(x)\,a_i(x) \qquad (i = 1, \ldots, n-1)\,,$$

where $\varkappa_i(x)$ stands for the eigenvalues of the linear transformation $\Lambda(x)$, which because of (3.8) is self-adjoint with respect to $G(x)\,h\,k$. For

$$h = \sum_{i=1}^{n-1} \xi^i\,a_i(x)\,,$$

therefore,

$$\Lambda(x)\,h = \sum_{i=1}^{n-1} \varkappa_i(x)\,\xi^i\,a_i(x)\,.$$

We shall return to the quantities $\varkappa_i(x)$ in another context.

3.7. The principal curvatures. In the following we investigate the curvature of a surface curve

$$y = y(x(\sigma)) \equiv \bar{y}(\sigma)\,,$$

at the point $y(x) = y(x(\sigma))$ of the surface. Here σ is the arc length of the surface curve; we assume that the preimage $x = x(\sigma)$ in the parameter space is twice continuously differentiable.

For this choice of the parameter the unit tangent becomes

$$e_1(\sigma) = e_1 = \bar{y}' = y'(x)\,x'\,,$$

and therefore

$$e_1' = y''(x)\,x'x' + y'(x)\,x''\,.$$

On the other hand, according to Frenet's formulas

$$e_1' = \varkappa\,e_2\,,$$

where $e_2 = e_2(\sigma)$ stands for the principal normal and $\varkappa = \varkappa(\sigma)$ for the principal curvature $\pm\,|\bar{y}''(\sigma)|$ of the surface curve.

In the theory of curves we have taken the principal normal in the direction of $\bar{y}''(\sigma)$, so that $\varkappa(\sigma)$ always turns out to be positive. In the present context the orientation is to be taken care of in a different way, namely, so that the angle enclosed by $e_2(\sigma)$ and the already fixed positive unit normal to the surface, $e(x)$, has a magnitude $\vartheta = \vartheta(\sigma) \leq \pi/2$, and thus $\cos\vartheta \geqq 0$.

From the above equations it now follows, because $\cos\vartheta = (e_2, e)$, that

$$\varkappa\cos\vartheta = (\varkappa\,e_2, e) = (e_1', e) = \big(y''(x)\,x'x', e\big) + \big(y'(x)\,x'', e\big)\,.$$

Here the second term on the right vanishes, while the first, according to the definition of the second fundamental form, is equal to $L(x)\,x'x'$. Therefore,

$$\varkappa(\sigma)\cos\vartheta(\sigma) = L\big(x(\sigma)\big)\,x'(\sigma)\,x'(\sigma)\,. \tag{3.9}$$

This *formula of Meusnier* shows that the curvature $\varkappa$ for the above normalization of the principal normal e_2 has the sign of $L(x)\, x'x'$ and otherwise depends only on the direction of the tangent $e_1 = y'(x)\, x'$ and on the angle ϑ. We can restrict ourselves to the case $\vartheta = 0$, where one is dealing with "normal sections", whose osculating plane contains the normal to the surface at the point in question.

In order to better understand the dependence of the curvature

$$\varkappa = L(x)\, x'x'$$

of such normal sections on the direction of the tangent it is advisable to once more metrize the parameter space at the point x in question by means of the first fundamental form (cf. 3.1), in order to be able to use the principal axes $a_1(x), \ldots, a_{n-1}(x)$ of the second fundamental form, introduced in the preceding section, as the coordinate system.

If

$$x' = \sum_{i=1}^{n-1} \xi^i\, a_i(x)\,,$$

where now

$$|x'| = \sqrt{G(x)\, x'\, x'} = |y'(x)\, x'| = |\bar{y}'| = 1\,,$$

and consequently $\sum_{i=1}^{n-1} (\xi^i)^2 = 1$, then (3.8)

$$\varkappa = L(x)\, x'x' = \sum_{i=1}^{n-1} L(x)\, a_i(x)\, a_i(x)\, (\xi^i)^2 = \sum_{i=1}^{n-1} \varkappa_i(x)\, (\xi^i)^2, \qquad (3.10)$$

where the $\varkappa_i(x)$ are the eigenvalues of the transformation $\Lambda(x)$ introduced earlier.

In this *formula of Euler* the eigenvalues

$$\varkappa_i(x) = G(x)\, a_i(x)\, \Lambda(x)\, a_i(x) = L(x)\, a_i(x)\, a_i(x)$$

are the curvatures in the directions

$$e_i(x) = y'(x)\, a_i(x) \qquad (i = 1, \ldots, n-1)\,,$$

which at the surface point $y(x)$ form an orthonormal system in the tangent space. These directions are called the *principal curvature directions* and the quantities $\varkappa_i(x)$ the *principal curvatures* of the surface at the point $y(x)$. Together with the unit normal $e(x)$ the principal curvature directions form an orthonormal coordinate system

$$e(x),\ e_1(x), \ldots, e_{n-1}(x)$$

for the embedding space R^n_x; this coordinate system is called the *n-frame* of the surface at the point $y(x)$.

Besides the principal curvatures $\varkappa_i(x)$, the elementary symmetric polynomials in these quantities, in particular the *Gaussian curvature*

$$K(x) \equiv \varkappa_1(x) \ldots \varkappa_{n-1}(x) = \prod_{i=1}^{n-1} G(x)\, a_i(x)\, \Lambda(x)\, a_i(x)$$

of the surface at the point $y(x)$, play a central role in the investigation of the "inner" (independent of the embedding in the surrounding $(m+1)$-dimensional space) geometry of the surface.

The Gaussian curvature vanishes if and only if one of the principal curvatures is equal to zero.

3.8. Exercises. 1. Prove that the projection $p(x)$ of a fixed vector $y = a$ on the tangent plane of a regular m-dimensional surface $y = y(x)$ in the space R^n_y $(m < n)$ is $(q-1)$ times continuously differentiable if $y(x)$ is continuously differentiable q times.

Hint. Let h_i $(i = 1, \ldots, m)$ be a (constant) linearly independent set of vectors. At the point x orthonormalize the tangent vectors $y'(x)\, h_i$ by the Schmidt process:

$$y_1(x) = \frac{y'(x)\, h_1}{|y'(x)\, h_1|}, \qquad y_i(x) = \frac{y'(x)\, h_i - p_i(x)}{|y'(x)\, h_i - p_i(x)|} \qquad (i = 1, \ldots, m),$$

where

$$p_i(x) = \sum_{j=1}^{i-1} \big(y'(x)\, h_i, y_j(x)\big)\, y_j(x).$$

From here one can see the claimed differentiability property for $y_1(x)$, $p_2(x)$, $y_2(x), \ldots, p_m(x)$, $y_m(x)$, and the assertion results from

$$p(x) = \sum_{i=1}^{m} \big(a, y_i(x)\big)\, y_i(x).$$

2. Prove: A regular m-dimensional surface in a euclidean space of dimension $m+1 = n$ which has nothing but *planar points*, so that the second fundamental form identically vanishes, is an m-dimensional subspace (or an m-dimensional portion) of the embedding space.

Hint. From Weingarten's formula (3.7) it follows that

$$e'(x)\, h = -\, y'(x)\, \Lambda(x)\, h \equiv 0;$$

for because $G(x)\, k\, \Lambda(x)\, h = L(x)\, h\, k = 0$, $\Lambda(x)\, h = 0$. Therefore, $e(x) \equiv e$ is constant, and

$$L(x)\, h\, k = \big(y''(x)\, h\, k, e_0\big) \equiv 0,$$

from which follows, first, $\big(y'(x)\, k, e_0\big) = \text{const.} = 0$ and then

$$\big(y(x) - y(x_0), e_0\big) \equiv 0.$$

3. Prove: A sphere

$$\big(y(x), y(x)\big) = \varrho^2$$

has only *umbilical points*, where all of the principal curvatures are equal.

Hint. We have $\big(y'(x)\, h, y(x)\big) = 0$, and thus $y(x) = \lambda\, e(x)$, $\lambda^2 = \varrho^2$, $\lambda = \pm\, \varrho$. We take, for example,

$$y(x) = -\, \varrho\, e(x).$$

It follows from Weingarten's differentiation formula that

$$y'(x)\,h = -\,\varrho\,e'(x)\,h = \varrho\,y'(x)\,\Lambda(x)\,h = y'(x)\,\big(\varrho\,\Lambda(x)\,h\big)\,,$$

and therefore

$$\Lambda(x)\,h = \frac{1}{\varrho}\,h = \varkappa\,h\,.$$

All of the principal curvatures are therefore equal to $1/\varrho$, and every point is consequently an umbilical point.

Conversely, it can be shown that such a surface is always a sphere (cf. 6.3, exercise 2).

§ 4. Vectors and Tensors

4.1. Parameter transformations. Before we develop differential geometry any further, the transformation character of the quantities introduced up until now is to be investigated as we change from one parameter x to another. For this the surface F^m is to be defined as a continuously differentiable (regular) m-dimensional manifold, in the sense of the discussion in § 1, so that we can disregard its embedding in the space R^n_y. Around the point p on F^m we mark off a portion G^m_p which is homeomorphic to a parameter region $G^m_x \subset R^m_x$, and here we introduce further admissible parameters $\bar{x}$, $\bar{\bar{x}}$, . . . The vector and the tensor calculus is concerned with the simplest kinds of transformations which quantities given on the manifold F^m can experience when the parameters are changed[1].

4.2. Invariants. From this standpoint the simplest structures are those which are *uniquely* defined in G^m_p. The "representatives" $F(x)$, $\bar{F}(\bar{x})$, . . . of such a quantity are obtained from one another through the law of *invariance*:

$$\bar{F}(\bar{x}) = F\big(x(\bar{x})\big)\,, \quad F(x) = \bar{F}\big(\bar{x}(x)\big)\,.$$

If, in particular, the range of F lies in the set of real numbers or, more generally, in a linear space, then such an invariant is also called a *scalar*.

4.3. Contravariant vectors. In the neighborhood G^m_p of the manifold F^m consider a point $p = (x, \bar{x}, \ldots)$. Let $dx = h$ be a differential, i. e. a vector in the space R^m_x, and form the differentials $d\bar{x} = \bar{h} = \frac{d\bar{x}}{dx}\,h$, etc., that correspond to h. The equivalence class

$$dp = (h, \bar{h}, \ldots)$$

[1] R. Nevanlinna [8].

defines a *differential* or a *contravariant vector* at the point p of the manifold.

Such a vector is determined by the pair $\{p, dp\}$, where dp is a class of vectors $h \in R_x^m$, $\bar{h} \in R_{\bar{x}}^m$, . . . , which are related by the law of *contravariance*

$$\bar{h} = \frac{d\bar{x}}{dx} h , \qquad h = \frac{dx}{d\bar{x}} \bar{h};$$

the derivatives here are to be taken at the point $p = (x, \bar{x}, \ldots)$.

The set of all vectors dp at the fixed point p form an m-dimensional linear space T_{dp}^m if the addition of two vectors $d_1p = (h_1, \bar{h}_1, \ldots)$ and $d_2p = (h_2, \bar{h}_2, \ldots)$ and the multiplication of $dp = (h, \bar{h}, \ldots)$ with a real number λ are defined by means of the equations

$$d_1p + d_2p = (h_1 + h_2, \bar{h}_1 + \bar{h}_2, \ldots) , \qquad \lambda\, dp = (\lambda h, \lambda \bar{h}, \ldots) .$$

T_{dp}^m is the *tangent plane* of the surface at the "point of tangency" p.

The representative in the parameter region G_x^m of a contravariant vector $\{p, dp\}$ is the pair $\{x, dx\}$ or also that vector in the affine space R_x^m that has x as its initial point and $x + dx$ as its end point (cf. I.1.5). This last interpretation corresponds to the elementary geometric view of a "vector on a surface".

On the *notation* let the following be remarked. Since a point $p = (x, \bar{x}, \ldots)$ is uniquely determined by any of its representatives and the same holds for a differential $dp = (dx, d\bar{x}, \ldots)$, the point and the differential can also be denoted by representatives x and dx in an arbitrary one of the admissible parameter spaces R_x^m. Thus if in what follows we speak simply of a contravariant vector h, say, we mean the entire equivalence class[1].

Let (p) be a point set on the surface. If a contravariant vector $dp = h = h(p)$ is associated with each point of (p), the set $(h(p))$ is called a contravariant *vector field* (more briefly a contravariant vector) on (p). The notions continuity, differentiability, etc., of such a field are defined in an obvious way with respect to the representatives $h(x)$, $\bar{h}(\bar{x})$, . . . of $h(p)$ in the admissible parameters $x, \bar{x}, \ldots$ for p.

4.4. Covariants. Now let a class of *linear operators* $A, \bar{A}, \ldots$ on the parameter spaces $R_x^m, R_{\bar{x}}^m, \ldots$ be given at the point $p = (x, \bar{x}, \ldots)$

[1] To distinguish between the admissible parameters x one can also use an index set (i). A point x on F_m is then defined by the equivalence class (x_i). Correspondingly, a contravariant vector h at the point x is given by a class (h_i), whereby $h_i \in R_{x_i}^m$ and the law of contravariance is satisfied: if i and j are two indices in the set (i), then

$$h_i = \frac{dx_i}{dx_j} h_j .$$

that transforms according to the law of *covariance*, i.e.,

$$\bar{A} = A \frac{dx}{d\bar{x}}, \qquad A = \bar{A} \frac{d\bar{x}}{dx},$$

where the derivatives are to be taken at the point p. The range of the invariant differential

$$A\, dp = (A\, dx, \bar{A}\, d\bar{x}, \ldots)$$

lies in an arbitrarily given, say n-dimensional, linear space. We call such an operator A a *covariant* of the manifold at the point p. One sees:

If a covariant operates on a contravariant vector, the result is an invariant.

This fact can be taken, conversely, as the definition of a covariant. For if the quantity $A\, h$ is invariant for *every* contravariant vector h, then, because $\bar{h} = \frac{d\bar{x}}{dx} h$, $A\, h = \bar{A}\, \bar{h} = \bar{A} \frac{d\bar{x}}{dx} h$; hence $A = \bar{A} \frac{d\bar{x}}{dx}$, i.e., A is covariant. On the other hand, the transformation law of contravariance follows from the definitions of invariance and covariance.

4.5. Covariant vectors. Let A be covariant and h contravariant. Because of the invariance of the differential $A\, h$, the covariant A can be thought of as a vector in an mn-dimensional operator space (I.3.5).

The case $n = 1$, where the invariant $A\, h$ can be taken to be a real number, merits special attention. Then

$$A = a^*$$

is a linear operator defined in a tangent space $T = T^m_{dp}$ and is thus a vector in the likewise m-dimensional space T^* *dual* to T. Because of the covariance of a^*,

$$\bar{a}^* = a^* \frac{dx}{d\bar{x}}, \qquad a^* = \bar{a}^* \frac{d\bar{x}}{dx},$$

a^* is called a *covariant vector*.

If a^* is a covariant vector and b a contravariant one, the expression $a^*\, b$ is a real invariant that is linear in both arguments $a^* \in T^*$ and $b \in T$. Here, as above, one can think of a^* as a linear operator from T^* defined on T, or (dually) of b as a linear operator from T defined on T^*. According to the notation we have adopted, the "contravariant argument" b stands to the right of the operator a^* in the invariant bilinear form, and the "covariant argument" a^* to the left of the "operator" b.

4.6. Gradient. The simplest example of a covariant vector is the derivative, the *gradient* a^* of a real invariant F, with representatives

$$a^* = \frac{dF}{dx}, \qquad \bar{a}^* = \frac{d\bar{F}}{d\bar{x}}, \ldots$$

Conversely, a covariant vector a^* in G_p^m is not always a gradient, i.e., the derivative of an invariant F. For that a^* must satisfy the integrability condition, i.e., in the neighborhood G_p^m one must necessarily have

$$\text{rot}\, a^* = \wedge \frac{da^*}{dx} = 0$$

(cf. III.3.3).

It is clear that this condition is *invariant* (independent of the choice of the admissible parameter x). For if one carries out a twice differentiable transformation of the parameter $x \leftrightarrow \bar{x}$, then for a covariant A and a contravariant h the invariance

$$A\,h = \bar{A}\,\bar{h}$$

holds, where $\bar{h} = \frac{d\bar{x}}{dx} h$. If this is differentiated, corresponding to a differential $dx = k, d\bar{x} = \bar{k} = \frac{d\bar{x}}{dx} k$, then

$$d(A\,h) = (dA)\,h = \frac{dA}{dx} k\,h = d(\bar{A}\,\bar{h}) = (d\,\bar{A})\,\bar{h} + \bar{A}\,(d\bar{h})$$

$$= \frac{d\bar{A}}{d\bar{x}} \bar{k}\,\bar{h} + \bar{A}\,\frac{d^2\bar{x}}{dx^2} k\,h\,.$$

Through a switch of h and k and a subsequent subtraction one obtains, because of the symmetry of the bilinear operator $d^2\bar{x}/dx^2$,

$$\frac{dA}{dx} k\,h - \frac{dA}{dx} h\,k = \frac{d\bar{A}}{d\bar{x}} \bar{k}\,\bar{h} - \frac{d\bar{A}}{d\bar{x}} \bar{h}\,\bar{k};$$

i.e., the bilinear form

$$\wedge \frac{dA}{dx} h\,k = \wedge \frac{d\bar{A}}{d\bar{x}} \bar{h}\,\bar{k}$$

is invariant, and the same is thus also true for the equation $\wedge\, dA/dx = 0$.

4.7. Tensors. For the bilinear operator $B = \wedge\, dA/dx$ the form $B\,h\,k$ is invariant, provided the vectors h and k are contravariant. By this property rot A is defined as a covariant *tensor* of rank 2. With this we come to the general notion of a tensor.

We consider a real $(\alpha + \beta)$-linear form, defined at the point p of the manifold F^m, that depends linearly on α contravariant vectors $h_1, \ldots, h_\alpha$ and β covariant vectors $k_1^*, \ldots, k_\beta^*$. Such a multilinear operator is to be denoted by A. In analogy with the above, we write the contravariant arguments to the right, the covariant arguments to the left.

Since the form is uniquely given at the point $p = (x, \bar{x}, \ldots)$, independently of the choice of the parameter, the multilinear form

$$k^*_\beta \ldots k^*_1 A\, h_1 \ldots h_\alpha$$

is thus an invariant (scalar).

Under this condition the operator is said to be an *α-covariant and β-contravariant tensor* at the point p. The sum $\alpha + \beta$ indicates the *rank* of the tensor. A vector is thus a tensor of rank one.

We shall also occasionally denote the tensor A by $\overset{\beta}{\underset{\alpha}{A}}$, where the upper index indicates the rank of its contravariance, the lower index the rank of its covariance. A somewhat more detailed notation is

$$\underbrace{\circ\cdots\circ}_{\beta}\ \overset{\beta}{\underset{\alpha}{A}}\ \overbrace{\circ\cdots\circ}^{\alpha}$$

with α contravariant "empty places" to the right and β empty covariant places to the left. If this empty form is saturated with α contravariant and β covariant vectors, the result is the above invariant multilinear form, which we can also write

$$\underset{1}{h^\beta} \ldots \underset{1}{h^1}\ \overset{\beta}{\underset{\alpha}{A}}\ \overset{1}{h_1} \ldots \overset{1}{h_\alpha}\,.$$

For the saturated form the sum of the upper indices is the same as the sum of the lower indices; both are equal to the rank $\alpha + \beta$ of the tensor.

The transformation law of the representatives $A, \bar{A}, \ldots$ of a tensor for the different parameters of the point $p = (x, \bar{x}, \ldots)$ results from the invariance

$$\underset{1}{\bar{h}^\beta} \ldots \underset{1}{\bar{h}^1}\ \overset{\beta}{\underset{\alpha}{\bar{A}}}\ \overset{1}{\bar{h}_1} \ldots \overset{1}{\bar{h}_\alpha} = \underset{1}{h^\beta} \ldots \underset{1}{h^1}\ \overset{\beta}{\underset{\alpha}{A}}\ \overset{1}{h_1} \ldots \overset{1}{h_\alpha}\,,$$

where

$$\overset{1}{\bar{h}_i} = \frac{d\bar{x}}{dx} \overset{1}{h_i}\,, \qquad \underset{1}{\bar{h}^j} = \underset{1}{h^j} \frac{dx}{d\bar{x}}\,.$$

Using the empty places this law is written

$$\underbrace{\circ \frac{dx}{d\bar{x}} \cdots \circ \frac{dx}{d\bar{x}}}_{\beta}\ \overset{\beta}{\underset{\alpha}{\bar{A}}}\ \overbrace{\frac{d\bar{x}}{dx}\circ \cdots \frac{d\bar{x}}{dx}\circ}^{\alpha} = \underbrace{\circ\cdots\circ}_{\beta}\ \overset{\beta}{\underset{\alpha}{A}}\ \overbrace{\circ\cdots\circ}^{\alpha}\,.$$

Conversely, the tensor character of the operator A could be defined by this transformation law for its representatives.

4.8. Transformation of the components. In the usual tensor calculus, the rule for the transformation of a tensor is given with the aid of its components, which are obtained by choosing a coordinate system in each of the parameter spaces. These rules result at once from the above coordinate-free definition of a tensor.

As is customary, we wish to indicate the contravariant components by upper, the covariant by lower indices. Further, we shall omit the summation signs whenever the summation is on an index that appears above as well as below (Einstein's summation convention).

Let $a_1, \ldots, a_m$ and $\bar{a}_1, \ldots, \bar{a}_m$ be two linear coordinate systems for the parameter spaces R_x^m and $R_{\bar{x}}^m$, and

$$dx = d\xi^i a_i , \qquad d\bar{x} = d\bar{\xi}^j \bar{a}_j$$

two corresponding parameter differentials, so that

$$d\bar{x} = \frac{d\bar{x}}{dx} dx , \qquad dx = \frac{dx}{d\bar{x}} d\bar{x} .$$

Then

$$d\bar{x} = \frac{\partial \bar{\xi}^j}{\partial \xi^i} d\xi^i \bar{a}_j = d\xi^i \frac{d\bar{x}}{dx} a_i$$

and consequently

$$\frac{d\bar{x}}{dx} a_i = \frac{\partial \bar{\xi}^j}{\partial \xi^i} \bar{a}_j , \qquad \frac{dx}{d\bar{x}} \bar{a}_j = \frac{\partial \xi^i}{\partial \bar{\xi}^j} a_i .$$

Now let l be an arbitrary contravariant vector at the point $(x, \bar{x}, \ldots)$ with the representatives

$$l = \lambda^i a_i , \qquad \bar{l} = \bar{\lambda}^j \bar{a}_j$$

in the parameter spaces R_x^m and $R_{\bar{x}}^m$. Then according to the above

$$\bar{l} = \frac{d\bar{x}}{dx} l = \lambda^i \frac{d\bar{x}}{dx} a_i = \lambda^i \frac{\partial \bar{\xi}^j}{\partial \xi^i} \bar{a}_j = \bar{\lambda}^j \bar{a}_j ,$$

and therefore

$$\bar{\lambda}^j = \frac{\partial \bar{\xi}^j}{\partial \xi^i} \lambda^i , \qquad \lambda^i = \frac{\partial \xi^i}{\partial \bar{\xi}^j} \bar{\lambda}^j .$$

These are the usual transformation formulas for the contravariant vector $l = (\lambda^1, \ldots, \lambda^m)$. Conversely, one obtains from this component representation the coordinate-free law of contravariance:

$$\bar{l} = \bar{\lambda}^j \bar{a}_j = \frac{\partial \bar{\xi}^j}{\partial \xi^i} \lambda^i \bar{a}_j = \lambda^i \frac{\partial \bar{\xi}^j}{\partial \xi^i} \bar{a}_j = \lambda^i \frac{d\bar{x}}{dx} a_i = \frac{d\bar{x}}{dx} l .$$

Second, let l^* be a covariant vector at the point $(x, \bar{x}, \ldots)$. Its representative l^* in the space R_x^m is a linear operator. For an $l = \lambda^i a_i$ of this space, therefore,

$$l^* l = \lambda^i l^* a_i = a^{*i} \lambda_i = \lambda^i \lambda_i ,$$

where we have set $a^{*i} = \lambda^i$ and $\lambda_i = l^* a_i$. The linear forms $a^{*i} = \lambda^i$ $(i = 1, \ldots, m)$ form the basis dual to the coordinate system a_i, and the λ_i are the coordinates of the vector l^* in this dual system. Because

$$\bar{l}^* = l^* \frac{dx}{d\bar{x}},$$

it turns out that

$$\bar{\lambda}_j = \bar{l}^* \bar{a}_j = l^* \frac{dx}{d\bar{x}} \bar{a}_j = l^* \frac{\partial \xi^i}{\partial \bar{\xi}^j} a_i = \frac{\partial \xi^i}{\partial \bar{\xi}^j} l^* a_i = \frac{\partial \xi^i}{\partial \bar{\xi}^j} \lambda_i .$$

According to this the law of covariance for the vector $l^* = (\lambda_1, \ldots, \lambda_m)$ is

$$\bar{\lambda}_j = \frac{\partial \xi^i}{\partial \bar{\xi}^j} \lambda_i , \qquad \lambda_i = \frac{\partial \bar{\xi}^j}{\partial \xi^i} \bar{\lambda}_j .$$

Finally, let $\overset{\beta}{\underset{\alpha}{A}}$ be a general mixed tensor of rank $\alpha + \beta$ at the point $(x, \bar{x}, \ldots)$ of the m-dimensional manifold F^m. With the above notations the invariant multilinear from

$$l^\beta \ldots l^1 \overset{\beta}{\underset{\alpha}{A}} l_1 \ldots l_\alpha$$

has the representative

$$a^{*j_\beta} \lambda^\beta_{j_\beta} \ldots a^{*j_1} \lambda^1_{j_1} \overset{\beta}{\underset{\alpha}{A}} \lambda^{i_1}_1 a_{i_1} \ldots \lambda^{i_\alpha}_\alpha a_{i_\alpha}$$

in the parameter space R^m_x, where the quantities

$$A^{j_1 \ldots j_\beta}_{i_1 \ldots i_\alpha} = a^{*j_\beta} \ldots a^{*j_1} \overset{\beta}{\underset{\alpha}{A}} a_{i_1} \ldots a_{i_\alpha}$$

are the $m^{\alpha+\beta}$ components of the representative $\overset{\beta}{\underset{\alpha}{A}}$.

By observing the transformation formulas for contra- and covariant vector components the customary transformation formulas for the components of the tensor A,

$$\bar{A}^{k_1 \ldots k_\beta}_{h_1 \ldots h_\alpha} = \frac{\partial \bar{\xi}^{k_\beta}}{\partial \xi^{j_\beta}} \ldots \frac{\partial \bar{\xi}^{k_1}}{\partial \xi^{j_1}} A^{j_1 \ldots j_\beta}_{i_1 \ldots i_\alpha} \frac{\partial \xi^{i_1}}{\partial \bar{\xi}^{h_1}} \ldots \frac{\partial \xi^{i_\alpha}}{\partial \bar{\xi}^{h_\alpha}},$$

result from the invariance of the above multilinear form.

From the coordinate-free tensor definition it follows at once that the transformation formulas for the components are invariant in form, independent of the choice of the coordinate system. If the tensor concept is defined by means of the formulas for the components, this invariance requires further verification.

4.9. Tensor algebra. If $\overset{\beta}{\underset{\alpha}{A}}$ and $\overset{\beta}{\underset{\alpha}{B}}$ are two tensors of equal rank, the sum of the saturated empty forms

$$\underbrace{\circ \cdots \circ}_{\beta} \overset{\beta}{\underset{\alpha}{A}} \overbrace{\circ \cdots \circ}^{\alpha} + \underbrace{\circ \cdots \circ}_{\beta} \overset{\beta}{\underset{\alpha}{B}} \overbrace{\circ \cdots \circ}^{\alpha}$$

is a real invariant. If

$$\underbrace{\bigcirc\cdots\bigcirc}_{\beta}\overset{\beta}{\underset{\alpha}{C}}\overbrace{\bigcirc\cdots\bigcirc}^{\alpha}$$

stands for this form, then $\overset{\beta}{\underset{\alpha}{C}}$ is an α-covariant, β-contravariant tensor, which one defines to be the sum of $\overset{\beta}{\underset{\alpha}{A}}$ and $\overset{\beta}{\underset{\alpha}{B}}$,

$$\overset{\beta}{\underset{\alpha}{C}} = \overset{\beta}{\underset{\alpha}{A}} + \overset{\beta}{\underset{\alpha}{B}} = \overset{\beta}{\underset{\alpha}{B}} + \overset{\beta}{\underset{\alpha}{A}}\,.$$

The product $\lambda\overset{\beta}{\underset{\alpha}{A}}$ (λ real) is defined correspondingly.

With these definitions the set of all α-covariant and β-contravariant tensors at the point p of the m-dimensional manifold F^m form a linear vector space of dimension $m^{\alpha+\beta}$. For $\alpha = 0$, $\beta = 1$ and $\alpha = 1$, $\beta = 0$ this space is the tangent space T or its dual space T^*, respectively, at the point $p \in F^m$.

Besides the above linear operations one can also introduce a commutative, associative and distributive *tensor product*

$$\overset{\beta}{\underset{\alpha}{A}}\overset{\delta}{\underset{\gamma}{B}} = \overset{\beta+\delta}{\underset{\alpha+\gamma}{C}}\,,$$

and, in fact, as that tensor whose saturated empty form is equal to the product of the corresponding empty forms of the factors.

If in particular the factors are *alternating*, then the tensor product will in general no longer possess this property. The *alternating part* of the saturated product

$$\wedge\left(\overset{\beta}{\underset{\alpha}{A}}\overset{\delta}{\underset{\gamma}{B}}\right) = \wedge\overset{\beta+\delta}{\underset{\alpha+\gamma}{C}}$$

is the Grassmann-Cartan exterior product of the tensors $\overset{\beta}{\underset{\alpha}{A}}$ and $\overset{\delta}{\underset{\gamma}{B}}$.

4.10. Contraction. With a tensor $\overset{\beta}{\underset{\alpha}{A}}$, where $\alpha, \beta \geqq 1$, one can associate a tensor $\overset{\beta-1}{\underset{\alpha-1}{A}}$ through the process of contraction, which can be defined in a coordinate-free way as follows.

Let us first consider the simplest case $\alpha = \beta = 1$, thus a "mixed" tensor $A = \overset{1}{\underset{1}{A}}$ of rank 2. In the corresponding invariant empty form $\bigcirc\overset{1}{\underset{1}{A}}\bigcirc$ fill only one of the empty places, for example, the contravariant one (on the right) with a contravariant vector $\overset{1}{h} = h \in T$. The "half-saturated" expression $\overset{1}{\underset{1}{A}}\overset{1}{h} = A\,h$ then defines A as a linear transformation of the tangent space T into itself, for the differential $A\,h$ is, because of the invariance of the "fully saturated" bilinear form $k^*(A\,h) = k^*A\,h$ (k^* a vector in the space T^*), again a contravariant vector.

If A is conceived of in this way as being a linear self-mapping of the tangent space $T = T^m_{dp}$, then according to exercise 4 in I.5.9 a real number α, namely the *trace* of A, can be associated with the latter transformation by means of an arbitrary, not identically vanishing, real, alternating form $D\,h_1 \ldots h_m$ ($h_i \in T$),

$$\alpha = Tr\,A = \frac{1}{D\,h_1 \ldots h_m} \sum_{i=1}^{m} D\,h_1 \ldots h_{i-1}\left(\overset{1}{\underset{1}{A}}\,h_i\right) h_{i+1} \ldots h_m\,.$$

This number is independent of the choice of the auxiliary form D and of the vectors $h_i \in T$. The construction can be repeated for each representative $A, \bar{A}, \ldots$ The corresponding traces are all *equal*:

$$Tr\,A = Tr\,\bar{A}\,.$$

The proof of this invariance is a simple consequence of the definition of trace according to exercise 4 in I. 5.9 (cf. 4.20, exercise 1).

The process of contraction, whereby there is associated with the tensor $\overset{1}{\underset{1}{A}}$ of rank 2 a real invariant $Tr\,A$ (a "tensor of rank 0"), consists in the formation of this trace.

The contraction of an arbitrary mixed tensor $\overset{\beta}{\underset{\alpha}{A}}$ now offers no difficulty. This procedure consists of the elimination of one argument on the right and one on the left of the form $\underbrace{\bigcirc \cdots \bigcirc}_{\beta}\,\overset{\beta}{\underset{\alpha}{A}}\,\overbrace{\bigcirc \cdots \bigcirc}^{\alpha}$, for example of that one with the index i on the right and j on the left. To this end one considers the expression

$$\underbrace{\bigcirc \cdots \bigcirc\,\underset{1}{h^j}\,\bigcirc \cdots \bigcirc}_{\beta}\,\overset{\beta}{\underset{\alpha}{A}}\,\overbrace{\bigcirc \cdots \bigcirc\,\overset{1}{h_i}\,\bigcirc \cdots \bigcirc}^{\alpha}\,,$$

which has only $\alpha + \beta - 2$ empty places. This invariant form, provided those $\alpha + \beta - 2$ empty places are filled with *fixed* vectors, defines $\overset{\beta}{\underset{\alpha}{A}}$ as a mixed tensor $\overset{1}{\underset{1}{B}} = \overset{\beta}{\underset{\alpha}{A}}$ of rank 2 with the two linear arguments $\underset{1}{h^j}$ and $\overset{1}{h_i}$. If $\overset{1}{\underset{1}{B}}$ is now contracted, the result is an invariant $(\alpha + \beta - 2)$-form

$$\underbrace{\bigcirc \cdots \bigcirc}_{\beta-1}\,\overset{\beta-1}{\underset{\alpha-1}{A}}\,\overbrace{\bigcirc \cdots \bigcirc}^{\alpha-1}\,,$$

and $\overset{\beta-1}{\underset{\alpha-1}{A}}$ is the sought-for contracted tensor of rank $\alpha + \beta - 2$.

4.11. The rotor. We have seen in 4.6 that when $A = A(p) = \underset{1}{A}$ is a differentiable covariant vector field defined in a portion G_p^m of the surface F^m and the equivalence relations which determine the points $p = (x, \bar{x}, \ldots)$ of G_p^m are twice differentiable, then the form

$$\wedge \frac{dA}{dx} h\, k = \frac{1}{2}\left(\frac{dA}{dx} h\, k - \frac{dA}{dx} k\, h\right),$$

where h and k are arbitrary contravariant vectors, is invariant. The bilinear operator rot $A = \wedge\, dA/dx$ is hence a covariant tensor of rank two.

The corresponding holds for the rotor of a covariant tensor $A = \underset{q}{A}$ of rank q. If one sets $\bar{h}_i = \frac{d\bar{x}}{dx} h_i$ $(i = 1, \ldots, q+1)$, where $h_1, \ldots, h_{q+1}$ are contravariant vectors, then

$$A\, h_1 \ldots \hat{h}_i \ldots h_{q+1} = \bar{A}\, \bar{h}_1 \ldots \hat{\bar{h}}_i \ldots \bar{h}_{q+1},$$

and through differentiation with the differential $dx = h_i$ one obtains

$$\begin{aligned} &\frac{dA}{dx} h_i\, h_1 \ldots \hat{h}_i \ldots h_{q+1} \\ &= \frac{d\bar{A}}{d\bar{x}} \bar{h}_i\, \bar{h}_1 \ldots \hat{\bar{h}}_i \ldots \bar{h}_{q+1} \\ &+ \sum_{\substack{j=1 \\ j \neq i}}^{q+1} \bar{A}\, \bar{h}_1 \ldots \bar{h}_{j-1} \left(\frac{d^2\bar{x}}{dx^2} h_i\, h_j\right) \bar{h}_{j+1} \ldots \hat{\bar{h}}_i \ldots \bar{h}_{q+1}, \end{aligned}$$

After multiplication with $(-1)^{i-1}/(q+1)$ and summation over $i = 1, \ldots, q+1$, the result, in view of the symmetry of the operator $d^2\bar{x}/dx^2$, is

$$\text{rot}\, A\, h_1 \ldots h_{q+1} = \wedge \frac{dA}{dx} h_1 \ldots h_{q+1} = \text{rot}\, \bar{A}\, \bar{h}_1 \ldots \bar{h}_{q+1},$$

which expresses the asserted covariance of the operator rot $A = \wedge\, dA/dx$.

From this it follows, in particular, that Stokes's formula is invariant with respect to parameter transformations.

4.12. The fundamental metric tensor. Lowering and raising indices. According to the definition of the fundamental symmetric operator $G(p) = (G(x), \bar{G}(\bar{x}), \ldots)$ the expression $G(x)\, h\, k$ is invariant:

$$G\, h\, k = G \frac{dx}{d\bar{x}} \bar{h} \frac{dx}{d\bar{x}} \bar{k} = \bar{G}\, \bar{h}\, \bar{k},$$

where h and k are arbitrary contravariant vectors. The operator G is thus a (symmetric) covariant tensor of rank two: $G = \underset{2}{G}$.

From here it follows in particular that the expression

$$h^* = G\,h = \underset{2}{G}\,\overset{1}{h}$$

is a covariant vector $h^* = \underset{1}{h}$. Conversely, one can, as a consequence of the Fréchet-Riesz theorem, associate with an arbitrary covariant vector $\underset{1}{h}$ a uniquely determined contravariant vector $\overset{1}{h}$ such that for every contravariant vector $\overset{1}{k}$

$$\underset{1}{h}\,\overset{1}{k} = \underset{2}{G}\,\overset{1}{h}\,\overset{1}{k}\,.$$

The dual vectors $\underset{1}{h}$ and $\overset{1}{h}$, which through G are in one-to-one correspondence, are said to be mutually *conjugate*; one also speaks of the "covariant or contravariant component" $\left(\underset{1}{h} \text{ or } \overset{1}{h}\right)$ of a vector h. The transition from the one component to the other corresponds in the usual tensor calculus to the process of raising and lowering of the indices.

In the equation $\underset{1}{h} = G\,\overset{1}{h}$, G can also be interpreted as a linear transformation that maps the "contravariant" parameter space into the dual "covariant" parameter space. This transformation is regular, the reciprocal transformation G^{-1} thus exists, and one has $\overset{1}{h} = G^{-1}\,\underset{1}{h}$.

4.13. Volume metric. At the point $p = (x, \bar{x}, \ldots)$ of the manifold F^m we consider m tangent vectors $d_i p = (h_i, \bar{h}_i, \ldots)$ $(i = 1, \ldots, m)$. In order to define the volume dv of the simplex $(d_1 p, \ldots, d_m p)$ spanned by these vectors take for a fixed p an arbitrary, real, nondegenerate, alternating, fundamental form $dv = D(p)\,d_1 p \ldots d_m p$. Since the differentials $d_i p$ are contravariant and the volume is to be invariant,

$$D(x)\,h_1 \ldots h_m = \bar{D}(\bar{x})\,\bar{h}_1 \ldots \bar{h}_m\,,$$

the operator $D(p) = \left(D(x), \bar{D}(\bar{x}), \ldots\right)$ is thereby defined as a covariant tensor of rank m. The volume element do is according to this definition endowed with a sign, corresponding to the orientation of the simplex $(d_1 p, \ldots, d_m p)$.

The differential dv is dependent on the arbitrary alternating tensor $D(p)$ and uniquely determined only up to an arbitrary real factor $\varrho(p)$. In order to determine this normalization factor in such a way that the volume metric at the point p is connected with the length metric determined by the fundamental tensor $G(p)$ in a euclidean fashion, one proceeds by I.6.10 as follows.

The determinant $\det(G(x)\, h_i\, k_j)$, where h_i, k_j $(i, j = 1, \ldots, m)$ are representatives of $2m$ tangent vectors, is nondegenerate and alternating in the differentials h_i as well as in the differentials k_j. For $k_j = h_i$ this determinant is nonnegative, and it vanishes if and only if the vectors $h_1, \ldots, h_m$ are linearly dependent. The magnitude of the "locally euclidean" volume element is then (cf. I.6.10) invariantly defined by the expression

$$|h_1, \ldots, h_m| = \frac{1}{m!}\sqrt{\det\left(G(x)\, h_i\, h_j\right)} \geqq 0 .$$

Now if $D(p) = \left(D(x), \bar{D}(\bar{x}), \ldots\right)$ is an arbitrary covariant alternating (nondegenerate) tensor of rank m, the volume element dv is normalized in a locally euclidean fashion by setting

$$dv = \varrho(p)\, D(p)\, d_1 p \ldots d_m p ,$$

where $\varrho(p)$ (> 0) is for linearly independent $d_i p$

$$\varrho(p) = \frac{|d_1 p \ldots d_m p|}{|D(p) d_1 p \ldots d_m p|} .$$

If one writes

$$\varepsilon(p) = \frac{D(p)\, d_1 p \ldots d_m p}{|D(p)\, d_1 p \ldots d_m p|} ,$$

then dv can also be expressed in the from

$$dv = \frac{\varepsilon(p)}{m!}\sqrt{\det\left(G(p)\, d_i p\, d_j p\right)} .$$

If the manifold F^m is embedded by means of a mapping $y = y(x)$ in a euclidean space R_y^n and one defines, as done in this chapter, the fundamental tensor $G(x)$, following Gauss, by $G(x)\, h\, k = \left(y'(x)\, h,\ y'(x)\, k\right)$, then the normalized volume element $|dv|$ coincides with the euclidean volume of the tangent simplex $(d_1 y, \ldots, d_m y)$ measured in the space R_y^n, whereby $d_i y = y'(x)\, d_i x$.

The above argument can be repeated without modification if one restricts the consideration to a q-dimensional subspace U^q of the tangent space T_{dp}^m $(1 \leqq q \leqq m)$.

4.14. The second fundamental form. We come to the question of the transformation character of the remaining fundamental quantities in the Gaussian theory of surfaces; the manifold F^m is thus given as an m-dimensional surface, defined by the mapping $y = y(p)$, embedded in a space R_y^n of dimension $n = m + 1$.

In the parameter space R_x^m the second fundamental form has the representation

$$L(x)\, h\, k = \left(y''(x)\, h\, k,\ e(x)\right) ,$$

where $e(x)$ is the representative of the (invariant) unit normal.

From the invariance $\frac{d\bar{y}}{d\bar{x}}\bar{k} = \frac{dy}{dx}k$, according to which

$$\frac{d\bar{y}}{d\bar{x}}\frac{d\bar{x}}{dx}k = \frac{dy}{dx}k,$$

it follows by means of differentiation with the differential $dx = h$ that

$$\frac{d^2\bar{y}}{d\bar{x}^2}\frac{d\bar{x}}{dx}h\frac{d\bar{x}}{dx}k + \frac{d\bar{y}}{d\bar{x}}\frac{d^2\bar{x}}{dx^2}h\,k = \frac{d^2y}{dx^2}h\,k,$$

and thus

$$\begin{aligned} \frac{d^2\bar{y}}{d\bar{x}^2}\bar{h}\,\bar{k} &= \left(\frac{d^2y}{dx^2} - \frac{d\bar{y}}{d\bar{x}}\frac{d^2\bar{x}}{dx^2}\right)h\,k, \\ \frac{d^2y}{dx^2}h\,k &= \left(\frac{d^2\bar{y}}{d\bar{x}^2} - \frac{dy}{dx}\frac{d^2x}{d\bar{x}^2}\right)\bar{h}\,\bar{k}. \end{aligned} \tag{4.1}$$

If the second term on the right in these transformation formulas were equal to zero, then d^2y/dx^2 as an operator would be a covariant of rank two. However, this is the case only when the relation $x \leftrightarrow \bar{x}$ is linear. The second derivative d^2y/dx^2 of the invariant y is therefore not a covariant.

But in view of the invariance $e(x) = \bar{e}(\bar{x})$ it does follow from the above transformation formulas that

$$L\,h\,k = \left(\frac{d^2y}{dx^2}h\,k,\,e\right) = \left(\frac{d^2\bar{y}}{d\bar{x}^2}\bar{h}\,\bar{k},\,\bar{e}\right) - \left(\frac{dy}{dx}\frac{d^2x}{d\bar{x}^2}\bar{h}\,\bar{k},\,\bar{e}\right)$$

and because the second term on the right does vanish,

$$L\,h\,k = \bar{L}\,\bar{h}\,\bar{k}.$$

Consequently, the second fundamental operator L is also a covariant tensor of rank two: $L = \underset{2}{L}$.

4.15. The operators $\Gamma(x)$ and $\Lambda(x)$. In order to ascertain the law of transformation for the Christoffel operator Γ, we start from the defining formula (3.5):

$$G(x)\,l\,\Gamma(x)\,h\,k = \frac{1}{2}\left(G'(x)\,h\,k\,l + G'(x)\,k\,l\,h - G'(x)\,l\,h\,k\right).$$

If the invariance $G\,k\,l = \bar{G}\,\bar{k}\,\bar{l} = \bar{G}\frac{d\bar{x}}{dx}k\frac{d\bar{x}}{dx}l$ is differentiated with the differential $dx = h$, then

$$\frac{dG}{dx}h\,k\,l = \frac{d\bar{G}}{d\bar{x}}\frac{d\bar{x}}{dx}h\frac{d\bar{x}}{dx}k\frac{d\bar{x}}{dx}l + \bar{G}\frac{d^2\bar{x}}{dx^2}h\,k\frac{d\bar{x}}{dx}l + \bar{G}\frac{d\bar{x}}{dx}k\frac{d^2\bar{x}}{dx^2}h\,l,$$

and therefore

$$\frac{dG}{dx}h\,k\,l = \frac{d\bar{G}}{d\bar{x}}\bar{h}\,\bar{k}\,\bar{l} + \bar{G}\,\bar{l}\,\frac{d^2\bar{x}}{dx^2}h\,k + \bar{G}\,\bar{k}\frac{d^2\bar{x}}{dx^2}h\,l.$$

Thus, according to the above formula for $G\,l\,\Gamma\,h\,k$,

$$G\,l\,\Gamma\,h\,k = \bar{G}\,\bar{l}\,\bar{\Gamma}\,\bar{h}\,\bar{k} + \bar{G}\,\bar{l}\,\frac{d^2\bar{x}}{dx^2}\,h\,k\,,$$

or, because $G\,l\,\Gamma\,h\,k = \bar{G}\,\bar{l}\,\dfrac{d\bar{x}}{dx}\,\Gamma\,h\,k$,

$$\bar{G}\,\bar{l}\left(\bar{\Gamma}\,\bar{h}\,\bar{k} - \frac{d\bar{x}}{dx}\,\Gamma\,h\,k + \frac{d^2\bar{x}}{dx^2}\,h\,k\right) = 0\,.$$

Since this holds for every $\bar{l}$, we have

$$\bar{\Gamma}\,\bar{h}\,\bar{k} = \left(\frac{d\bar{x}}{dx}\,\Gamma - \frac{d^2\bar{x}}{dx^2}\right)h\,k\,. \tag{4.2}$$

As in the transformation formula (4.1) for the second derivative d^2y/dx^2 a "disturbing term" also turns up here on the right: the operator Γ is not a covariant. We shall presently return to the similarity of both of these transformations.

The operator Λ was uniquely determined using both of the fundamental operators G and L by means of formula (3.8),

$$G(x)\,k\,\Lambda(x)\,h = L(x)\,h\,k\,.$$

It follows from here that

$$\bar{G}\,\bar{k}\,\bar{\Lambda}\,\bar{h} = \bar{L}\,\bar{h}\,\bar{k} = L\,h\,k = G\,k\,\Lambda\,h = \bar{G}\,\bar{k}\,\frac{d\bar{x}}{dx}\,\Lambda\,h\,,$$

and hence

$$\bar{G}\,\bar{k}\left(\bar{\Lambda}\,\bar{h} - \frac{d\bar{x}}{dx}\,\Lambda\,h\right) = 0\,,$$

which yields the transformation formula

$$\bar{\Lambda}\,\bar{h} = \frac{d\bar{x}}{dx}\,\Lambda\,h$$

for Λ. With a contravariant vector $\overset{1}{h}$ and a covariant vector $\underset{1}{h}$, therefore,

$$\underset{1}{h}\,\Lambda\,\overset{1}{h}$$

is invariant and real: Λ is a mixed tensor of rank two: $\Lambda = \overset{1}{\underset{1}{\Lambda}}$.

4.16. Invariance of the principal curvatures. The principal curvature directions $e_1, \ldots, e_{n-1}$ and the corresponding principal curvatures $\varkappa, \ldots, \varkappa_{n-1}$ of the surface at a point $p = (x, \bar{x}, \ldots)$ of the surface are as a consequence of their geometric significance invariant quantities. This can also be verified using the above transformation formulas. Let e_j be one of these unit vectors and $\varkappa_j$ the associated principal curvature, then in the parameter space R_x^m, provided it is metrized at the point x by means of the first fundamental form,

$$e_j(x) = y'(x)\, a_j(x)$$

and

$$\Lambda(x)\, a_j(x) = \varkappa_j(x)\, a_j(x) \qquad (|a_j(x)| \equiv 1)$$

(cf. 3.6—7). If one goes by means of the mapping $\bar{x} = \bar{x}(x)$, $x = x(\bar{x})$ over to the parameter space $\bar{R}^m_{\bar{x}}$ and if one sets $\frac{d\bar{x}}{dx}\, a_j = \bar{a}_j$, then the transformation formula for the operator Λ yields

$$\bar{\Lambda}\, \bar{a}_j = \frac{d\bar{x}}{dx} \Lambda\, a_j = \varkappa_j \frac{d\bar{x}}{dx} a_j = \varkappa_j\, \bar{a}_j\, ,$$

from which it is clear that the eigenvectors a_j are transformed covariantly and that the eigenvalues $\varkappa_j$ remain invariant. Then also

$$\bar{e}_j = \frac{dy}{d\bar{x}} \bar{a}_j = \frac{dy}{dx} a_j = e_j$$

is invariant.

Conversely, it would have been possible to deduce the transformation law for Λ from the invariance of e_j and $\varkappa_j$ and the contravariance of a_j.

As a consequence of the invariance of the principal curvatures $\varkappa_j$, the elementary symmetric polynomials in the $\varkappa_j$, thus in particular the Gaussian curvature

$$K(p) = \varkappa_1(p) \ldots \varkappa_{n-1}(p)\, ,$$

are likewise invariant.

4.17. The covariant derivative. We turn back to the transformation formulas (4.1) and (4.2) derived above for the second derivative d^2y/dx^2 and for the operator Γ.

By (4.1) we have for d^2y/dx^2

$$\frac{d^2y}{d\bar{x}^2} \bar{h}\, \bar{k} = \frac{d^2y}{dx^2} h\, k - \frac{dy}{d\bar{x}} \frac{d^2\bar{x}}{dx^2} h\, k\, ,$$

and it follows from the transformation formula (4.2) for Γ, when the operator $dy/d\bar{x}$ ixs applied to both sides, that

$$\frac{dy}{d\bar{x}} \bar{\Gamma}\, \bar{h}\, \bar{k} = \frac{dy}{dx} \Gamma\, h\, k - \frac{dy}{d\bar{x}} \frac{d^2\bar{x}}{dx^2} h\, k\, .$$

Subtracting these equations therefore yields the invariance

$$\left(\frac{d^2y}{d\bar{x}^2} - \frac{dy}{d\bar{x}} \bar{\Gamma}\right) \bar{h}\, \bar{k} = \left(\frac{d^2y}{dx^2} - \frac{dy}{dx} \Gamma\right) h\, k\, ,$$

which also follows directly from Gauss's derivative formula (3.4)

$$y''(x)\, h\, k - y'(x)\, \Gamma(x)\, h\, k = L(x)\, h\, k\, e(x)\, ,$$

in view of the invariance of the right hand side. Hence the operator

$$\frac{d^2y}{dx^2} - \frac{dy}{dx} \Gamma$$

is a covariant tensor of rank two.

This holds not only for the derivative operator dy/dx, but for *any* simply covariant operator A on the manifold F^m. In fact, if k stands for the representative of a contravariant vector in the parameter space R^m_x, then

$$A\,k = \bar{A}\,\bar{k} = \bar{A}\frac{d\bar{x}}{dx}k\,,$$

from which it follows by means of differentiation with the differential $dx = h$ that

$$\frac{dA}{dx}h\,k = \frac{d\bar{A}}{d\bar{x}}\frac{d\bar{x}}{dx}h\frac{d\bar{x}}{dx}k + \bar{A}\frac{d^2\bar{x}}{dx^2}h\,k = \frac{d\bar{A}}{d\bar{x}}\bar{h}\,\bar{k} + \bar{A}\frac{d^2\bar{x}}{dx^2}h\,k\,.$$

On the other hand, the transformation formula (4.2) for Γ yields

$$A\,\Gamma\,h\,k = \bar{A}\,\bar{\Gamma}\,\bar{h}\,\bar{k} + \bar{A}\frac{d^2\bar{x}}{dx^2}h\,k\,,$$

and the asserted invariance

$$\left(\frac{d\bar{A}}{d\bar{x}} - \bar{A}\,\bar{\Gamma}\right)\bar{h}\,\bar{k} = \left(\frac{dA}{dx} - A\,\Gamma\right)h\,k$$

follows by subtraction.

We call the covariant operator of rank two

$$'A \equiv A' - A\,\Gamma \tag{4.3}$$

the *covariant derivative* $'A = {}'A(x)$ *of the covariant* A (with respect to an operator Γ which obeys the transformation law (4.2)).

If A is in particular a covariant vector field $l^* = l^*(x)$, then

$$'(l^*)\,h\,k = \big((l^*)' - l^*\,\Gamma\big)\,h\,k$$

is a real invariant. The operator $'(l^*)$ is thus a covariant tensor of rank two.

The covariant derivative of a contravariant vector can be defined in a corresponding fashion. If $k = k(x)$ is the representative of such a field, then with a covariant vector l^*

$$l^*\,k = \bar{l}^*\,\bar{k} = l^*\frac{dx}{d\bar{x}}\bar{k}\,.$$

If this equation is differentiated with the differential $dx = h$, then

$$l^*\frac{dk}{dx}h = l^*\frac{d^2x}{d\bar{x}^2}\bar{h}\,\bar{k} + \bar{l}^*\frac{d\bar{k}}{d\bar{x}}\bar{h}\,.$$

On the other hand, according to the transformation formula (4.2) for Γ

$$l^*\,\Gamma\,k\,h = \bar{l}^*\,\bar{\Gamma}\,\bar{k}\,\bar{h} - l^*\frac{d^2x}{d\bar{x}^2}\bar{k}\,\bar{h}\,,$$

and adding these equations yields the invariance

$$l^*\left(\frac{dk}{dx}+\Gamma k\right)h=\bar{l}^*\left(\frac{d\bar{k}}{d\bar{x}}+\bar{\Gamma}\bar{k}\right)\bar{h}\,,$$

by which the operator $'k = 'k(x)$, the *covariant derivative of the contravariant vector k defined by*

$$'k\equiv k'+\Gamma k\,, \tag{4.4}$$

is a mixed tensor of rank two.

The covariant derivative of an invariant is by definition equal to the ordinary derivative.

The above definitions can be generalized to tensors of arbitrary rank, and we shall come back to this in exercise 1 of 7.7.

Thus, for example, the covariant derivatives of the tensors G and L are covariant tensors of rank three, and the corresponding linear forms

$$'G(x)\,h\,k\,l=G'(x)\,h\,k\,l-G(x)\,k\,\Gamma(x)\,h\,l-G(x)\,l\,\Gamma(x)\,h\,k\,,$$

$$'L(x)\,h\,k\,l=L'(x)\,h\,k\,l-L(x)\,k\,\Gamma(x)\,h\,l-L(x)\,l\,\Gamma(x)\,h\,k\,,$$

are real invariant forms, where, by the way, the first as a consequence of (3.5) vanishes identically. The second, as we shall see in 6.1, is symmetric not only in k and l, but also in h and k.

As a second example, we consider the mixed tensor Λ of rank two. From the transformation law for this operator, according to which for each contravariant k and covariant l^*

$$l^*\,\Lambda\,k=\bar{l}^*\,\bar{\Lambda}\,\bar{k}=l^*\frac{dx}{d\bar{x}}\bar{\Lambda}\frac{d\bar{x}}{dx}k\,,$$

it follows by differentiation with the differential $dx = h$ that

$$l^*\frac{d\Lambda}{dx}h\,k=l^*\frac{d^2x}{d\bar{x}^2}\bar{h}\,\bar{\Lambda}\,\bar{k}+\bar{l}^*\frac{d\bar{\Lambda}}{d\bar{x}}\bar{h}\,\bar{k}+\bar{l}^*\,\bar{\Lambda}\,\frac{d^2\bar{x}}{dx^2}h\,k\,.$$

But on the other hand, by (4.2)

$$l^*\,\Gamma\,h\,\Lambda\,k=-\,l^*\frac{d^2x}{d\bar{x}^2}\bar{h}\,\bar{\Lambda}\,\bar{k}+\bar{l}^*\,\bar{\Gamma}\,\bar{h}\;\bar{\Lambda}\,\bar{k}\,,$$

$$-\,l^*\,\Lambda\,\Gamma\,h\,k=-\,l^*\,\bar{\Lambda}\,\frac{d^2\bar{x}}{dx^2}h\,k-\bar{l}^*\,\bar{\Lambda}\,\bar{\Gamma}\,\bar{h}\,\bar{k}\,,$$

and the addition of these three equations yields the invariance of the form

$$l^*\,'\Lambda(x)\,h\,k=l^*\,\Lambda'(x)\,h\,k+l^*\,\Gamma(x)\,h\,\Lambda(x)\,k-l^*\,\Lambda(x)\,\Gamma(x)\,h\,k\,,$$

where the covariant derivative

$$\circ\,'\Lambda\circ\circ\equiv\circ\,\Lambda'\circ\circ+\circ\,\Gamma\circ\Lambda\circ-\circ\,\Lambda\,\Gamma\circ\circ$$

is a doubly covariant and simply contravariant tensor of rank three. This operator is also symmetric in the contravariant arguments h and k.

It is easily verified that the above coordinate-free definitions of covariant differentiation coincide, after the introduction of coordinates, with the usual definitions.

4.18. The divergence. Let $\overset{1}{u}(p) = u(p)$ be a contravariant vector field which is differentiable on the surface $G_p^m \subset F^m$. As in 4.13, we make use of a real, differentiable, alternating and invariant fundamental form $D(p)\, d_1p \ldots d_mp$, whereby the differentials d_ip vary in the tangent space $T = T_{dp}^m$. Because of the contravariance of $u(p)$ the alternating differential form of rank $m - 1$

$$D(p)\, u(p)\, d_1p \ldots d_{m-1}\, p$$

is an invariant. From that it follows that its rotor

$$\operatorname{rot} \big(D(p)\, u(p)\big)\, d_1p \ldots d_mp$$

is an invariant alternating differential form of rank m. Referring to the definition of the divergence (cf. III. 2.10) we now define this quantity to be the rotor density

$$\operatorname{div} u(p) \equiv m \frac{\operatorname{rot} \big(D(p)\, u(p)\big) d_1p \ldots d_mp}{D(p)\, d_1p \ldots d_mp} .$$

According to this definition div u is given as a *scalar*.

In particular, according to 4.13, the fundamental form D can be chosen in a "locally euclidean" way by means of the fundamental metric form G.

Using the extended definition of the operator rot (cf. III.2.6) one also arrives at an extension of div u that does not necessarily presume the differentiability of u.

4.19. The Laplace operator. If the differentiable vector field is covariant, $\underset{1}{u}(p) = u^*(p)$, the technique of 4.18 cannot be used directly for the formation of the divergence. But if the manifold F^m is metrized by a Gaussian tensor $G(p)$, then by the procedure of 4.12 the index can be raised and the given field u^* replaced by the dual contravariant field $u(p)$, which is uniquely determined by the equation

$$u^*(p) = G(p)\, u(p) , \qquad u(p) = G^{-1}(p)\, u^*(p) .$$

One then defines

$$\operatorname{div} u^*(p) \equiv \operatorname{div} u(p) = \operatorname{div} \big(G^{-1}(p)\, u^*(p)\big) .$$

This technique can be used for the definition of the *Laplace differential operator*. Let $f(p)$ be a real twice differentiable invariant. The gradient of this quantity, which is defined by the class of representatives

$$\frac{df}{dx}$$

as a covariant vector, has by the above definition a well-defined divergence, and this quantity

$$\Delta f \equiv \operatorname{div} \frac{df}{dx}$$

is defined to be the generalized Laplace operator (Beltrami-operator) of f. According to this definition Δf is an invariant (in this regard cf. exercise 3 in 4.20).

By the extended definition of the rotor and of the divergence, the operator Δ can also be defined without assuming f to be twice differentiable.

4.20. Exercises. 1. Prove the invariance of the trace for a mixed tensor $\overset{1}{A}$ of rank 2.

Hint. If $\overset{1}{A} = \overset{1}{A}$ is interpreted as a linear transformation of the tangent space $\overset{1}{T}$, then its trace is defined (cf. I.5.9, exercise 4) as

$$Tr\, A = \frac{1}{D h_1 \ldots h_m} \sum_{i=1}^{m} D\, h_1 \ldots h_{i-1}\, A\, h_i\, h_{i+1} \ldots h_m;$$

this expression is independent of the vectors $h_1, \ldots, h_m$ as well as of the choice of the real nondegenerate alternating form D. We choose $h_1, \ldots, h_m$ to be contravariant and fix D arbitrarily.

In order to form the trace, observe first that A follows the transformation law $\bar{A}\,\bar{h} = \frac{d\bar{x}}{dx} A\, h$ (h a contravariant vector). To form the trace we use the nondegenerate alternating form

$$\bar{D}\,\bar{h}_1 \ldots \bar{h}_m \equiv D \frac{dx}{d\bar{x}} \bar{h}_1 \ldots \frac{dx}{d\bar{x}} \bar{h}_m = D\, h_1 \ldots h_m\,,$$

which is permitted. With this we have

$$\begin{aligned} Tr\, \bar{A} &= \frac{1}{\bar{D}\,\bar{h}_1 \ldots \bar{h}_m} \sum_{i=1}^{m} \bar{D}\,\bar{h}_1 \ldots \bar{h}_{i-1}\, \bar{A}\, \bar{h}_i\, \bar{h}_{i+1} \ldots \bar{h}_m \\ &= \frac{1}{D\, h_1 \ldots h_m} \sum_{i=1}^{m} \bar{D} \frac{d\bar{x}}{dx} h_1 \ldots \frac{d\bar{x}}{dx} h_{i-1} \frac{d\bar{x}}{dx} (A\, h_i) \frac{d\bar{x}}{dx} h_{i+1} \ldots \frac{d\bar{x}}{dx} h_m \\ &= \frac{1}{D\, h_1 \ldots h_m} \sum_{i=1}^{m} D\, h_1 \ldots h_{i-1}\, A\, h_i\, h_{i+1} \ldots h_m = Tr\, A\,. \end{aligned}$$

2. Show that the covariant derivative of the first fundamental tensor vanishes:

$$'G(x) \equiv 0\,.$$

3. Show that

$$\Delta f \equiv \operatorname{div} \frac{df}{dx}$$

is invariant with respect to transformations of the parameter.

§ 5. Integration of the Derivative Formulas

5.1. Formulation of the problem. We turn back to the theory of surfaces and in particular to the differentiation formulas of Gauss and Weingarten. For a twice differentiable m-dimensional surface F^m: $y = y(x)$ $(x \in G_x^m,\ y \in R_y^n,\ n = m + 1)$ it follows from (3.4) and (3.7) that in every admissible parameter space

$$y''(x)\, h\, k = y'(x)\, \Gamma(x)\, h\, k + L(x)\, h\, k\, e(x)\,, \tag{5.1}$$

$$e'(x)\, h = -\, y'(x)\, \Lambda(x)\, h\,, \tag{5.2}$$

where the operators Γ and Λ are uniquely determined by means of the formulas (cf. (3.5), (3.8))

$$G(x)\, l\, \Gamma(x)\, h\, k = \frac{1}{2}\left(G'(x)\, h\, k\, l + G'(x)\, k\, l\, h - G'(x)\, l\, h\, k\right), \tag{5.3}$$

$$G(x)\, k\, \Lambda(x)\, h = L(x)\, h\, k\,, \tag{5.4}$$

and by the fundamental Gaussian forms

$$G(x)\, h\, k = \left(y'(x)\, h,\, y'(x)\, k\right), \tag{5.5}$$

$$L(x)\, h\, k = \left(y''(x)\, h\, k,\, e(x)\right); \tag{5.6}$$

h, k, l stand for representatives of three arbitrary contravariant vectors.

As in the theory of curves, we shall now reverse the problem and ask:

Suppose two sufficiently differentiable *covariant and symmetric tensors G and L of rank two* are given on a sufficiently differentiable m-dimensional $(m > 1)$ manifold F^m, whereby G is to be positive definite. Under which additional necessary and sufficient conditions does there then exist a regular m-dimensional surface in a euclidean space R_y^n of dimension $n = m + 1$ with the prescribed fundamental tensors G and L?

That certain additional integrability conditions are necessary is clear, based on the theory of differential equations given in Chapter IV, since now $m > 1$, and we are dealing with a system of partial differential equations. We shall in what follows set up these necessary and sufficient conditions, integrate the differentiation formulas and show that up to a translation and an orthogonal transformation of the space R_y^n the surface we seek is uniquely determined by G and L.

5.2. Summary of the derivative formulas. In order to obtain a direct connection to the existence theorem in IV.2. we first summarize the differentiation formulas in one single differential equation.

For this we consider the linear space R_z^{mn} (of dimension $mn = m\,(m + 1)$) of all linear operators z that map the m-dimensional manifold F^m into the space R^n_y and then form the product space $R^{mn}_z \times R^n_y$, whose elements consist of all ordered pairs of vectors

$$u = [z, y] .$$

Here

$$u_1 = [z_1, y_1] = [z_2, y_2] = u_2$$

precisely when $z_1 = z_2$ and $y_1 = y_2$; further

$$u_1 + u_2 = [z_1 + z_2, y_1 + y_2]$$

and

$$\lambda u = [\lambda z, \lambda y] .$$

With these definitions for the linear relations the product space is linear and its dimension equal to $m n + n = (n - 1) n + n = n^2$ (cf. I.1.6, exercises 6—7); for this reason we denote it by $R^{n^2}_u$.

Assuming there exists an invariant $y(p)$ on F^m with the prescribed fundamental tensors $G(p)$ and $L(p)$, which we now assume to be *twice continuously differentiable*, let $y'(x) \equiv z(x)$ and $e(x)$ be the representatives of the derivative $y'(p)$, which is a covariant, and of the invariant unit normal $e(p)$ in the parameter region G^m_x of the manifold F^m. Then the vector function

$$u(x) \equiv [z(x), e(x)] , \tag{5.7}$$

according to the differentiation formulas (5.1) and (5.2), satisfy the differential equation

$$u'(x) h = [z'(x) h, e'(x) h]$$
$$= [z(x) \Gamma(x) h + e(x) L(x) h, - z(x) \Lambda(x) h] .$$

Here for each $x \in G^m_x$ the right hand side is linear in h as well as in u, and $u = u(x)$ therefore satifsies the linear differential equation

$$du = B(x) dx u , \tag{5.8}$$

where

$$B(x) h u = B(x) h [z, y] \equiv [z \Gamma(x) h + e L(x) h, - z \Lambda(x) h] . \tag{5.9}$$

Further, since the prescribed tensors G and L are twice continuously differentiable, the operators Γ and Λ, because of the formulas (5.3) and (5.4), and hence also the bilinear operator B, are continuously differentiable once.

Conversely, if $u(x)$ is a solution of the differential equation (5.8), it is twice differentiable, and the same holds for the quantities $z(x)$ and $e(x)$ uniquely determined from (5.7), which according to the definition of $B(x)$ then satisfy the differentiation formulas

$$z'(x) h = z(x) \Gamma(x) h + e(x) L(x) h , \quad e'(x) h = - z(x) \Lambda(x) h . \tag{5.10}$$

The integration of the differentiation formulas (5.1) and (5.2) is now reduced to the solution of the above linear differential equation (5.8) for $u = u(x)$.

The existence theorem in IV. 2.1 provides necessary and sufficient conditions for the integration, which can then be related by means of the definition of the operator B to the operators Γ, Λ and L and thus ultimately to G and L. We shall return to these integrability conditions later and in this connection extract the following from the existence theorem:

If the integrability conditions hold in the region G_x^m, then the differential equation (5.8) for $u = u(x)$ has one and only one continuously differentiable solution

$$u(x) \equiv [z(x), e(x)]$$

which at an arbitrarily given point $x_0 \in G_x^m$ assumes an arbitrarily prescribed value $u(x_0) = u_0 = [z_0, e_0]$.

Transferred to the linear operator z and the vector e this means that the equations (5.10) for the derivatives then have solutions $z(x)$ and $e(x)$ that are uniquely determined if one arbitrarily prescribes the linear operator $z(x_0) = z_0$ and the vector $e(x_0) = e_0$.

5.3. Construction of the surface from $z(x)$ and $e(x)$. If there exists a surface on the manifold F^m with the derivative $y'(x) = z(x)$ and the unit normal $e(x)$ in the parameter region G_x^m, then for each x in the region G_x^m and for each h in the space R_x^m we must have

$$\tilde{\zeta}(x)\, h \equiv (z(x)\, h, e(x)) = 0\,, \qquad \tilde{\varepsilon}(x) \equiv (e(x), e(x)) = 1\,. \tag{5.11}$$

Since the first fundamental form of the surface is prescribed and equals $G(x)\, h\, k$, we further have

$$\tilde{G}(x)\, h\, k \equiv (z(x)\, h, z(x)\, k) = G(x)\, h\, k\,. \tag{5.11'}$$

Because

$$\begin{aligned}(y''(x)\, h\, k, e(x)) &= (z'(x)\, h\, k, e(x))\\ &= (z(x)\, \Gamma(x)\, h\, k + e(x)\, L(x)\, h\, k, e(x)) = L(x)\, h\, k\,,\end{aligned}$$

the surface also has the prescribed second fundamental form.

Hence the initial operator $z(x_0) = z_0$ and the initial vector $e(x_0) = e_0$ must be chosen so that

$$(e_0, e_0) = 1\,, \qquad (z_0\, h, e_0) = 0\,, \qquad (z_0\, h, z_0\, k) = G(x_0)\, h\, k\,. \tag{5.12}$$

These initial conditions are satisfied if first an arbitrary unit vector in the euclidean space R_y^n is taken for e_0 and then, for z_0, an arbitrary operator that maps the parameter space R_x^m, metrized with $G(x_0)\, h\, k$, orthogonally onto the m-dimensional subspace of the space R_y^n orthogonal to e_0 ($n = m + 1$).

We claim that the quantities $\tilde{\zeta}(x)$, $\tilde{\varepsilon}(x)$ and $\tilde{G}(x)$ then satisfy the above identities (5.11) and (5.11') not only for $x = x_0$, but for every $x \in G_x^m$.

In fact, according to the differentiation formulas (5.10),

$$\begin{aligned}\tilde{G}' h k l &= (z' h k, z l) + (z k, z' h l)\\ &= \tilde{G} l \Gamma h k + \tilde{G} k \Gamma h l + L h k \tilde{\zeta} l + L h l \tilde{\zeta} k,\\ \tilde{\zeta}' h k &= (z' h k, e) + (z k, e' h)\\ &= \tilde{\zeta} \Gamma h k + L h k \tilde{\varepsilon} - \tilde{G} k \Lambda h,\\ \tilde{\varepsilon}' h &= 2 (e' h, e) = -2 \tilde{\zeta} \Lambda h.\end{aligned}$$

But the functions $G(x)\, h\, k$, $\zeta(x) \equiv 0$, $\varepsilon(x) \equiv 1$ also satisfy the same system of linear differential equations. For according to equations (5.3) and (5.4), which determine the operators Γ and Λ,

$$G' h k l = G l \Gamma h k + G k \Gamma h l, \qquad L h k - G k \Lambda h = 0,$$

and consequently

$$\begin{aligned}G' h k l &= G l \Gamma h k + G k \Gamma h l + L h k \zeta l + L h l \zeta k,\\ \zeta' h k &= \zeta \Gamma h k + L h k \varepsilon - G k \Lambda h,\\ \varepsilon' h &= -2 \zeta \Lambda h.\end{aligned}$$

This system can now, just as in 5.2, be summarized in one single linear differential equation if one goes over to the product space of the real ε-axis, of the ζ-space dual to R_x^m and of the space of the symmetric tensor G. In this product space one obtains as the equivalent of the above system one single linear differential equation for the quantity $[\tilde{\varepsilon}(x), \tilde{\zeta}(x), \tilde{G}(x)]$ or for $[\varepsilon(x), \zeta(x), G(x)]$, to which the existence theorem in IV.2.1 can be applied.

In particular, the *uniqueness* of the solution for a given initial value at x_0 follows from this theorem. Now since according to the choice of e_0 and z_0

$$\tilde{G}(x_0) = G(x_0), \qquad \tilde{\zeta}(x_0) = \zeta(x_0) = 0, \qquad \tilde{\varepsilon}(x_0) = \varepsilon(x_0) = 1,$$

then in G_x^m

$$\big(z(x)\, h, z(x)\, k\big) = G(x)\, h\, k, \qquad \big(z(x)\, h, e(x)\big) = 0, \qquad \big(e(x), e(x)\big) = 1,$$

and the claim is proved. As already mentioned it follows from here that

$$\big(z'(x)\, h\, k, e(x)\big) = L(x)\, h\, k.$$

The construction of a surface $y = y(x)$ with the prescribed fundamental forms now offers no difficulties. It only remains to integrate the simple differential equation

$$y'(x)\, h = z(x)\, h. \tag{5.13}$$

Because of the symmetry of the operators Γ and L,

$$z' h k = z \Gamma h k + e L h k = z' k h,$$

and the integrability condition is consequently satisfied. Since $z(x)$ was continuously differentiable twice, it follows from the general theory (cf. III.3.3) that equation (5.13) has a solution $y(x)$ which is continuously differentiable three times and which is uniquely determined if the point $y(x_0) = y_0$ in R^n_y is prescribed arbitrarily.

Then according to the above, for $h \in R^m_x$

$$(y' h, e) = (z h, e) = 0 , \qquad (e, e) = 1 ,$$

and $e = e(x)$ is therefore the unit normal to the surface $y = y(x)$ with the prescribed fundamental forms

$$G h k = (z h, z k) = (y' h, y' k) ,$$

$$L h k = (z' h k, e) = (y'' h k, e) .$$

Observe that this surface, because of the regularity of the operator $z(x_0) = z_0$, is regular at least in a sufficiently small neighborhood of the point $y_0 = y(x_0)$.

The above consideration was restricted to a region G^m_x in a given parameter space R^m_x. One obtains the invariant surface $y = y(p)$ on the m-dimensional manifold F^m if for every admissible parameter transformation $x = x(\bar{x})$, $\bar{x} = \bar{x}(x)$, following the law of invariance, $\bar{y}(\bar{x})$ and $\bar{e}(\bar{x})$ are *defined* by

$$\bar{y}(\bar{x}) = y(x(\bar{x})) , \qquad \bar{e}(\bar{x}) = e(x(\bar{x})) .$$

Since G and L were given as covariant tensors of rank two, the relations

$$(y'(x) h, y'(x) k) = G(x) h k , \qquad (y''(x) h k, e(x)) = L(x) h k ,$$

then hold however the parameter x is chosen.

5.4. Discussion of uniqueness. We refer to what was said in 3.6—7. Let

$$\varkappa_i = \varkappa_i(x_0) \qquad (i = 1, \ldots, n - 1)$$

stand for the eigenvalues of the linear transformation $\Lambda(x_0)$ and let $a_i = a_i(x_0)$ be the corresponding orthonormal eigenvectors with respect to $G(x_0) h k$ as fundamental metric form. They are thus quantities which are uniquely determined by the prescribed fundamental forms

$$G(x) h k , \qquad L(x) h k = G(x) h \Lambda(x) k .$$

The initial operator $z_0 = y'(x_0)$ maps the eigenvectors a_i to the principal curvature directions

$$e_i = e_i(x_0) = z_0 a_i \qquad (i = 1, \ldots, n - 1)$$

at the point $y_0 = y(x_0)$ of the surface which has been constructed. Together with the arbitrarily prescribed unit normal $e_0 = e(x_0)$ these principal curvature directions constitute an orthonormal coordinate system

$$e_0, e_1, \ldots, e_{n-1}$$

at the point y_0, the n-frame of the constructed surface at this point.

Now if besides the unit normal e_0 the principal curvature directions $e_1, \ldots, e_{n-1}$ are arbitrarily prescribed in a sequence corresponding to the somehow ordered eigenvalues $\varkappa_i$, the initial operator z_0 is thereby uniquely determined; for there exists precisely one linear mapping z_0 of R_x^m into the orthogonal complement of e_0 such that $z_0 a_i = e_i$ for $i = 1, \ldots, n-1$.

If in addition the point $y_0 = y(x_0)$ is fixed arbitrarily, then the solutions of the integrated differential equations, and therefore also the surface constructed in the previous section, are uniquely determined by y_0 and by the n-frame given here. We conclude:

A surface is uniquely determined by its fundamental tensors G and L up to a translation and an orthogonal transformation of the embedding space.

5.5. The main theorems of the theory of surfaces. The integrability conditions for the differentiation formulas (5.10):

$$z'(x)\,h = z(x)\,\Gamma(x)\,h + e(x)\,L(x)\,h\,, \qquad e'(x)\,h = -\,z(x)\,\Lambda(x)\,h$$

or for the linear differential equation (5.8) equivalent to this system:

$$u'(x)\,h = B(x)\,h\,u(x)$$

are still to be set up. Here $u = u(x)$ varies in the n^2-dimensional product space $R_u^{n^2} = R_z^{mn} \times R^n$, and the bilinear operator $B(x)$ is defined in G_x^m by (5.9):

$$B(x)\,h\,u = B(x)\,h[z, y] \equiv [z\,\Gamma(x)\,h + y\,L(x)\,h,\ -z\,\Lambda(x)\,h]\,.$$

If, as above, the fundamental tensors G and L are taken to be twice continuously differentiable, then Γ and Λ, and as a consequence also B, are continuously differentiable once in G_x^m.

The existence theorem in IV.2.1 then states the following:

In order that the linear differential equation (5.8) possess a solution which is uniquely determined by the initial value

$$u_0 = [z_0, e]\,,$$

which is arbitrarily prescribed at the arbitrary point x_0 of the region G_x^m, it is necessary and sufficient that the equation

$$R(x)\,h\,k\,u$$

$$\equiv \frac{1}{2}\left((B'(x)\,h\,k - B(x)\,h\,B(x)\,k) - (B'(x)\,k\,h - B(x)\,k\,B(x)\,h)\right)u = 0 \tag{5.14}$$

be satisfied in G_x^m for each pair of tangent vectors h, k and each u in the product space $R_u^{n^2}$.

Here according to the definition of the linear operator B

$$B'\,h\,k\,u = [z\,\Gamma'\,h\,k + y\,L'\,h\,k,\ -z\,\Lambda'\,h\,k]$$

and

$$B\,h\,B\,k\,u = [z\,\Gamma k\,\Gamma h + y\,L\,k\,\Gamma h - z\,\Lambda\,k\,L\,h,\ -z\,\Gamma k\,\Lambda\,h - y\,L\,k\,\Lambda\,h].$$

Thus if the integrability condition is to hold for *every* u, i.e., for every linear operator $z \in R_z^{m\,n}$ and every vector $y \in R_y^n$, then necessarily

$$\Gamma' h\,k\,l - \Gamma k\,\Gamma h\,l + \Lambda\,k\,L\,h\,L - \Gamma' k\,h\,l + \Gamma h\,\Gamma k\,l - \Lambda\,h\,L\,k\,l = 0, \tag{5.15a}$$

$$L'\,h\,k\,l - L\,k\,\Gamma\,h\,l - L'\,k\,h\,l + L\,h\,\Gamma\,k\,l = 0\,, \tag{5.15b}$$

$$\Gamma\,k\,\Lambda\,h - \Lambda'\,h\,k - \Gamma\,h\,\Lambda\,k + \Lambda'\,k\,h = 0\,, \tag{5.15c}$$

$$L\,k\,\Lambda\,h - L\,h\,\Lambda\,k = 0\,, \tag{5.15d}$$

where h, k, l are representatives of contravariant vectors.

Before we go on to the analysis of these integrability conditions, we wish to show directly by means of the differentiation formulas (5.1) and (5.2) that they are in any case necessary.

If our problem has a solution, assuming the prescribed tensors G and L to be twice continuously differentiable, then $z(x)$ will be twice and $y(x)$ therefore three times continuously differentiable. From Gauss's differentiation formula (5.1):

$$y''(x)\,k\,l = y'(x)\,\Gamma(x)\,k\,l + L(x)\,k\,l\,e(x)$$

it then follows by differentiation with the parameter differential $dx = h$ that

$$y'''\,h\,k\,l = y''\,h\,\Gamma\,k\,l + y'\,\Gamma'\,h\,k\,l + L\,k\,l\,e'\,h + L'\,h\,k\,l\,e\,,$$

which in view of the two differentiation formulas yields the following decomposition of $y'''\,h\,k\,l$ into tangential and normal components:

$$y'''\,hkl = y'(\Gamma'\,h\,k\,l + \Gamma\,h\,\Gamma\,k\,l - \Lambda\,h\,L\,k\,l) + (L\,h\,\Gamma\,k\,l + L'\,h\,k\,l)\,e\,.$$

Here the left hand side is symmetric in h and k, and hence also is the right. Further, since $y'(x)$ is regular, this yields the above equations (5.15a) and (5.15b).

If one differentiates Weingarten's differentiation formula (5.2):

$$e'(x)\,k = -\,y'(x)\,\Lambda(x)\,k\,,$$

the result is

$$e''\,h\,k = -\,y''\,h\,\Lambda\,k - y'\,\Lambda'\,h\,k\,,$$

which in view of Gauss's differentiation formula (5.1) implies that

$$e''\,h\,k = -\,y'(\Gamma\,h\,\Lambda\,k + \Lambda'\,h\,k) - L\,h\,\Lambda\,k\,e\,.$$

Both sides here are symmetric in h and k, and this gives equations (5.15c) and (5.15d).

By the above, there are apparently four integrability conditions imposed on the tensors G and L, which by means of formulas (5.3) and (5.4) define the operators Γ and Λ. If these conditions are to be compatible and our problem to have any solution at all, these must reduce to at most two independent conditions. That is in fact the case.

First, by formula (5.4)

$$L\,h\,\Lambda\,k = G\,\Lambda\,k\,\Lambda\,h\,,$$

and condition (5.15d) is as a consequence of the hypothesized symmetry satisfied with no further ado.

According to the same formula (5.4)

$$G\,l\,\Lambda\,k = L\,k\,l\,,$$

from which

$$G\,l\,\Lambda'\,h\,k = L'\,h\,k\,l - G'\,h\,l\,\Lambda\,k = L'\,h\,k\,l - G'\,h\,\Lambda\,k\,l$$

follows by differentiation.

Further, formula (5.3) with $\Lambda\,k$ instead of k yields

$$G\,l\,\Gamma\,h\,\Lambda\,k = \frac{1}{2}\,(G'\,h\,\Lambda\,k\,l + G'\,\Lambda\,k\,l\,h - G'\,l\,h\,\Lambda\,k)\,,$$

and by adding these equations one obtains

$$G\,l(\Lambda'\,h\,k + \Gamma\,h\,\Lambda\,k) = L'\,h\,k\,l - \frac{1}{2}(G'\,h\,\Lambda\,k\,l + G'\,l\,h\,\Lambda\,k - G'\,\Lambda\,k\,l\,h)\,.$$

Here the subtrahend on the right is, according to formulas (5.3) and (5.4), equal to

$$G\,\Lambda\,k\,\Gamma\,h\,l = L\,k\,\Gamma\,h\,l\,,$$

and therefore

$$G\,l\,(\Lambda'\,h\,k + \Gamma\,h\,\Lambda\,k) = L'\,h\,k\,l - L\,k\,\Gamma\,h\,l\,.$$

This relation shows at once that equations (5.15b) and (5.15c) are equivalent.

According to this, the necessary and sufficient integrability conditions are reduced to two, namely

$$\Gamma'(x)\,h\,k\,l + \Gamma(x)\,h\,\Gamma(x)\,k\,l - \Gamma'(x)\,k\,h\,l - \Gamma(x)\,k\,\Gamma(x)\,h\,l$$
$$= L(x)\,k\,l\,\Lambda(x)\,h - L(x)\,h\,l\,\Lambda(x)\,k\,, \tag{5.16a}$$

$$L'(x)\,h\,k\,l - L(x)\,k\,\Gamma(x)\,h\,l = L'(x)\,k\,h\,l - L(x)\,h\,\Gamma(x)\,k\,l\,. \tag{5.16b}$$

These are the fundamental equations of the theory of surfaces. The first is the *Gauss-Codazzi formula*, the second the *Codazzi-Mainardi formula*.

§ 6. Theorema Egregium

6.1. The curvature tensors. We are going to analyze the fundamental equations (5.16a) and (5.16b) more carefully, and start with formula (5.16b), where we now write h_1, h_2, h_3 instead of h, k, l. If one subtracts

$$L(x)\, h_3\, \Gamma(x)\, h_1\, h_2 = L(x)\, h_3\, \Gamma(x)\, h_2\, h_1\,,$$

the covariant derivative can be used to write this equation briefly as

$$'L(x)\, h_1\, h_2\, h_3 = {}'L(x)\, h_2\, h_1\, h_3\,. \tag{6.1}$$

Since the real form on the left

$$'L(x)\, h_1\, h_2\, h_3 \equiv L'(x)\, h_1\, h_2\, h_3 - L(x)\, h_2\, \Gamma(x)\, h_1\, h_3 - L(x)\, h_3\, \Gamma(x)\, h_1\, h_2$$

is obviously also symmetric in h_2 and h_3, the Codazzi-Mainardi equation is equivalent to the following statement:

The covariant derivative of the second fundamental tensor L is *symmetric.*

We go on to formula (5.16a), and for short we write it

$$R(x)\, h_1\, h_2\, h_3 = L(x)\, h_2\, h_3\, \Lambda(x)\, h_1 - L(x)\, h_1\, h_3\, \Lambda(x)\, h_2\,, \tag{6.2}$$

where

$$R(x)\, h_1\, h_2\, h_3 = 2 \wedge \left(\Gamma'(x)\, h_1\, h_2 + \Gamma(x)\, h_1\, \Gamma(x)\, h_2\right) h_3\,. \tag{6.3}$$

From the transformation formulas for Λ and L it follows at once that the operator

$$R(x) = \underset{3}{\overset{1}{R}}(x)$$

is a triply covariant and simply contravariant tensor of rank four; this is the *mixed Riemannian curvature tensor.*

Observe that this tensor is uniquely determined by the first fundamental tensor G alone. For the expression (6.3) for $\underset{3}{\overset{1}{R}}$ contains only the operators Γ and Γ'', and these can, based on formula (5.3) which uniquely determines Γ, be computed from G, G' and G''.

We introduce a fourth arbitrary contravariant vector h_4. If one then sets

$$\underset{4}{R}(x)\, h_1\, h_2\, h_3\, h_4 \equiv G(x)\, h_4\, \underset{3}{\overset{1}{R}}(x)\, h_1\, h_2\, h_3\,, \tag{6.4}$$

the Gauss-Codazzi formula can, as a consequence of the equations

$$G(x)\, h_4\, \Lambda(x)\, h_1 = L(x)\, h_1\, h_4\,, \quad G(x)\, h_4\, \Lambda(x)\, h_2 = L(x)\, h_2\, h_4\,,$$

be brought into the equivalent form

$$\underset{4}{R}(x)\, h_1\, h_2\, h_3\, h_4 = L(x)\, h_1\, h_4\, L(x)\, h_2\, h_3 - L(x)\, h_1\, h_3\, L(x)\, h_2\, h_4\,. \tag{6.5}$$

Here, according to its definition (6.4), $\underset{4}{R}$ is a covariant tensor of rank four, the *covariant Riemannian curvature tensor.*

Like $\underset{3}{\overset{1}{R}}$, $\underset{4}{R}$ is also uniquely determined by the first fundamental tensor G alone, and it can be computed from G, G' and G''.

Observe further that the above two curvature tensors vanish simultaneously, so that the equations

$$\underset{3}{\overset{1}{R}}(x) = 0\,, \qquad \underset{4}{R}(x) = 0$$

are equivalent.

Certain symmetry properties of the tensor $\underset{4}{R}$ result from the right side of formula (6.5).

The symmetric group of the 24 permutations of the indices 1, 2, 3, 4 has as normal subgroup the "four-group", which consists of the identity permutation and the permutations

$$(12)\,(34)\,, \qquad (13)\,(24)\,, \qquad (14)\,(23)\,.$$

The corresponding quotient group is isomorphic to the symmetric permutation group on three elements. One sees immediately that the form on the right in (6.5) is *invariant* for the permutations in the four-group and, corresponding to the permutations in the quotient group, assumes altogether six different forms which differ pairwise with respect to sign.

The above situation, according to which $\underset{4}{R}$ is uniquely determined by G, together with these symmetry properties contain the essence of the Gauss-Codazzi formula.

6.2. Theorema Egregium. Among other things, Gauss's classical "theorema egregium" follows from the Gauss-Codazzi formula (6.5).

That coset of the four-group whose permutations change the sign of $\underset{4}{R}(x)\, h_1\, h_2\, h_3\, h_4$ contains the permutations (12), (12) (12) (34) = (34), (12) (13) (24) = (1423) and (12) (14) (23) = (1324). According to this there are precisely two transpositions, (12) and (34), which change the sign of the form named. For fixed h_3 and h_4, $\underset{4}{R}(x)\, h_1\, h_2\, h_3\, h_4$ is alternating in h_1 and h_2, for fixed h_1 and h_2, in h_3 and h_4.

But the form

$$\underset{4}{C}(x)\, h_1\, h_2\, h_3\, h_4 \equiv G(x)\, h_1\, h_4\, G(x)\, h_2\, h_3 - G(x)\, h_1\, h_3\, G(x)\, h_2\, h_4$$

also has the same property. This immediately implies: If a_1, a_2 and likewise a_3, a_4 are linearly independent vectors of the parameter space R_x^m, then

$$\frac{\underset{4}{R}(x)\, h_1\, h_2\, h_3\, h_4}{\underset{4}{C}(x)\, h_1\, h_2\, h_3\, h_4} = \frac{\underset{4}{R}(x)\, a_1\, a_2\, a_3\, a_4}{\underset{4}{C}(x)\, a_1\, a_2\, a_3\, a_4}$$

for each pair of vectors h_1, h_2 from the two-dimensional subspace spanned by a_1, a_2 and each pair of vectors h_3, h_4 from the subspace spanned by a_3, a_4. If in particular one takes $a_1 = a_3 = h$, $a_2 = a_4 = k$, then

$$\frac{\underset{4}{R}(x)\, h_1\, h_2\, h_3\, h_4}{\underset{4}{C}(x)\, h_1\, h_2\, h_3\, h_4} = \frac{\underset{4}{R}(x)\, h\, k\, h\, k}{\underset{4}{C}(x)\, h\, k\, h\, k},$$

provided h_1, h_2, h_3, h_4 vary in the two-dimensional subspace of R_x^m spanned by h and k.

Now if in particular $m = 2$, $n = 3$, the above holds with no restrictions on h_1, h_2, h_3, h_4, however the linearly independent coordinate axes h and k are taken for the parameter space R_x^2. Now bring $L(x)$ into the principal axis form with respect to $G(x)$ and take the principal axis directions $a_1(x)$ and $a_2(x)$ for h and k (cf. 3.6—7). Then

$$L(x)\, a_i(x)\, a_i(x) = \varkappa_i(x) \quad (i = 1, 2)\,, \quad L(x)\, a_1(x)\, a_2(x) = 0\,,$$

and therefore according to (6.5)

$$\underset{4}{R}(x)\, h\, k\, h\, k = -\varkappa_1(x)\, \varkappa_2(x)\,,$$

and

$$G(x)\, a_i(x)\, a_i(x) = 1 \qquad (i = 1, 2)\,, \qquad G(x)\, a_1(x)\, a_2(x) = 0\,.$$

Consequently

$$\underset{4}{C}(x)\, h\, k\, h\, k = -1\,.$$

Therefore

$$K(x) \equiv \varkappa_1(x)\, \varkappa_2(x) = \frac{\underset{4}{R}(x)\, h_1\, h_2\, h_3\, h_4}{\underset{4}{C}(x)\, h_1\, h_2\, h_3\, h_4}$$

$$= \frac{\underset{4}{R}(x)\, h_1\, h_2\, h_3\, h_4}{G(x)\, h_1\, h_4\, G(x)\, h_2\, h_3 - G(x)\, h_1\, h_3\, G(x)\, h_2\, h_4}\,. \tag{6.6}$$

Here on the left stands the *Gaussian curvature* of the surface at the point $y(x)$, and on the right an expression that does not depend on h_1, h_2, h_3, h_4 and that therefore *can be computed from* $G(x)$, $G'(x)$ *and* $G''(x)$ *alone*. That proves the *Theorema egregium of Gauss*.

6.3. Exercises. 1. Let $\underset{2}{R}\, h_2\, h_3$ be the differential form that arises by contracting the Riemannian differential form $h^1\, \underset{3}{\overset{1}{R}}\, h_1\, h_2\, h_3$ with respect to h^1 and h_1, and $\underset{1}{\overset{1}{R}}\, h_2$ the linear transformation that results from $\underset{2}{R}\, h_2$ through raising an index. Finally let R be the real scalar that comes from the contraction of $h^2\, \underset{1}{\overset{1}{R}}\, h_2$. $\underset{2}{R}$ and $\underset{1}{\overset{1}{R}}$ are the so-called covariant and mixed *Ricci tensors*, respectively, and R is the scalar *Riemannian curvature*. Show:

If $e_1, \ldots, e_m$ stands for an arbitrary coordinate system orthonormalized at the point x with respect to the fundamental metric tensor, then

$$\underset{2}{R}\, h_2\, h_3 = \sum_{i=1}^{m} \underset{4}{R}\, h_2\, e_i\, e_i\, h_3 = \sum_{i=1}^{m} \underset{4}{R}\, h_3\, e_i\, e_i\, h_2 = \underset{2}{R}\, h_3\, h_2\,,$$

$$\overset{1}{\underset{1}{R}}\, h_2 = \sum_{i=1}^{m} \overset{1}{\underset{3}{R}}\, h_2\, e_i\, e_i\,,$$

$$R = -\sum_{i=1}^{m} \sum_{j=1}^{m} \underset{4}{R}\, e_i\, e_j\, e_i\, e_j = \sum_{\substack{i,j=1\\ i \neq j}}^{m} \varkappa_i\, \varkappa_j\,,$$

where $\varkappa_1, \ldots, \varkappa_m$ stand for the principal curvatures at the point x.

2. Show, inverting exercise 3 in 3.8, that a surface with nothing but umbilical points is a sphere.

Hint. For each $x \in G_x^m$ all principal curvatures are equal, and consequently $\Lambda(x)\, h = \varkappa(x)\, h$ and

$$L(x)\, k\, l = G(x)\, k\, \Lambda(x)\, l = \varkappa(x)\, G(x)\, k\, l\,.$$

Because $'G(x)\, h\, k\, l = 0$, covariant differentiation yields

$$'L(x)\, h\, k\, l = \varkappa'(x)\, h\, G(x)\, k\, l\,,$$

and therefore according to the Codazzi-Mainardi formula

$$\varkappa'(x)\, h\, G(x)\, k\, l = \varkappa'(x)\, k\, G(x)\, h\, l\,.$$

If for an arbitrary h the vectors k and l are taken so that $l = k \neq 0$ and $G(x)\, h\, k = 0$, then $\varkappa'(x)\, h \equiv 0$, and $\varkappa$ is therefore independent of x. The claim is then an immediate consequence of Weingarten's formula.

§ 7. Parallel Translation

7.1. Definition. In a neighborhood G^m on the m-dimensional manifold F^m let a piecewise regular arc $p = p(t)$ be given, where t varies in an interval of some one-dimensional parameter space R_t^1. At each point p of this arc let a differentiable *contravariant* vector $u = u(p) = \overset{1}{u}(p)$ be defined. Then the covariant derivative $'u(p)$ is a mixed tensor of rank 2. One says the vector field $u(p)$ has come into being along the curve $p = p(t)$ by means of *parallel trnslation* if the former derivative vanishes on the curve.

In the parameter space R_x^m, corresponding to the quantities $p = p(t)$ and $u = u(p)$, there is an arc $x = x(t)$ and a contravariant vector field $u = u(x)$. The condition for parallelism of the field along $x = x(t)$ is thus, provided $dx = x'(t)\, dt$,

$$'u\, dx \equiv du + \Gamma\, u\, dx = 0\,. \tag{7.1}$$

The parallel translation of a differentiable *covariant* vector field $u = u(p) = \underset{1}{u}(p)$ along the curve $p = p(t)$ is defined correspondingly. The covariant derivative $'u(p)$ is in this case a covariant tensor of rank two and parallelism is expressed through the equation

$$'u\, dx \equiv du - u\, \Gamma\, dx = 0\,. \tag{7.1'}$$

The left hand side of the last equation is a covariant vector.

The condition for the parallelism of a vector field $u(p)$ is thus given by means of a normal linear homogeneous differential equation which is invariant with respect to parameter transformations. Conversely, if the curve $p = p(t)$ is prescribed, a parallel vector field can be constructed by integrating the defining differential equation along the curve. According to the general theory of normal systems (cf. IV.1), the field is uniquely determined if one prescribes the initial value $u_0 = u(p_0)$ of the field vector in an arbitrary way at an arbitrarily chosen point $p_0 = p(t_0)$. The integration then certainly succeeds if the Christoffel operator $\Gamma(p)$ is continuous or, equivalently, if the fundamental metric tensor is continuously differentiable once. In the "embedding theory" it suffices to assume the embedding mapping $y = y(x)$ to be twice continuously differentiable.

7.2. The translation operator. According to the general theory of linear homogeneous differential equations (cf. IV.3.8) there is associated with equation (7.1) (or (7.1')) a family (T) of regular linear transformations of the tangent space (or of the space dual to the tangent space) to the manifold F^m with the following properties:

1. Corresponding to each oriented piecewise differentiable path l in a (sufficiently small) neighborhood on the manifold F^m there is a well-determined linear transformation $T = T_l$.

2. For the product $l = l_2 l_1$ of two paths l_1 and l_2 one has $T_l = T_{l_2} T_{l_1}$.

3. For the path l^{-1} reciprocal to l, $T_l T_{l^{-1}} = T_{l^{-1}} T_l = I$ (the identity transformation).

4. If the path l joins the points $p = p_1$ and $p = p_2$ on the manifold, then the vectors $u_1 = u(p_1)$ and $u_2 = u(p_2)$ of the field $u(p)$ which is parallel along l are connected through the relations[1]

$$u_2 = T_l u_1\,, \qquad u_1 = T_l^{-1} u_2 = T_{l^{-1}} u_2\,.$$

In the parameter space R_x^m, corresponding to the operator T_l there is a transformation T_l of this space, and one has for an increment dx of the path $l = l_x \subset R_x^m$ at the point x the relation

$$du = T_{dx} u - u = (dT)\, u = -\Gamma\, dx\, u\,.$$

[1] For the case of a covariant field $u(p)$ it is advisable, in conformity with our presentation of the tensor calculus, to write the operator T_l to the right of the argument u: $u_2 = u_1 T_l$, etc.

Thus T is differentiable at the point $x \leftrightarrow p$, and the derivative is $T' = -\Gamma$. Conversely, the theory of parallelism can be most simply constructed based on a given operator group (T) and postulates 1—4[1].

7.3. Metric properties. Since the Christoffel operator Γ is uniquely determined by the fundamental metric tensor G and its first derivative, the translation operator is determined by G and its derivatives G', G''. To more carefully investigate this connection, we consider in the parameter space R_x^m a piecewise regular arc $x = x(t)$, and along it take two parallel contravariant, say, vector fields $u(x)$ and $v(x)$. Then the expression $G(x)\,u(x)\,v(x)$ is an invariant, and its derivative is thus the same as its covariant derivative. If one differentiates it along the curve $x = x(t)$, the result is thus

$$(G\,u\,v)'\,dx = '(G\,u\,v)\,dx = 'G\,dx\,u\,v + G('u\,dx)\,v + G\,u('v\,dx)\,.$$

Here $'G = 0$ (cf. 4.20, exercise 2), and since, because of the parallelism, $'u = 'v = 0$, the entire above expression vanishes. It follows from this that $G\,u\,v$ is constant along the curve $x = x(t)$, and one concludes that *with respect to the local euclidean metric* $G(x)$ *the translation operator* T_l *is an orthogonal transformation of the space* R_x^m.

Now if, as is the case in the Gaussian theory, the manifold F^m is embedded in the space R_y^{m+1}, so that the metric G is induced by the euclidean metric on the latter space, it turns out that parallel translation on the embedding surface F^m along an arc joining two surface points y_1 and y_2 maps the tangent planes to the surface at these points orthogonally (euclideanly) onto one another. This mapping can be completed to an orthogonal transformation of the entire space R_y^{m+1} by requiring that the unit normals at the points y_1 and y_2 correspond.

7.4. Geodesic lines. We now set ourselves the task of determining the "straightest" lines on the manifold, i.e., those paths whose tangents are parallel. If the equation for the lines we are seeking is written in the form $x = x(\tau)$ (τ real), then a tangent vector has the form $u = u(\tau) = \lambda(\tau)\,x'(\tau)$, where $\lambda(\tau)$ (> 0) is a real multiplier. If this vector $u = u(\tau)$ is to now define a parallel field along $x = x(\tau)$, then by 7.3 its length in the metric $G(x)$ is constant:

$$G\,\lambda \frac{dx}{d\tau}\,\lambda \frac{dx}{d\tau} = \lambda^2 \left(\frac{d\sigma}{d\tau}\right)^2 = \text{const.}\,,$$

where $d\sigma$ stands for the length of the arc differential $dx = x'(\tau)\,d\tau$. Thus, up to a *constant* multiplier, we must have $\lambda = d\tau/d\sigma$, and the tangent vector which is to be translated in a parallel fashion will be equal to

$$\lambda \frac{dx}{d\tau} = \frac{dx}{d\sigma}\,.$$

[1] For this, cf. W. Graeub and R. Nevanlinna [1].

Now if the arc length σ of the path to be determined is chosen as the parameter, substitution of $u = dx/d\sigma$ in the equation (7.1) for parallel translation yields the condition

$$\frac{d^2x}{d\sigma^2} + \Gamma \frac{dx}{d\sigma}\frac{dx}{d\sigma} = 0 \tag{7.2}$$

for the "straightest" or *geodesic* line $x = x(\sigma)$.

In order to integrate this second order differential equation, one again introduces an arbitrary parameter τ. In this way a second order normal differential equation is obtained for the geodesic line $x = x(\tau)$, which according to the theory of normal systems (cf. IV.1) can be integrated (cf. 7.7, exercise 5). Through each point x there goes a one parameter family of straightest line arcs, which are uniquely determined if the direction of the tangent is fixed at the former point.

The geodesic lines are also characterized by a metric condition, namely they are the shortest lines joining two points (which are sufficiently close to one another) on the manifold (cf. 7.7, exercise 6).

7.5. Integrability of the parallel translation equation. Until now the differential equation for the parallel translation of a vector $u = u(x)$ has been integrated along a prescribed path $x = x(t)$. The question now arises, under which conditions can the partial differential equations for the parallel translation of a (for example, contravariant) vector u,

$$du + \Gamma\, u\, dx = 0\,,$$

be integrated in a full m-dimensional neighborhood on the manifold F^m. This is the case if and only if the translation operator T_l is independent of the course of the path l which joins its fixed beginning and end points. For this, by IV.3.,9 it is necessary and sufficient that the relation

$$T_\gamma = I$$

holds for the boundary $\gamma = \partial s^2$ of every two-dimensional simplex s^2 on the manifold. If Γ is continuously differentiable (G is hence twice continuously differentiable) this condition is equivalent with the trilinear differential form (cf. IV.3.12)

$$\wedge\, (\Gamma'\, h\, k + \Gamma\, h\, \Gamma\, k)\, l$$

being zero for every h, k, $l \in R^m_x$. But the operator in this form is (up to a factor of 1/2) nothing other than the mixed Riemannian curvature tensor $\overset{1}{\underset{3}{R}}$ of the manifold (cf. (6.3)), and therefore:

For the parallel translation equation to be integrable on the manifold F^m it is necessary and sufficient that its curvature vanish.

7.6. Determining manifolds of curvature zero. We assume that the curvature $\overset{1}{\underset{3}{R}}(x) = 0$ in a certain parameter neighborhood in the space R_x^m and wish to show that for a suitable choice of the parameter x the Christoffel operator $\Gamma(x)$ can then be made to vanish.

To this end we first fix the parameter space R_x^m arbitrarily. If the Christoffel operator does not yet vanish we try to determine a new admissible parameter $\bar{x} = \bar{x}(x)$ so that (cf. (4.2) in 4.15)

$$\bar{\Gamma}\,\bar{h}\,\bar{k} = \frac{d\bar{x}}{dx}\,\Gamma\,h\,k - \frac{d^2\bar{x}}{dx^2}\,h\,k = 0\,. \tag{7.3}$$

To solve this second order differential equation for $\bar{x}(x)$, we introduce the regular operator $z = d\bar{x}/dx$ as the new variable. The first order differential equation which thus results,

$$\frac{dz}{dx}\,dx - z\,\Gamma\,dx = 0 \tag{7.3'}$$

can according to IV.3.12 be completely integrated provided for each h and k

$$z \wedge (\Gamma'\,h\,k + \Gamma\,h\,\Gamma\,k) = 0\,.$$

Because of the regularity of z this means that

$$(\Gamma'\,h\,k + \Gamma\,h\,\Gamma\,k)\,l = \frac{1}{2}\,\overset{1}{\underset{3}{R}}\,h\,k\,l = 0\,.$$

By hypothesis this integrability condition is satisfied, and the operator $z = d\bar{x}/dx$ is hence uniquely determined, if it is arbitrarily fixed at an initial point $x = x_0$.

To then integrate the equation

$$d\bar{x} = z\,dx\,, \tag{7.3''}$$

observe that by (7.3′) the rotor of the operator z is equal to

$$\wedge\, z'\,h\,k = z \wedge \Gamma\,h\,k = 0\,.$$

Therefore, the condition of integrability for the differential $d\bar{x} = z\,dx$ is satisfied, and by III.3.3 $\bar{x}(x)$ is determined by means of $z(x)$ up to an additive constant. The sought-for parameter space $\bar{R}_{\bar{x}}^m$, in which $\bar{\Gamma}(\bar{x}) \equiv 0$, is therewith constructed; it is uniquely determined if one associates with an arbitrary element $\{p, dp\}$ of the tangent space of F^m an arbitrary line element $\{\bar{x}, d\bar{x}\}$.

In this distinguished parameter space the equation for parallel translation is simply $d\bar{u} = 0$, $\bar{u}(\bar{x}) = \text{const}$. Thus parallel translation coincides with the elementary translations of the space $\bar{R}_{\bar{x}}^m$. Actually, the geometry of a manifold F^m of curvature zero is *euclidean*, for it follows from $\bar{\Gamma} = 0$ that the derivative $'G(x) = 0$ (cf. 4.20, exercise 2).

The metric tensor $\bar{G}(\bar{x})$ is thus independent of the point $\bar{x}$, which implies the euclideanness of the geometry on $R^m_{\bar{x}}$.

If conversely a manifold F^m admits a parametric representation where $G = \text{const.}$, $\Gamma = 0$, then its curvature is obviously zero, and one concludes:

For a manifold to be euclidean it is necessary and sufficient that its curvature vanish.

7.7. Exercises. 1. Let F^m_p be a differentiable manifold of dimension m and $x \in R^m_x$ the representative of a point $p \in F^m_p$. Further, suppose $A(x)\, h_1 \dots h_q$ is an invariant q-linear form in the contravariant arguments h_i. The covariant derivative $'A$ is defined as a covariant tensor $\underset{q+1}{B}$ by taking the arguments $h_1, \dots, h_q$ to be vectors which are parallel along an arc emanating from x that has the tangent vector $dx = h$ at the initial point x and then differentiating the form $\underset{q}{A}\, h_1 \dots h_q$ at the point x with the differential $dx = h$. One then sets $\underset{q+1}{B}\, h\, h_1 \dots h_q \equiv d(\underset{q}{A}\, h_1 \dots h_q)$.

According to this, what is the general form of the covariant derivative $'\underset{q}{A}$? Also define analoguously the covariant derivative of a contravariant tensor $\underset{q}{A}$.

2. Determine the alternating part of the second covariant derivative of a contravariant vector field $u(x)$.

3. Let $D_0\, h_1\, h_2$ ($h_1, h_2 \in R^2_x$) be a nondegenerate alternating real form. The alternating fundamental form

$$D(x)\, h_1\, h_2 \equiv \pm \sqrt{\det\left(G(x)\, h_i\, h_j\right)}\,,$$

where the sign is chosen to be the same as the sign of $D_0\, h_1\, h_2$, satisfies the relation

$$'D\, h\, h_1\, h_2 = D'\, h\, h_1\, h_2 - D\, h_1\, \Gamma\, h_2\, h + D\, h_2\, \Gamma\, h_1\, h = 0\,.$$

4. Prove the so-called *Bianchi identity*

$$'\underset{4}{R}\, h_1\, h_2\, h_3\, h_4\, h_5 + {}'\underset{4}{R}\, h_2\, h_3\, h_1\, h_4\, h_5 + {}'\underset{4}{R}\, h_3\, h_1\, h_2\, h_4\, h_5 = 0\,.$$

5. Show that the geodesic line emanating from the point p_0 of the twice continuously differentiable m-dimensional manifold F^m is uniquely given if one prescribes the direction of the tangent at the initial point p_0. These geodesic lines form a field which covers a certain neighborhood of p_0 simply.

Hint. In the parameter space R^m_x, where p_0 has the representative x_0, the equation for the geodesic lines is

$$x'' + \Gamma\, x'\, x' = 0\,, \tag{a}$$

where $x' = dx/d\sigma$ (σ arc length). For an arbitrary parameter τ one has ($\dot{x} = dx/d\tau$)

$$x' = \frac{\dot{x}}{\dot{\sigma}}, \quad x'' = \frac{\dot{\sigma}\,\ddot{x} - \ddot{\sigma}\,\dot{x}}{\dot{\sigma}^3},$$

and equation (a) becomes

$$\ddot{x} + \Gamma\,\dot{x}\,\dot{x} = \frac{\ddot{\sigma}}{\dot{\sigma}}\,\dot{x}\,. \tag{b}$$

Equation (a) is thus invariant in its form if $\ddot{\sigma} = 0$, $\sigma = \alpha\,\tau + \beta$, i.e., when the parameter is, up to a trivial normalization, equal to the arc length σ.

On the other hand, if $x = x(\tau)$ satisfies the equation

$$\ddot{x} + \Gamma\,\dot{x}\,\dot{x} = 0\,, \tag{c}$$

then the parameter τ (up to an affine transformation) is equal to the arc length of the curve $x = x(\tau)$. For $\dot{\sigma}^2 = G\,\dot{x}\,\dot{x}$, and by (c)

$$\frac{d(\dot{\sigma})^2}{d\tau} = \frac{dG}{dx}\,\dot{x}\,\dot{x}\,\dot{x} + 2\,G\,\dot{x}\,\ddot{x} = \frac{dG}{dx}\,\dot{x}\,\dot{x}\,\dot{x} - 2\,G\,\dot{x}\,\Gamma\,\dot{x}\,\dot{x} = 0\,,$$

and consequently $\dot{\sigma} = \alpha$, $\sigma = \alpha\,\tau + \beta$.

Equation (c) is equivalent to the normal system

$$\dot{x} = u\,, \qquad \dot{u} = -\,\Gamma\,u\,u\,,$$

whose solution is uniquely determined if the initial values $x_0 = x(\tau_0)$, $\dot{x}_0 = \dot{x}(\tau_0)$ are given. The first part of the claim follows from that.

For the proof of the field property of the solutions in the vicinity of the point x_0, set $\tau_0 = \sigma_0 = 0$, and take a unit vector e $(G(x_0)\,e\,e = 1)$ for the initial tangent $\dot{x}(0)$. If $x = x(\tau)$ stands for the solution of equation (c) that satisfies the initial conditions $x(0) = x_0$, $\dot{x}(0) = e$, then $\beta = 0$ and $\alpha = 1$, hence $\tau = \sigma$.

Corresponding to each e with $G(x_0)\,e\,e = 1$ there is for sufficiently small $\sigma < \sigma^*$ a well-determined point $x = x(\sigma; e)$. Thus if one sets $\sigma\,e = t$, then

$$x = x(t) \qquad (x(0) = x_0)$$

is a self-mapping of the space R_x^m metrized with $G(x)$ which is well-defined in the sphere $|t| < \sigma^*$ of the space R_t^m metrized with $G(x_0)$. Further, since the solutions of equation (c) $(= (a))$, assuming sufficient differentiability of the tensor $G(x)$, are differentiable with respect to σ and e, the derivative dx/dt exists. For $t = 0$ it reduces to the identity transformation and is therefore regular. It then follows from the inversion theorem in II.4.2 that

$$t = t(x) \qquad (t(x_0) = 0)$$

is single-valued (and differentiable) in a certain neighborhood of the point p_0: Through each point x of this neighborhood there consequently goes precisely one geodesic line which emanates from x_0, namely that one with the unit tangent e at the point x_0, where $e = t(x)/\sqrt{G(x_0)\, t(x)\, t(x)} = t(x)/\sigma(x)$.

6. The shortest line joining two points which lie sufficiently close to one another on the manifold F^m is geodesic.

Hint. Without drawing upon general principles from the calculus of variations, the assertion can be proved in the following direct way. Let p_0 be a point on F^m and x_0 its representative in R_x^m. The geodesic lines emanating from x_0, according to the above, form a field: thus if τ denotes an arbitrary *common* parameter for this field of curves, then for each x in a certain neighborhood of x_0 the initial tangents $\dot{x}(0)$ $(\dot{x}(\tau) = dx/d\tau)$ can be taken in a unique way so that the geodesic line $x = x(\tau; \dot{x}(0))$ joins the point $x_0 = x(0; \dot{x}(0))$ with x.

Let $dx = h$ be a fixed differential. If one takes into account that the differential

$$d\dot{x} = \frac{d}{dx}\left(\frac{dx}{d\tau}\right) h = \frac{d}{d\tau}\left(\frac{dx}{dx}h\right) = \frac{dh}{d\tau} = 0$$

and that

$$\frac{d}{d\tau}(d\sigma) = \frac{d}{d\tau}\left(\frac{d\sigma}{dx}h\right) = \frac{d}{dx}\left(\frac{d\sigma}{d\tau}\right) h = d\dot{\sigma}\,,$$

then differentiation of the equation $\dot{\sigma}^2 = G\,\dot{x}\,\dot{x}$ with the differential dx yields

$$2\,\dot{\sigma}\frac{d}{d\tau}(d\sigma) = d(\dot{\sigma}^2)$$

$$= \frac{dG}{dx}dx\,\dot{x}\,\dot{x} + 2\,G\,\dot{x}\,d\dot{x} = \frac{dG}{dx}dx\,\dot{x}\,\dot{x} = \dot{\sigma}^2\frac{dG}{dx}dx\,x'\,x'$$

$(x' = dx/d\sigma)$, and hence

$$2\frac{d}{d\tau}(d\sigma) = \dot{\sigma}\frac{dG}{dx}dx\,x'\,x'\,.$$

Here (cf. 3.5))

$$\frac{dG}{dx}dx\,x'\,x' = 2\frac{dG}{dx}x'\,x'\,dx - 2\,G\,dx\,\Gamma\,x'\,x'\,,$$

so that finally

$$\frac{d}{d\tau}(d\sigma) = \dot{\sigma}\left(\frac{dG}{dx}x'\,x'\,dx - G\,dx\,\Gamma\,x'\,x'\right).$$

This result is valid for a given G, assuming sufficient differentiability, for any field of curves. If in particular the curves are geodesic, then $\Gamma\,x'\,x' = -\,x''$, and the above equation yields

$$\frac{d}{d\tau}(d\sigma) = \dot{\sigma}\left(\frac{dG}{dx}x'\,x'\,dx + G\,x''\,dx\right) = \frac{d\sigma}{d\tau}\frac{d}{d\sigma}(G\,x'\,dx) = \frac{d}{d\tau}(G\,x'\,dx)\,,$$

from which it can be seen that the difference $d\sigma - G\,x'\,dx$ is constant on the geodesic arc that joins the points x_0 and x. But for $\tau \to 0$, $\sigma \to 0$ and $x \to x_0$

$$G(x)\,x'(\sigma)\,dx \to G(x_0)\,e\,dx\,,$$

where $e = x'(0)$ stands for the unit tangent to the arc at the initial point x_0. Further, since near x_0 $(\sigma(x))^2$ can be replaced by $G(x_0)\,x\,x$, as $x \to x_0$ we also have

$$d\sigma(x) = \frac{d\sigma}{dx}\,dx \to G(x_0)\,e\,dx\,,$$

and the above difference is therefore $= 0$ on the entire geodesic arc, and consequently

$$d\sigma(x) = G(x)\,x'(\sigma)\,dx\,,$$

where $x'(\sigma)$ stands for the unit tangent to the geodesic line through x at the latter point and dx is an arbitrary differential, $d\sigma(x)$ being the corresponding differential of the field function.

From this it follows by means of Schwarz's inequality that

$$|d\sigma|^2 \leqq G(x)\,x'\,x'\,G(x)\,dx\,dx = G(x)\,dx\,dx = |dx|^2\,,$$

and therefore $|d\sigma| \leqq |dx|$, where $|dx|$ is the length of the line element dx at the point x measured in the metric $G(x)$. The claim is an immediate result of this inequality.

§ 8. The Gauss-Bonnet Theorem

8.1. The geodesic curvature vector. Suppose a regular, twice differentiable arc on the manifold F^m has the equation $x = x(\sigma)$ in the parameter region G_x^m ($\subset R_x^m$), where σ stands for the arc length. The contravariant vector

$$g(x) \equiv x'' + \Gamma\,x'\,x' = {}'(x')\,x' \qquad \left(x' = \frac{dx}{d\sigma}\right) \tag{8.1}$$

vanishes when the arc is geodesic and thus gives a measure for the curvature of the are in the metric $G(x)$. One calls it the *geodesic curvature vector* of the arc at the point $x = x(\sigma)$.

It follows by differentiation of the identity $G\,x'\,x' = 1$ with respect to σ that

$$0 = \frac{dG}{dx}\,x'\,x'\,x' + 2\,G\,x'\,x'' = 2\,G\,x'(x'' + \Gamma\,x'\,x') = 2\,G\,x'\,g\,,$$

from which it can be seen that the geodesic curvature vector is a normal to the arc $x = x(\sigma)$.

8.2. The total geodesic curvature. In the following the dimension m of the manifold F^m is assumed to be equal to 2.

We consider on the surface F^2 a neighborhood that corresponds to the region G_x^2 in the parameter space R_x^2. For the orientation we introduce in R_x^2 an arbitrary nondegenerate real alternating form $D_0\, h_1\, h_2$ $(h_i \in R_x^2)$. Then if one sets

$$\det\big(G(x)\, h_i\, h_j\big) = \big(D(x)\, h_1\, h_2\big)^2 , \tag{8.2}$$

where $D(x)\, h_1\, h_2$ is to have the sign of $D_0\, h_1\, h_2$, $D(x)\, h_1\, h_2$ is a bilinear form, defined for each $x \in G_x^2$, which is alternating and nondegenerate. In the following we let $[h_1\, h_2]$ stand for the angle formed by the vectors h_1 and h_2, which in the locally euclidean metric defined by the fundamental tensor $G(x)$ is uniquely determined modulo 2π by the relations

$$\begin{aligned} |h_1|\, |h_2| \sin\, [h_1\, h_2] &= D\, h_1\, h_2 \\ |h_1|\, |h_2| \cos\, [h_1\, h_2] &= G\, h_1\, h_2\, . \end{aligned} \tag{8.3}$$

$(|h_i|^2 = G\, h_i\, h_i)$.

Now let $x = x(\sigma)$ be a twice differentiable arc in G_x^2 that joins the points $x_1 = x(\sigma_1)$ and $x_2 = x(\sigma_2)$. Since the geodesic curvature vector $g(x)$ is at each point of the curve normal to the curve's tangent $dx = x'(\sigma)\, d\sigma$, according to the first formula (8.3)

$$D\, dx\, g = |dx|\, |g| \sin\, [dx\, g] = \pm\, |g|\, d\sigma\, ,$$

where $|g|^2 = G\, g\, g$ and the sign $\pm$ at each point of the arc is fixed by the sign of $D_0\, dx\, g$. The integral

$$\int_{x_1 x_2} D(x)\, dx\, g(x) = \int_{\sigma_1}^{\sigma_2} \pm |g(x(\sigma))|\, d\sigma \tag{8.4}$$

is called the *total geodesic curvature* of the arc $x = x(\sigma)$ with respect to the metric $G(x)$ and the orientation D_0 of the plane R_x^2.

8.3. Computation of the total geodesic curvature. We wish to derive an expression for the total geodesic curvature (8.4) which is important for what follows.

For this we consider on the arc $x = x(\sigma)$ two arbitrary contravariant and continuously differentiable vector fields $u(x)$ and $v(x)$ which we normalize to length one relative to the metric $G(x)$,

$$G\, u\, u = G\, v\, v = 1\, , \tag{8.5}$$

and compute the derivative of the angle $[u\, v]$ formed by the vectors u and v, which according to (8.3) and (8.5) is determined at each point of the arc $x = x(\sigma)$ (modulo 2π) by the relations

$$\sin\, [u\, v] = D\, u\, v\, , \qquad \cos\, [u\, v] = G\, u\, v\, .$$

If the first of these formulas is differentiated with respect to σ, then in view of exercise 3 in 7.7 we have

$$G\, u\, v\, \frac{d[u\, v]}{d\sigma} = D\, u('v\, x') - D\, v('u\, x')\, .$$

Here, as a consequence of (8.5), u and the contravariant vector

$$'u\,x' = \frac{du}{d\sigma} + \Gamma \frac{dx}{d\sigma} u$$

obtained by covariant differentiation are mutually perpendicular, and the same is true of v and $'v\,x'$ $\left(G\,u('u\,x') = G\,v('v\,x') = 0\right)$, and consequently

$$u = G\,u\,v\,v + \frac{G\,u('v\,x')}{|'v\,x'|^2}\,'v\,x', \quad v = G\,v\,u\,u + \frac{G\,v('u\,x')}{|'u\,x'|^2}\,'u\,x' .$$

If this is substituted in the above equation, the result is the formula

$$\frac{d[u\,v]}{d\sigma} = D\,v('v\,x') - D\,u('u\,x') . \tag{8.6}$$

If in this formula one makes the choise

$$v = x' = \frac{dx}{d\sigma},$$

so that according to (8.1) $D\,v\,('v\,x') = D\,x'\,g$, then for the total geodesic curvature of the curve $x = x(\sigma)$ we find the expression we seek

$$\int_{x_1 x_2} D\,dx\,g = \int_{x_1 x_2} D\,u('u\,dx) + \int_{x_1 x_2} d[u\,x'] , \tag{8.7}$$

which forms the foundation for the discussion to follow.

8.4. Case of a closed curve. We shall now apply formula (8.7) to a twice continuously differentiable closed curve $\gamma: x = x(\sigma)\ \left(x(\sigma_1) = x(\sigma_2)\right)$. Since the angle $[u\,x']$ is well-determined modulo 2π, because of the continuity of the vector u and of the tangent x', its increase on the closed path γ is a multiple of 2π, and one has

$$\int_{\gamma} D\,dx\,g = \int_{\gamma} D\,u('u\,dx) + 2\pi\nu , \tag{8.8}$$

where ν is an integer.

This result presumes that the curve γ is twice continuously differentiable. If this is the case only piecewise, then a modification enters in. We indicate this in the special case where γ is the boundary of a triangle $s^2 = s^2(x_1, x_2, x_3)$ under the additional assumption that the vector field u is *continuously differentiable* not only on $\gamma = \partial s^2$, but *on the entire closed simplex* s^2. We wish to determine the sequence of vertices x_1, x_2, x_3 so that the orientation of ∂s^2 induced by D_0 is positive.

We start from formula (8.7) and apply it for the three edges $x_i\,x_{i+1}$ ($i = 1, 2, 3$; $x_4 = x_1$). By summation one obtains

$$\int_{\partial s^2} D\,dx\,g = \int_{\partial s^2} D\,u('u\,dx) + \sum_{i=1}^{3} \int_{x_i x_{i+1}} d[u\,x'] . \tag{8.9}$$

The angle here, $[u\,x']$, has a jump at the vertices x_i which (modulo 2π) is equal to the angle of rotation experienced by the tangent vector x' at x_0 measured in the metric $G(x)$. If the corresponding interior angle of the triangle is equal to ω_i, then the former angle is $\pi - \omega_i$, and thus one has

$$\sum_{i=1}^{3} \int_{x_i x_{i+1}} d[u\, x'] = \int_{\partial s^2} d[u\, x'] - 3\pi + \Omega\,, \qquad (8.9')$$

where Ω gives the sum of the angles in the triangle s^2 in the metric $G(x)$ and the boundary integral on the right is to be taken in the sense of Stieltjes, taking into account the jumps $\pi - \omega_i$ at the vertices. Since u and x' after a complete trip around ∂s^2 return to their original positions, this Stieltjes integral is in every case an integral multiple of 2π,

$$\int_{\partial s^2} d[u\, x'] = 2\pi\nu\,. \qquad (8.10)$$

We shall show that here $\nu = 1$.

For the proof we decompose s^2 into four triangles s_j^2 $(j = 1, 2, 3, 4)$ by drawing through the midpoints of each side of s^2 the parallels to the two remaining sides. Suppose the integral (8.10) over the boundary ∂s_j^2 has the value $2\pi\nu_j$, so that

$$\sum_{j=1}^{4} \int_{\partial s_j^2} d[u\, x'] = 2\pi \sum_{j=1}^{4} \nu_j\,.$$

At a midpoint of a side the three adjoining angles have the sum π, and the corresponding contribution from these three vertex jumps of the small triangles to the sum on the left is therefore $3(3\pi - \pi) = 6\pi$. Since the contributions of the interior angles cancel out in the summation, the above sum is, according to this, larger than the integral (8.10) by 6π, and thus $\nu = \sum_{j=1}^{4} \nu_j - 3$ and

$$\nu - 1 = \sum_{j=1}^{4} (\nu_j - 1)\,.$$

From this it follows that

$$|\nu - 1| \leqq \sum_{j=1}^{4} |\nu_j - 1| \leqq 4\,|\nu_1 - 1|\,,$$

if $|\nu_s - 1|$ stands for the largest of the numbers $|\nu_j - 1|$. By repeating the thus started "Goursat procedure", one obtains a sequence of nested triangles s_n^2 which converge to a point x_0 of the closed triangle s^2, and

$$|\nu - 1| \leqq 2^{2n}\,|\nu_n - 1|\,,$$

where

$$\int_{\partial s_n^2} d[u\, x'] = 2\pi\nu_n\,.$$

But for a sufficiently large $n \geqq n_0$ this integral has the value 2π. For ultimately the triangle s_n^2 lies in an arbitrarily small neighborhood of the point x_0 and because of the continuity of $u(x)$ and of the fundamental metric form $G(x)$ the above integral differs arbitrarily little from the integral

$$\int_{\partial s_n^2} d[u_0\, x']_0 \,,$$

where $u_0 = u(x_0)$ and the angle $[u_0\, x']_0$ is measured in the *constant* euclidean metric $G(x_0)$. This integral is obviously equal to 2π.

According to this $\nu_n - 1 = 0$ for $n \geqq n_0$, and it therefore follows from the above inequality for $|\nu - 1|$ that $\nu = 1$, which was to be proved.

If the value 2π of the Stieltjes integral (8.10) is substituted in (8.9′) and (8.9) the relation (8.9) assumes the form

$$\int_{\partial s^2} D\, dx\, g = \int_{\partial s^2} D\, u('u\, dx) + \Omega - \pi \,. \tag{8.11}$$

8.5. The Gauss-Bonnet theorem. We now come to the computation of the integral

$$\int_{\partial s^2} D\, u('u\, dx)$$

on the right in formula (8.11). For this we use the Stokes transformation formula, whose application to the linear form

$$A\, dx \equiv D\, u('u\, dx)$$

is permitted if, for example, we hypothesize the given unit vector field $u = u(x)$ to be twice continuously differentiable on the simplex s^2. Observe that this hypothesis implies no restriction, for it is a remarkable fact that the value of the above boundary integral in no way depends on the choice of the vector field u, provided only that it is continuously differentiable once, as can be read off immediately from formula (8.11).

For the computation of the operator rot A we again use the formula in exercise 3 in 7.7, according to which for $h, k \in R_x^2$

$$A'\, h\, k - A\, \Gamma\, h\, k = 'A\, h\, k = D\, u(''u\, h\, k) + D('u\, h)\, ('u\, k);$$

$''u$ stands for the second covariant derivative of the contravariant vector u.

Now for the three arbitrary vectors $a, b, c \in R_x^2$ the formula[1]

$$G\, c\, c\, D\, a\, b = D\, c\, b\, G\, c\, a - D\, c\, a\, G\, c\, b$$

---

[1] If one sets $G\, a\, a = |a|^2$, $G\, b\, b = |b|^2$, $G\, c\, c = |c|^2$, then according to (8.3) $D\, c\, b\, G\, c\, a - D\, c\, a\, G\, c\, b = |a|\, |b|\, |c|^2 (\sin [c\, b] \cos [c\, a] - \sin [c\, a] \cos [c\, b])$ $= |a|\, |b|\, |c|^2 \sin([c\, b] - [c\, a]) = |a|\, |b|\, |c|^2 \sin [a\, b] = G\, c\, c\, D\, a\, b.$

holds, from which, because $G\,u\,u = 1$ and $G\,u('u\,dx) = 0$, it follows, with $a = 'u\,h$, $b = 'u\,k$, $c = u$, that

$$D('u\,h)\,('u\,k) = 0\,.$$

Thus in view of exercise 2 in 7.7 we have

$$2\,\mathrm{rot}\,A\,h\,k = 2\,D\,u \wedge {''u}\,h\,k = D\,u\,\overset{1}{\underset{3}{R}}\,h\,k\,u\,,$$

where $\overset{1}{\underset{3}{R}}$ is the mixed Riemannian tensor defined by (6.3).

In order to go on, we now take a unit vector v orthogonal to u, so that $G\,u\,v = 0$, $D\,u\,v = 1$. Then

$$2\,\mathrm{rot}\,A\,h\,k = |\overset{1}{\underset{3}{R}}\,h\,k\,u|\sin\,[u\,\overset{1}{\underset{3}{R}}\,h\,k\,u] = |\overset{1}{\underset{3}{R}}\,h\,k\,u|\cos\,[v\,\overset{1}{\underset{3}{R}}\,h\,k\,u]$$
$$= G\,v\,\overset{1}{\underset{3}{R}}\,h\,k\,u = \underset{4}{R}\,h\,k\,u\,v\,,$$

where $\underset{4}{R}$ is the covariant Riemannian tensor (cf. (6.4)). Since this tensor in alternating in h and k, the quotient $\underset{4}{R}\,h\,k\,u\,v/D\,h\,k$ is independent of h and k, and in view of the expression (6.6) for the Gaussian curvature $K(x)$ one finds that

$$\underset{4}{R}\,h\,k\,u\,v = \frac{\underset{4}{R}\,h\,k\,u\,v}{D\,h\,k}\,D\,h\,k = \frac{\underset{4}{R}\,u\,v\,u\,v}{D\,u\,v}\,D\,h\,k$$
$$= \underset{4}{R}\,u\,v\,u\,v\,D\,h\,k = -\,K\,D\,h\,k\,.$$

Further, since the oriented area element df (the area of the simplex spanned by h and k) is according to the first of formulas (8.3) equal to $D\,h\,k/2$, Stokes's formula finally yields

$$\int_{\partial s^2} D\,u('u\,dx) = \int_{\partial s^2} A\,dx = \int_{s^2} \mathrm{rot}\,A\,d_1x\,d_2x = -\int_{s^2} K\,df\,.$$

When this result is combined with formula (8.11) we obtain the *Gauss-Bonnet theorem*:

$$\int_{\partial s^2} D(x)\,dx\,g(x) + \int_{s^2} K(x)\,df = \Omega - \pi\,. \tag{8.12}$$

The terms on the left are the total geodesic curvature of the boundary ∂s^2 and the total Gaussian curvature of the simplex s^2. The sum of these two curvatures is equal to the *angular* excess $\Omega - \pi$ of the triangle measured in the metric $G(x)$.

8.6. Extensions. The Gauss-Bonnet formula yields a corresponding general relation for a polygon $\pi^2 \subset R_x^2$. For this one has to triangulate the polygon and to add the formulas (8.12) for the individual subsimplexes s^2 in the decomposition. In this way one obtains

$$\int_{\partial\pi^2} D\,dx\,g + \int_{\pi^2} K\,df = \sum_{s^2} (\Omega - \pi)\,.$$

To evaluate the last sum, let α_0, α_1, α_2 stand for the number of vertices, edges and triangles in the polyhedron that results from the triangulation of π^2. If α_{01} and α_{02} stand for the number of interior and boundary vertices, respectively, of the polyhedron, then one has

$$\alpha_0 = \alpha_{01} + \alpha_{02} \quad \text{and} \quad 3\,\alpha_2 - 2\,\alpha_1 + \alpha_{02} = 0\,.$$

We now obtain

$$\sum_{s^2} (\Omega - \pi) = \sum_{s^2} \Omega - \pi\,\alpha_2 = 2\,\pi\,\alpha_{01} + \sum \omega - \pi\,\alpha_2\,,$$

where the ω are the angles of the polygon π^2. If the angles $\varphi = \pi - \omega$ supplementary to the angles ω are then introduced,

$$\sum \omega = \pi\,\alpha_{02} - \sum \varphi\,,$$

and the above expression is therefore equal to

$$\begin{aligned} &2\,\pi\,\alpha_{01} + \pi\,\alpha_{02} - \pi\,\alpha_2 - \sum \varphi = 2\,\pi\,\alpha_0 - \pi\,\alpha_2 - \pi\,\alpha_{02} - \sum \varphi \\ &= 2\,\pi(\alpha_0 - \alpha_1 + \alpha_2) - \sum \varphi - \pi\,(3\,\alpha_2 - 2\,\alpha_1 + \alpha_{02}) \\ &= -\,2\,\pi\,\chi - \sum \varphi\,, \end{aligned}$$

where

$$\chi = -\,\alpha_0 + \alpha_1 - \alpha_2$$

stands for the *Euler characteristics* of the polyhedral surface π^2.

We thus have, in summary, the Gauss-Bonnet formula for the polygron π^2

$$\int_{\partial\pi^2} D\,dx\,g + \int_{\pi^2} K\,df + 2\,\pi\,\chi + \Phi = 0\,, \qquad (8.12')$$

where χ is the characteristic and $\Phi = \sum \varphi$ is the sum of the polygon's supplementary angles.

It is an important property of the Gauss-Bonnet theorem that all of the four terms which appear are invariant with respect to twice continuously differentiable transformations of the variable x. From this it follows that the theorem holds unchanged for a curve polygon π^2 whose bounding sides $x = x(\sigma)$ are twice continuously differentiable. And with that the validity of the theorem is established for arbitrary triangulable polygons π^2 on the manifold F^2. One only has to give a decomposition of π^2 so fine that the triangles each lie in a parameter neighborhood, and the summation of the Gauss-Bonnet triangle formulas yields the theorem for π^2.

The above discussion takes on an especially simple form for a *closed* triangulable surface F^2. In this case one finds

$$\int_{F^2} K\,df = -\,2\,\pi\,\chi\,.$$

VI. Riemannian Geometry

Gaussian surface theory, treated in Chapter V, is *relative* insofar as the surface metric, which changes from point to point, is induced by the metric of the surrounding higher dimensional euclidean space. For Gauss, however, the crucial point was to construct a theory of surfaces "from the inside out", so to speak, using only concepts that relate to the surface itself, ignoring its embedding in a higher dimensional space.

This guiding principle of Gauss's *inner absolute geometry* was suggested by a practical task, the geodesic survey of the Kingdom of Hannover, which was entrusted to Gauss in the years 1821–1825. Geodesic cartography rests fundamentally upon local observations and measurements of the topological-metric structure of the surface of the earth undertaken on the latter *surface itself.*

But here another point enters in that occupies a central position in the Gaussian theory: Any attempt to represent a compact surface (say a sphere) by a two-dimensional, planar map cannot, for topological reasons, succeed "globally" (as already emphasized in V.1.4) using one single "chart". The surface is covered with a set of local neighborhoods (H) each of which can be mapped onto a planar chart K. If two neighborhoods (H) have a nonempty intersection D, any point P of D has an image point P_1 and P_2 on each of the corresponding charts K_1 and K_2. Conversely, one again obtains the entire surface by identifying corresponding points ($P_1 \leftrightarrow P_2$, etc.) on the individual charts. In this way the individual charts are joined together into a *global* map ("atlas") of the surface. This idea is basic in practical geodesy, where the surface is triangulated and the individual triangular maps are then joined together, using the mapping correspondences ($P_1 \leftrightarrow P_2$, etc.), into a global entity.

Starting from this idea, which Gauss developed in his general theory of surfaces (and which was also decisive for the discussion in Chapter V), Riemann was able three decades later to construct his general theory of space, the theory of n-dimensional manifolds. Together with the notion of a "Riemann surface", which Riemann introduced in his fundamental investigations of abelian integrals of a complex variable, the ideas of Gauss and Riemann are basic in the later investigations of topology, differential geometry and geometric function theory.

In the present concluding chapter, the basic features of Riemannian geometry are presented. The reader who wishes to skip Chapter V can begin reading the differential geometric portion of this book directly with the present Chapter VI. In the following we shall, as the occasion demands, indicate at which points it is necessary to refer to the discussion in Chapter V.

§ 1. Affine Differential Geometry

1.1. Elementary affine geometry. In Chapter I we discussed affine vector spaces and the *parallel translation* of vectors in such a space (I.1.5). If a vector u_0 is given at the point x_0 and $x = x(t)$ designates some arc $\big(x_0 = x(t_0)\big)$, the vector is moved parallel along the curve with a family of *translations* $T\big(x(t)\big)$ (cf. I.3.9) that form an abelian group.

This elementary notion can be conceptually generalized in *affine differential geometry*. This has already been done within the context of the embedding theory (V, § 7). To facilitate the reading we shall briefly summarize the ideas basic to the theory of parallel displacement.

1.2. Manifolds[1]**.** A set R_p of objects ("points") p is called a *topological space* if the following axioms are satisfied:

I. In the point set R_p certain subsets (H) of points (called "*open*" sets) are distinguished.

I.1. The union of arbitrarily many and the intersection of finitely many open sets (H) is again an open set H.

I.2. The union of all open sets H is the entire set R_p.

The open set H is said to be a "neighborhood" of each point p contained in H.

A topological space is called *Hausdorff* if the following *separation axiom* is satisfied:

I.3. Two different points p of R_p have two disjoint neighborhoods.

A Hausdorff space R_p is called an *m-dimensional manifold* (R_p^m) if a system of covering neighborhoods (H_p) exists in R_p each of which is homeomorphic to an (open) set H_x of an m-dimensional linear space R_x^m [2].

1.3. Chart relations. Let H_p and $\overline{H}_{\bar{p}}$ be two neighborhoods on the manifold R_p^m and H_x and $\overline{H}_{\bar{x}}$ their homeomorphic images ("charts", "maps" or "parameter regions") in the linear spaces R_x^m and $R_{\bar{x}}^m$.

---

[1] Cf. V.1.3.

[2] That is, there exists a mapping $H_p \leftrightarrow H_x$ which is one-t-one and continuous in both directions, so that each open set in H_x corresponds to an open set in H_p, and conversely.

Provided H_p and $\bar{H}_{\bar{p}}$ are not disjoint, corresponding to their intersection D are two open domains G_x and $\bar{G}_{\bar{x}}$ in the charts H_x and $\bar{H}_{\bar{x}}$ that are related to one another through a topological mapping $x \leftrightarrow \bar{x}$ (Fig. 6). This chart relation is reflexive, symmetric and transitive.

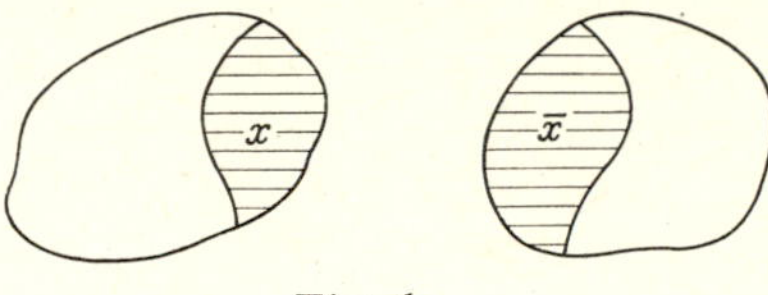

Fig. 6

It defines an equivalence between the chart points $(x, \bar{x}, \ldots)$ that correspond to the same point (p) of the manifold R_p^m [1].

1.4. Differentiable manifolds[2]. If the chart relations $x \leftrightarrow \bar{x}$ are regularly differentiable

$$d\bar{x} = \frac{d\bar{x}}{dx} dx \,, \qquad dx = \frac{dx}{d\bar{x}} d\bar{x} \,, \tag{1.1}$$

then the manifold R_p^m is said to be *regular* or *differentiable*. If the chart relations are differentiable m-times, this designation is likewise carried over to the manifold.

Let $p = (x, \bar{x}, \ldots)$ be a point on the differentiable manifold R_p^m. Let the parameters $(x, \bar{x}, \ldots)$ be increased by the differentials $(dx, d\bar{x}, \ldots)$, which are connected through the transformations (1.1). The classes (p fixed)

$$dp \equiv (dx, d\bar{x}, \ldots)$$

obtain in a clear fashion a linear structure (as carefully demonstrated in V.1.2). They form an m-dimensional linear space R_{dp}^m, the *tangent space* of the manifold R_p^m at the point p.

1.5. Invariants, vectors, tensors. The basic concepts of the affine tensor calculus are developed in Chapter V (4.1—4.12). We refer to this exposition.

1.6. Parallel displacement. Affine differential geometry concerns itself with differential manifolds on which is given a *parallel displacement* or *linear translation*. While in elementary affine geometry two

[1] Conversely, an abstract manifold R_p^m can be *generated* by means of the system of charts $H_x, \bar{H}_{\bar{x}}, \ldots$, which are provided with the given chart relations, by associating a "point" p with each equivalence class $(x, \bar{x}, \ldots)$.

Besides the system of charts $(H_x, \bar{H}_{\bar{x}}, \ldots)$ used to define the manifold one can by means of additional topological parameter transformations of open subsets of the former charts admit new maps. A corresponding *admissible extension of the charts* is also to be allowed for more special manifolds (regular, differentiable).

[2] This notion (already introduced in V.1.3) is briefly recapitulated here.

vectors at points p_1 and p_2 are either parallel or not parallel, in differential geometry the parallelism relation is not given as a simple "distant parallelism", but placed in relation to the paths along which the vectors are transported. It is customary to define this translation by means of the Christoffel operators Γ^i_{hk} (Christoffel symbols of the first kind) and the associated linear differential equation, as has been shown in Chapter V, § 7. However, instead of using this equation, it is advisable to proceed more directly from its integrals and to base the translation on their group structure[1]. This procedure, which has already been indicated in V.7.2 and whose characteristics proceed from the discussion in IV.8.3, is to serve as the foundation of parallelism.

1.7. The translations operator T. The manifold R^m_p is now assumed to be continuously differentiable[2]. An arc $p = p(\tau)$, where τ is a real parameter, is called regular when its chart projections are regular[3].

We consider a connected open region G on the manifold R^m_p and fix on it a piecewise regular arc l with initial point p_1 and final point p_2. To each such path let there be assigned a regular linear self-transformation of a linear m-dimensional space R^m_u.

By means of this *displacement operator* (T_l) an "affine translation" or "parallelism" between the contravariant vectors $u = u(p)$ $(p \in G)$ is defined in the following way:

If u_1 is a tangent vector at the point p_1, then one assigns to the end point p_2 of the arc l $(= p_1 p_2)$ the vector

$$u_2 = T_l u_1 .$$

1.8. Axioms of parallel displacement. The parallel displacement of vectors in *elementary* affine geometry is *symmetric* and *transitive*[4]. These two properties are to be retained for the generalized parallelism concept. The displacement operator (T_l) is therefore to satisfy the following special axioms:

1. Let $l_1 = p_1 p_2$, $l_2 = p_2 p_3$ be two paths on G and $l_2 l_1$ the composite path $p_1 p_2 p_3$. Then

$$T_{l_1 l_1} = T_{l_2} T_{l_1} .$$

---

[1] Indeed, this point of view ought to be clearly emphasized in the theory of differential equations.

[2] I.e., the equivalence relations $x \leftrightarrow \bar{x}$ are continuously differentiable.

[3] More precisely: Let $\tau_1 \leqq \tau \leqq \tau_2$ be an interval on which the arc $p(\tau)$ has the projection $x = x(\tau)$ on a chart H_x; this arc is to be regular (the derivative $x' = dx/d\tau$ is continuous and $\neq 0$). Because of the hypothesized differentiability of the manifold, this definition is invariant with respect to transformations $x \leftrightarrow \bar{x}$.

[4] The transitivity of parallelism in affine geometry is guaranteed by Desargues's theorem.

2. If $l^{-1} = p_2 p_1$ is the reorientation of the path $l = p_1 p_2$, then it is required that

$$T_{l^{-1}} = T_l^{-1} .$$

Later (1.10) a kind of *continuity axiom* is to be added to these group theoretical axioms.

Let us assume the path l $(= p_1 p_2)$ is so short that it lies in the intersection of two charts H_x and $\bar{H}_{\bar{x}}$. Let $l = l_x$, $\bar{l} = l_{\bar{x}}$ be the representatives of l in the latter charts. Corresponding to the operator T_l are then two regular linear transformations $T = T_l$ and $\bar{T} = \bar{T}_{\bar{l}}$ of representatives u and $\bar{u}$, respectively, of the tangent space. One has

$$u(x_2) = T_l(u(x_1)) , \quad \bar{u}(\bar{x}_2) = \bar{T}_{\bar{l}}(\bar{u}(\bar{x}_1)) ,$$

and, because of the contravariance of the vector u,

$$u(x_1) = \frac{dx(\bar{x}_1)}{d\bar{x}} \bar{u}(\bar{x}_1) , \qquad u(x_2) = \frac{dx(\bar{x}_2)}{d\bar{x}} \bar{u}(\bar{x}_2) ,$$

and hence

$$\frac{dx(x_2)}{d\bar{x}} \bar{u}(\bar{x}_2) = T_l\, u(x_1) = T_l \frac{dx(\bar{x}_1)}{d\bar{x}} \bar{u}(\bar{x}_1) .$$

Consequently,

$$\bar{u}(\bar{x}_2) = \frac{d\bar{x}(x_2)}{dx} T_l \frac{dx(\bar{x}_1)}{d\bar{x}} \bar{u}(\bar{x}_1) ,$$

and the operator T is transformed in the transition $x \leftrightarrow \bar{x}$ according to

$$\bar{T}_{\bar{l}} = \frac{d\bar{x}(x_2)}{dx} T_l \frac{dx(x_1)}{d\bar{x}} . \tag{1.2}$$

1.9. Path independence of parallel displacement. In general the parallel displacement which is defined by a translation operator T_l $(l = p_1 p_2)$ depends on the choice of the path l joining the points p_1 and p_2 on the manifold R_p^m. This leads us to ask, under which conditions is parallel translation *independent* of the path? If the initial point p_0 is fixed, the translation operator $T_l = T_{p_0 p}$ becomes a well-defined function of the end point p_2.

The same problem has already been treated in the framework of the Gaussian embedding theory (cf. IV.3.7—3.9 and V.7.2), and in what follows we can rely upon this discussion.

1.10. The operator $U_l = T_l - I$. If parallel displacement is path independent on a connected subregion $G_p^m \subset R_p^m$ of the manifold, then for any two paths l_1 and l_2 between two prescribed points p_1 and p_2 on G_x^m, $T_{l_1} = T_{l_2}$. For the closed path $\gamma = l_2^{-1} l_1$, by postulate 2,

$$T_\gamma = I , \qquad U_\gamma = T_\gamma - I = 0 . \tag{1.3}$$

If, on the other hand, $U_\gamma = 0$ for each closed path in G_p^m, then $T_\gamma = I$, from which the path independence of parallel displacement follows, because of postulate 2.

In order to further analyze the condition $U_\gamma = 0$, we restrict our attention to a neighborhood of the point $p = p_0$ lying within a chart H_x $(x_0 \leftrightarrow p_0)$. On this chart we introduce an arbitrary euclidean metric in which each regular arc $l \subset H_x$ obtains a length $|l| = \int_l |dx|$. To investigate the path independence of the operator T, we assume that a kind of Lipschitz condition is valid:

3. $|U_l| = |T_l - I| \leqq M\,|l|$,

where M is a fixed finite constant.

Path independence can then be established within a convex (or starlike) subregion $G_x \subset H_x$ by following the procedure of IV.3.8. One first considers the closed boundary $\gamma = \partial s^2$ of a triangle $s^2 \subset G_x$ and proves (cf. IV.3.10):

In order that $U_\gamma = 0$, it is necessary and sufficient that

$$\lim \frac{U_{\partial s^2}}{\Delta} = 0$$

for each point x of the closed simplex s_0, when the simplex s converges to the point x in the two-dimensional plane E_0 spanned by s_0[1]. *Here Δ is the area of the triangle s^2.*

By successive applications of this theorem we are able to conclude that parallel displacement is path independent in G_x for polygonal paths l_x. If we restrict ourselves to such paths and assign an arbitrary (contravariant) vector $u_0 = u(x_0)$ to some fixed point $x_0 \in G_x$, a *parallel* field $u = u(x)$ is defined in G_x by transporting u_0 along the segment $l = x_0 x$ to the point $x \in G_x$. This restriction to polygonal paths in G_x suffers from a defect, however, for the class of such paths π is not invariant with respect to transformations of the parameter. In order to extend the above discussion to arbitrary piecewise regular paths l, one must require that

4. $T_\pi \to T_l$, when the distribution on l of successive vertices of the polygonal path π inscribed on l is refined without bound.

1.11. Distant parallelism. These remarks solve the problem of path independence of parallel translation *locally* on the manifold M_p^m. The corresponding global problem of so-called "distant parallelism" can be solved, as a consequence of the above local result, only under more special topological conditions relating to the manifold:

---

[1] It even suffices to assume that the convergence $s \to x$ is regular, i.e., that the quotient δ^2/Δ remains bounded in the process, where δ stands for the longest side of s.

Let G_p^m be a simply connected region on the manifold R_p^m. If p_1 and p_2 are two arbitrarily chosen points on G_p^m, two paths l_1 and l_2 $(\subset G_p^m)$ having p_1 as initial point and p_2 as end point can be continuously deformed on G_p^m into one another[1]. Under this condition the paralellism is path independent in G_p^m.

Remark. There exists an interesting relationship between the notion of distant parallelism and the theory of the Lie groups (cf. W. Greub [2]).

1.12. Differentiability of the operator *T*. Up to this point the theory of parallel displacement has been constructed based solely upon the four postulates 1—4. The continuity of T follows from 3: $T_h \to I$ if the segment h with the fixed end point x tends to zero.

We now assume that T is differentiable. For this one denotes the segment $x\, x_0$ by$l$ and, keeping the point x_0 fixed, sets

$$T_l = T(x_0, x) .$$

T is thereby defined in the neighborhood of $x = x_0$ as a linear operator function of x. We now suppose that T is differentiable (in the sense of the definition of II.1.9). For $x = x_0$, if we set

$$\left(\frac{dT}{dx}\right)_{x=x_0} = \Gamma(x_0) ,$$

$$dT = \Gamma(x_0)\, dx . \tag{1.4}$$

This expression (like T itself) is a linear operator. Writing x instead of x_0, $\Gamma(x)$ is a *bilinear* operator, the so-called *Christoffel operator*.[2]

1.13. Covariant derivative. In a neighborhood of the point $x = x_0 \in H_x$ let a differentiable contravariant vector field $x \to u = u(x)$ be given. If $u(x)$ is transported parallel along the segment $x\, x_0$ to the point x_0, there results a contravariant vector $T(x_0, x)\, u(x)$. Hence for an arbitrary covariant vector u_0^* at the point x_0 the expression $u_0^*\, T(x_0, x)\, u(x)$ is scalar, and so is its differential

$$u_0^*(dT)\, u(x) + u_0^*\, T(x_0, x)\, du(x)$$
$$= u_0^* \left[\left(\frac{dT}{dx}\, dx\right) u(x) + T(x_0, x)\, u'(x)\, dx\right].$$

---

[1] The notion of "continuous deformation" can (more simply than when making use of the more customary "deformation rectangles") be reduced to successive "elementary deformations". If $a\, b$ is a segment on $l = p_1\, a\, b\, p_2$ represented in a convex region D_x of a chart H_x by the path l_x, then l_x is deformed "elementarily" by replacing the segment $a_x\, b_x$ by another arbitrary path that runs within D_x and has the same end points a_x and b_x. The two paths l_1 and l_2 are then said to be continuously deformable into one another provided they can be transformed into one another by means of a finite number of elementary deformations.

[2] In this connection, cf. V.3.4, where the operator Γ was derived from the Riemannian tensor. We shall return to this question once more in § 2 of this chapter.

When $x \to x_0$ it follows from the above that the sum $\Gamma(x)\, dx\, u(x) + u'(x)\, dx$ is *contravariant.*

Thus, if one defines the operator $'u(x)$ by

$$'u\, dx = \Gamma\, dx\, u + u'\, dx\,, \tag{1.5}$$

$'u(x)$, the *covariant derivative* of the contravariant vector $u(x)$, becomes a mixed tensor of rank 2 (cf. V.4.17).

1.14. Parallel translation of covariant vectors. This is defined with the operator T_l as follows. If u_0^* is a covariant vector at the initial point x_0 of $l = x_0\, x$, then the vector $u^* = u^*(x)$, which has been translated parallel along l to the end point x, is given by

$$u^* = u_0^*\, T_l\,.$$

When this expression is differentiated with respect to x, one obtains, if $x \to x_0$,

$$du^*(x) = u_0^*(x)\, dT - u^*(x)\, \Gamma(x)\, dx\,,$$

so that the parallel tr anslation is now given by

$$du^* = u^*\, \Gamma\, dx = 0 \tag{1.6}$$

(cf. V.4.17).

The *covariant derivative* of a covariant vector $u^*(x)$ is, correspondingly,

$$'u^* = u^{*\prime} - u^*\, \Gamma\,. \tag{1.7}$$

It is a covariant tensor of rank two.

1.15. The curvature tensor. Assuming the Christoffel operator to be *differentiable,* we are going to examine more carefully the expression

$$U_{\partial s} = T_{\partial s} - I\,, \tag{1.8}$$

which is decisive for path independence. We fix a point $p \leftrightarrow x = x_0$ and consider a small simplex $s = (x_0, x_1, x_2)$ and the expression

$$U_{\partial s}\, u_0 = \int_{\partial s} du(x) = -\int_{\partial s} \Gamma(x)\, dx\, u(x)\,, \tag{1.9}$$

where $\partial s = x_0\, x_1\, x_2\, x_0$; $u(x_0) = u_0$ is the value of the transported vector u at the initial point x.

By integration twice, starting with the constant value $u(x) = u_0$, one finds, upon setting $h = x_1 - x_0$,

$$u(x_1) - u_0 = -\int_{x_0}^{x_1} (\Gamma(x_0) + \Gamma'(x_0)\,(x - x_0) + \ldots)\, dx\, u(x) \tag{1.10'}$$

$$= (-\,\Gamma(x_0)\, h + \frac{1}{2}\, \Gamma(x_0)\, h\, \Gamma(x_0)\, h - \frac{1}{2}\, \Gamma'(x_0)\, h\, h)\, u_0 + \cdots\,.$$

In a similar fashion, integration along the segment $x_1 x_2$ yields

$$\int_{x_1}^{x_2} du = u(x_2) - u(x_1) = (-\Gamma(x_0)\,(k-h) + \frac{1}{2}\,\Gamma(x_0)\,(k-h)$$

$$\Gamma(x_0)\,(k+h) - \frac{1}{2}\,\Gamma'(x_0)\,(k+h)\,(k-h)\,u_0 + \cdots, \qquad (1.10'')$$

where $k = x_2 - x_0$.

Finally, we have

$$\int_{x_2}^{x_0} du = (\Gamma(x_0)\,k - \frac{1}{2}\,\Gamma(x_0)\,k\,\Gamma(x_0)\,k + \frac{1}{2}\,\Gamma'(x_0)\,k\,k)\,u_0 + \cdots. \qquad (1.10''')$$

Addition of the three above integrals gives the desired result:

$$\int_{\partial s} du = \frac{1}{2}\,(\Gamma(x_0)\,k\,\Gamma(x_0)\,h - \Gamma(x_0)\,h\,\Gamma(x_0)\,k$$

$$+ \Gamma'(x_0)\,k\,h - \Gamma'(x_0)\,h\,k)\,u_0 + \cdots. \qquad (1.10)$$

Here the remainder term denoted by $\cdots$ is of the order of magnitude $\delta^2(\delta)$, where δ stands for the larger of the norms $|h|$, $|k|$ (in some local euclidean metric in the parameter space).

Expansion (1.10) can be derived somewhat more simply with the aid of Stokes's theorem (cf. III.2.7).

The expression

$$R(x)\,h\,k\,l = \Big(\bigwedge_{h\,k} \Gamma(x)\,h\,\Gamma(x)\,k + \Gamma'(x)\,h\,k\Big)\,l \qquad (1.11)$$

is a contravariant vector, and the operator $R(x)$ is therefore a tensor of rank four with signature

$$R = \overset{1}{\underset{3}{R}}\,.$$

R is called the *curvature* tensor.

From expansion (1.12) it follows that the limit

$$\lim \frac{U_{\partial s}}{D\,k\,h} = \varrho(x)$$

exists when the triangle $(x, x+h, x+k)$ converges regularly to the point x. This quantity is a tensor density (cf. III.2.5) which carries the signature $\overset{1}{\underset{1}{\varrho}}$.

In the above derivation we assumed that the simplex s has the point $x = x_0$ as a vertex. This condition is not essential in order that expansion (1.10) hold. For if one considers a simplex (x_1, x_2, x_3) which lies in a plane through the point $x = x_0$ in a neighborhood $|x - x_0| \leqq \varrho$ of x_0, the expansion (1.10) can be applied to the three simplexes

(x_0, x_1, x_2), (x_0, x_1, x_3) and (x_0, x_2, x_3). Then, because of the additivity of alternating forms, addition yields an expansion which is again of the form (1.10), where now $h = x_2 - x_1$, $k = x_3 - x_1$ (where a cyclical change in the indices 1, 2, 3 is permitted).

In summary we therefore have this result:

If the Christoffel operator

$$\Gamma(x) = T'(x)$$

is differentiable, then the limit

$$\frac{1}{D\,k\,h}\,U_{\partial s} = \frac{1}{D\,h\,k}\,(T_{\partial s} - I) \to \varrho(x)$$

exists when the simplex s, which lies in a plane that goes through the point x and is spanned by the vectors h and k, converges regularly to the point x.

Here $D\,h\,k$ is an arbitrary alternating real form ($\neq 0$). The trilinear form

$$R(x)\,h\,k\,l \equiv D\,h\,k\,\varrho(x)\,l\,,$$

where h, k, l are arbitrary contravariant vectors, is contravariant, and

$$R = \overset{1}{\underset{3}{R}}\,,$$

the curvature operator of the parallelism given by T, is a tensor of rank four.

Observe that the existence and differentiability is sufficient (but not necessary) for the existence of the operator $\Gamma(x)$ of the limit $\varrho(x)$ and (at the same time) of the curvature tensor $R(x)$.

1.16. Local path independence of linear translation. From the theorem in 1.1 it follows that the condition $R(x) = 0$ is necessary and sufficient for the independence of parallelism, provided one restricts oneself to simply connected neighborhoods on the manifold. There parallelism is defined by means of the translation operator T. If the latter is assumed to be differentiable and its derivative, the Christoffel operator, is differentiable, then the curvature tensor (p. 210)

$$R = \wedge\,(\Gamma 0\,\Gamma 0 + \Gamma 0 6')$$

exists, and the integrability condition is

$$\underset{h\,k}{\wedge}\,(\Gamma\,h\,\Gamma\,k + \Gamma'\,h\,k) = 0\,,$$

where h and k are arbitrary contravariant vectors.

If, conversely, one defines parallel displacement, as is customary in classical differential geometry, by giving the Christoffel operator $\Gamma(x)$ directly and integrating the differential equation

$$'u\,dx = du + \Gamma\,dx\,u = 0\,, \tag{1.12}$$

then the translation operator can, by integrating this differential equation along a path $l = x_0 x$, be determined from

$$T_l u_0 = u(x) ,$$

where $u(x)$ stands for the final value and u_0 the initial value of the integral. In this regard, cf. IV, 3.8.

1.17. Elementary affine geometry. Parallel displacement is characterized in elementary geometry by the condition $T = I$, $U = 0$. We set ourselves the task of investigating under which conditions the operator T, for a suitable choice of the parameter x, can be transformed into the identity I.

First, because of the independence of parallelism from the path, the curvature R vanishes. Assuming this, the operator T_l in an x-chart where l corresponds to a path $x_0 x$ is then a well-defined function $T(x_0, x)$ of x_0 and x in a (simply connected) neighborhood of $x = x_0$. The transformation law for T under a parameter change $x \leftrightarrow \bar{x}$ is (cf. VI.1.8)

$$\bar{T}(\bar{x}_0, \bar{x})\, d\bar{x} = \frac{d\bar{x}(x_0)}{dx}\, T(x_0, x)\, dx .$$

Now if $\bar{T}$ is to be the identity, we deduce the necessary and sufficient condition

$$d\bar{x} = A_0\, T(x_0, x)\, dx .$$

This differential equation, provided $A_0 = \bar{x}'(x_0)$ is prescribed, can be solved if and only if the bilinear operator $d/dx\, (A_0\, T(x_0, x)) = A_0\, T(x_0, x)\, \Gamma(x)$, and (because of the regularity of the operator $A_0\, T$) also $\Gamma(x)$ is symmetric:

$$\Gamma(x)\, h\, k = \Gamma(x)\, k\, h .$$

Under this condition the integral

$$\bar{x}(x) = \bar{x}(x_0) + \bar{x}'(x_0)\,(x - x_0) + \cdots$$

is uniquely determined in a neighborhood of the point $x = x_0$, so that $\bar{T} = I$. This solves the problem.

We have found

$$R = 0 \quad \text{and} \quad \wedge\, \Gamma = 0$$

to be necessary and sufficient conditions for the existence in a suitable parametric chart of an elementary affine translation.

Remark. In general the Christoffel operator is not symmetric. The *alternating* part $\wedge\, \Gamma$ of Γ is called the *torsion* of the manifold.

§ 2. Riemannian Geometry

Riemann, generalizing the Gaussian surface theory, developed in his famous inaugural dissertation, "Über die Hypothesen, welche der

Geometrie zu Grunde liegen", the theory of manifolds, where the geometry is determined by a euclidean metric which varies from point to point. The metric at a point $p \leftrightarrow x$ is therefore determined by means of a symmetric, positive definite invariant bilinear form

$$G(x)\, h\, k \tag{2.1}$$

in the tangent vectors h, k. From this operator $G(x)$, one can derive the symmetric Christoffel operator $\Gamma(x)$ and thereby define parallel translation. Conversely, under certain conditions an affine differential geometry can be completed to a Riemannian manifold.

2.1. Local metrization of an affine manifold. As already done in the context of the theory of surfaces (V.7.3), the connection between metric and affine differential geometry is effected by requiring the translation operator T_l to be an orthogonal transformation of the tangent space, in the sense of the metric $G(x)$, at each point x of the path $l = x_0\, x$. Thus, provided T is differentiable and $u(x)$ and $v(x)$ are two contravariant vectors which are transported parallel along the path l, $G(x)\, u(x)\, v(x)$ is constant along l. If the operator $G(x)$ is also assumed to be differentiable, then it follows that

$$0 = d(G\, u\, v) = G'\, dx\, u\, v + G(du)\, v + G\, u(dv)\,.$$

Here dx is initially the tangent vector to the path l at the point x. But since the path l, which was only assumed to be piecewise differentiable, can be continued in an arbitrary direction starting from the point x, the above equation holds for every contravariant vector $dx = k$ at the point x. If in addition one observes that according to (1.5)

$$du = -\,\Gamma(x)\, k\, u\,, \qquad dv = -\,\Gamma(x)\, k\, v\,,$$

then

$$G'\, k\, u\, v = G\, \Gamma\, k\, u\, v + G\, \Gamma\, k\, v\, u\,, \tag{2.2}$$

where now k, u, v are *arbitrary* contravariant vectors.

2.2. Determination of Γ through G. Henceforth we assume that the torsion vanishes and that the operator Γ is therefore *symmetric*. Under this assumption one obtains by permuting the vectors k, u, v cyclically and adding the corresponding equations (2.2) (where the first is to be multiplied by -1)

$$\frac{1}{2}\,(G'\, u\, v\, k + G'\, v\, k\, u - G'\, k\, u\, v) = G\, k\, \Gamma\, u\, v\,. \tag{2.3}$$

This same formula has already been found in the Gaussian embedding theory For a given fundamental metric form $G\, h\, k$ the Christoffel operator Γ is uniquely determined by equation (2.3) (cf. V.3.5).

2.3. Integration of the differential equation (2.2). Conversely, the Riemannian operator $G(x)$ for the given $\Gamma(x)$ is determined through integration of equation (2.2), where u and v are assumed to be constant and $k = dx$ designates the differential of x. Interpreted in this way, the differential equation can be written more briefly as

$$'G = 0\,, \qquad (2.3')$$

where $'G$ stands for the covariant derivative of the tensor G (cf. VI.1.13)

For a given Γ and fixed contravariant vectors u and v, the relation (2.2) provides a linear homogeneous differential equation for the determination of the linear operator $G(x)$. It can be solved locally as follows by means of the technique developed in IV.

Let B_x be a convex region on a chart of the manifold M^n. We assume that the operator $\Gamma(x)$, which determines parallel displacement on B_x, is differentiable. Then let $x_0 \in B_x$, and suppose u and v are two arbitrarily fixed constant (contravariant) vectors in the parameter space R^n_x. In equation (2.2) we write $k = dx$, choose an arbitrary point $x \in B_x$ and integrate (2.2) along the *segment* $x_0\,x$. This normal system then yields as its solution a radial integral $G = G^*(x)$. The latter is uniquely determined at each point $x \in B_x$, provided the initial value $G(x_0)\,u\,v$ is given. If this is chosen arbitrarily as a positive definite symmetric form $G(x_0)\,u\,v$ in the vectors u and v, then the integral $\overline{G}$ also defines a similar form $G^*(x)\,u\,v$ in the vicinity of x_0.

By means of radial integration of equation (2.2) we have so determined a positive definite bilinear form $G^*(x)u\,v$ in a neighborhood of x_0 that $d(G^*\,u\,v) = 0$, and $G^*(x)\,u\,v$ therefore remains constant when the vectors u and v are transported parallel along the segment $x_0\,x$. This form is further invariant with respect to a parameter change $x \leftrightarrow \bar{\ }$, which is verified by applying the formula for transforming the operator.

2.4. The Riemannian curvature tensor. The radial metric tensor G^* does not in general satisfy the requirement $G^*\,u\,v = \text{const.}$ for parallel displacement along paths l that are not radial. For arbitrary piecewise differentiable paths (with $x = x_0$ as initial point) this is true only if $G^*(x)$ satisfies the differential equation for *arbitrarily* directed differentials $dx = h$. According to IV this is the case if and only if the expression G'' defined by equation (2.2) is symmetric for $G = G^*$. In order to show this, differentiate equation (2.2), for two arbitrarily fixed (constant) vectors u, v and once again for an arbitrary differential $dx = h$. In this fashion one first obtains

$$G''\,h\,k\,u\,v = G'\,h\,\Gamma\,k\,u\,v + G'\,h\,\Gamma\,k\,v\,u + G\,\Gamma'\,h\,k\,u\,v + G\,\Gamma'\,h\,k\,v\,u\,.$$

If the first terms on the right are replaced by the values determined from (2.2), then in view of the symmetry of the tensor G and of the operator Γ, the result is

$$G''\,h\,k\,u\,v = G\,(\Gamma\,h\,\Gamma\,k\,u + \Gamma'\,h\,k\,u)\,v + G\,(\Gamma\,h\,\Gamma\,k\,v + \Gamma'\,h\,k\,v)\,u$$
$$+\, G\,\Gamma\,h\,v\,\Gamma\,k\,u + G\,\Gamma\,k\,v\,G\,h\,u\,.$$

Hence, the integrability condition is

$$\bigwedge_{hk} G\,(\Gamma\,h\,\Gamma\,k\,u + \Gamma'h\,k\,u)\,v + \bigwedge_{hk} G\,(\Gamma\,h\,\Gamma\,k\,v + \Gamma'\,h\,k\,v)\,u = 0\,,$$

or

$$G\,v\,\overset{1}{\underset{3}{R}}\,h\,k\,u + G\,u\,\overset{1}{\underset{3}{R}}\,h\,k\,v = 0\,, \tag{2.4}$$

where $\overset{1}{\underset{3}{R}}$ is the curvature tensor of the affine manifold M^n determined by (1.11).

The *Riemannian curvature tensor* is defined as a covariant tensor of rank four by the invariant form

$$\underset{4}{R}\,h_1\,h_2\,h_3\,h_4 = G(\overset{1}{\underset{3}{R}}\,h_1\,h_2\,h_3)\,h_4\,, \tag{2.5}$$

where $h_1, \ldots, h_4$ are contravariant vectors.

Condition (2.4) for the integrability of the differential equation (2.2) can therefore be stated:

The Riemannian curvature tensor (2.5) *is alternating in the last two arguments* h_3 *and* h_4.

By definition $\underset{4}{R}$ is also alternating in the first two arguments h_1 and h_2.

2.5. Summary. Provided a given parallel translation is metrizable by means of a Riemannian tensor in a neighborhood of the point x_0, this solution is uniquely determined in the neighborhood of x_0 by prescribing the tensor G at the point $x = x_0$. It is equal to the tensor $G^*(x)$ constructed by means of radial integration.

In order for the problem to be solvable in the neighborhood of point x_0, it is necessary and sufficient that the Riemannian tensor $\underset{4}{R}$ be alternating in its last two arguments.

Bibliography

BÄCHLI, G.: [1] Über die Integrierbarkeit von Systemen partieller, nichtlinearer Differentialgleichungen erster Ordnung. Comment. Math. Helv. **36**, 245–264 (1961/1962).

BARTLE, R. G.: [1] Implicit functions and solutions of equations in groups. Math. Z. **62**, 335–346 (1955). — [2] On the openness and inversion of differentiable mappings. Ann. Acad. Sci. Fenn. A I **257** (1958).

BOURBAKI, N.: [1] Éléments de mathématique. VII. Algèbre multilinéaire. Actualités Sci. Ind. **1044**, Paris: Hermann (1948).

CARTAN, E.: [1] Leçons sur les invariants intégraux. Paris: Hermann (1922). — [2] Les systèmes différentiels extérieurs et leur applications géométriques. Actualités Sci. Indust. **994**, Paris: Hermann (1945).

DOMBROWSKI, P., and F. HIRZEBRUCH: [1] Vektoranalysis. [Hectographed lecture notes.] Bonn: Univ. Bonn, Math. Inst. (1962).

DUNFORD, N., u. J. T. SCHWARTZ (in collaboration with W. G. BADE and R. G. BARTLE): [1] Linear operators. I. General theory. Pure Appl. Math. **7**, New York/London: Interscience (1958).

FISCHER, H. R.: [1] Differentialkalkül für nicht-metrische Strukturen. Ann. Acad. Sci. Fenn. A I **247** (1957). — [2] Differentialkalkül für nicht-metrische Strukturen. II. Differentialformen. Arch. Math. **8**, 438–443 (1957).

FRÉCHET, M.: [1] Sur quelques points du calul fonctionnel. Thèse. Rend. Circ. Mat. Palermo XXII (1906). — [2] La notion de differentielle dans l'analyse générale. Ann. sec. E'c. Norm. sup. XLII (1925).

FREUDENTHAL, H.: [1] Simplizialzerlegungen von beschränkter Flachheit. Ann. of Math. **43**, 580–582 (1942).

GILLIS, P.: [1] Sur les formes différentielles et la formule de Stokes. Acad. Roy. Belg. Cl. Sci. Mém. Coll. in-8° (2) **20**:**3** (1942).

GRAEUB, W., and R. NEVANLINNA: [1] Zur Grundlegung der affinen Differentialgeometrie. Ann. Acad. Sci. Fenn. A I **224** (1956).

GRAUERT, H., and W. FISCHER: [1] Differential- und Integralrechnung. II. Heidelberger Taschenbücher **36**, Berlin/Heidelberg/New York: Springer (1968).

GRAUERT, H., and I. LIEB: [1] Differential- und Integralrechnung. III. Heidelberger Taschenbücher **43**. Berlin/Heidelberg/New York: Springer (1968).

GREUB, W. H.: [1] Linear algebra. [Third edition.] Grundlehren math. Wiss. **97**. Berlin/Heidelberg/New York: Springer (1967). — [2] Multilinear algebra. Grundlehren math. Wiss. **136**. Berlin/Heidelberg/New York: Springer (1967).

HAAHTI, H.: [1] Über konforme Abbildungen eines euklidischen Raumes in eine Riemannsche Mannigfaltigkeit. Ann. Acad. Sci. Fenn. A I **287** (1960).

HAAHTI, H., and T. KLEMOLA: [1] Zur Theorie der vektorwertigen Differentialformen. Ann. Acad. Sci. Fenn. A I **318** (1962).

HEIKKILÄ, S.: [1] On the complete integrability of the first order total differential equation. Ann. Acad. Sci. Fenn. A I **495** (1971).

HERMANN, P.: [1] Über eine Verallgemeinerung der alternierenden Ableitung von Differentialformen. Univ. Jyväskylä, Math. Inst., Bericht 12 (1971). — [2] Über eine Verallgemeinerung der alternierenden Ableitung von Differentialformen. I. Math. Nachr. 52, 85—99 (1972). — [3] Über eine Verallgemeinerung der alternierenden Ableitung von Differentialformen. II. Math. Nachr. 52, 101—111 (1972). — [4] Eine Übertragung der Differentialgleichung erster Ordnung auf Differentialformen in normierten linearen Räumen. To appear in: Colloquium on mathematical analysis, Jyväskylä 1970. Lecture Notes in Mathematics. Berlin/Heidelberg/New York: Springer. — [5] Über den Satz von Stokes. To appear in Ann. Acad. Sci. Fenn. A I.

HILLE, E., and R. S. PHILLIPS: [1] Functional analysis and semi-groups. Colloquium Publ. 31, Providence (R. I.): Amer. Math. Soc. (1957).

KELLER, H. H.: [1] Über die Differentialgleichung erster Ordnung in normierten linearen Räumen. Rend. Circ. Mat. Palermo (2) 8, 117—144 (1959).

KLEMOLA, T.: [1] Über lineare partielle Differentialgleichungen erster Ordnung. Ann. Acad. Sci. Fenn. A I 260 (1958). — [2] Reguläre Mengen von Simplexen und der Satz von Stokes. Ann. Acad. Sci. Fenn. A I 295 (1961).

KRICKEBERG, K.: [1] Über den Gaußschen und den Stokesschen Integralsatz. III. Math. Nachr. 12, 341—365 (1954).

LAUGWITZ, D.: [1] Differentialgeometrie ohne Dimensionsaxiom. I. Tensoren auf lokal-linearen Räumen. Math. Z. 61, 100—118 (1954). — [2] Differentialgeometrie ohne Dimensionsaxiom. II. Riemannsche Geometrie in lokal-linearen Räumen. Math. Z. 61, 134—149 (1954).

LICHNEROWICZ, A.: [1] Algèbre et analyse linéaires. Collection d'ouvrages de Mathématiques à l'usage des Physiciens, Paris: Masson (1947).

LONKA, H.: [1] Über lineare vektorielle Differentialgleichungen zweiter Ordnung. Ann. Acad. Sci Fenn. A I 350 (1964).

LOUHIVAARA, I. S.: [1] Über die Differentialgleichung erster Ordnung in normierten linearen Räumen. Rend. Circ. Mat. Palermo (2) 10, 45—58 (1961). — [2] Bemerkungen zur Vektoranalysis. Studia logico-mathematica et philosophica in honorem Rolf Nevanlinna die natali eius septuagesimo 22. X. 1965. Acta Philos. Fenn. 18, 95—115 (1965).

MANNINEN, J.: [1] Zur Charakteristikentheorie von Systemen partieller Differentialgleichungen erster Ordnung. Ann. Acad. Sci. Fenn. A I 283 (1960),.

MICHAL, A. D., and V. ELCONIN: [1] Completely integrable differential equations in abstract spaces. Acta Math. 68, 71—107 (1937).

MÜLLER, CL.: [1] Über die Grundoperationen der Vektoranalysis. Math. Ann. 124, 427—449 (1952). — [2] Grundprobleme der mathematischen Theorie elektromagnetischer Schwingungen. Grundlehren math. Wiss. 88. Berlin/Göttingen/Heidelberg: Springer (1957). — [3] Über einen neuen Zugang zur mehrdimensionalen Differential- und Integralrechnung. To appear.

NEVANLINNA, F.: [1] Über die Umkehrung differenzierbarer Abbildungen. Ann. Acad. Sci. Fenn. A I 245 (1957). — [2] Über absolute Analysis. Treizième Congrès des Mathématiciens Scandinaves à Helsinki 1957, 178—197, Helsinki (1958).

NEVANLINNA, F. and R.: [1] Über die Integration eines Tensorfeldes. Acta Math. 98, 151—170 (1957).

NEVANLINNA, R.: [1] Bemerkung zur Funktionalanalysis. Math. Scand. 1, 104—112 (1953). — [2] Bemerkung zur absoluten Analysis. Ann. Acad. Sci. Fenn. A I 169 (1954). — [3] Über die Umkehrung differenzierbarer Abbildungen. Ann. Acad. Sci. Fenn. A I 185 (1955). — [1] Über den Satz von Stokes. Ann. Acad. Sci. Fenn. A I 219 (1946). — [5] Zur Theorie der Normalsysteme von gewöhnlichen Differentialgleichungen. Hommage à S. Stoïlow pour son 70e anniversaire. Rev.

Math. Pures Appl. **2**, 423—428 (1957). — [*6*] Sur les équations aux dérivées partielles du premier ordre. C. R. Acad. Sci. Paris **247**, 1953—1954 (1958). — [*7*] Application d'un principe de E. Goursat dans la théorie des équations aux dérivées partielles du premier ordre. C. R. Acad. Sci. Paris **247**, 2087—2090 (1958). — [*8*] Über Tensorrechnung. Rend. Circ. Mat. Palermo (2) 7, 285—302 (1958). — [*9*] Über fastkonforme Abbildungen. Proceedings of the International colloquium on the theory of functions., Helsinki 1957. Ann. Acad. Sci. Fenn. A I **251/7** (1958). — [*10*] Über die Methode der sukzessiven Approximationen. Ann. Acad. Sci. Fenn. A I **291** (1960). — [*11*] On differentiable mappings. Analytic functions, Princeton Math. Ser. 24, Princeton (N.J.): Princeton Univ. Press, 3—9 (1960). — [*12*] Remarks on complex and hypercomplex systems. Soc. Sci. Fenn. Comment. Phys.-Math. 26:3B (1962). — [*13*] Calculus of variation and partial differential equations. J. Analyse Math. **19**, 273—281 (1967).

NIEMINEN, T.: [*1*] On decompositions of simplexes and convex polyhedra. Soc. Sci. Fenn. Comment. Phys.-Math. **20**:5 (1957).

PÓLYA, G.: [*1*] Über die Funktionalgleichung der Exponentialfunktion im Matrizenkalkül. S.-B. Preuß. Akad. Wiss. Phys.-Math. Kl., 96—99 (1928).

REICHARDT, H.: [*1*] Vorlesungen über Vektor- und Tensorrechnung. Hochschulbücher für Mathematik **34**. Berlin: VEB Deutscher Verlag der Wissenschaften (1957).

DE RHAM, G.: [*1*] Über mehrfache Integrale. Abh. Math. Sem. Univ. Hamburg **12**, 313—339 (1937/1938).

RIKKONEN, H.: [*1*] Zur Einbettungstheorie. Ann. Acad. Sci. Fenn. A I **300** (1961).

ROTHE, E. H. : [*1*] Gradient mappings. Bull. Amer. Math. Soc. **59**, 5—19 (1953).

SCRIBA, C.-H.: [*1*] Über die Differenzierbarkeit der radialen Lösung einer expliziten vektoriellen Differentialgleichung erster Ordnung. Math. Nachr. **32**, 25—40 (1966).

SEBASTIÃO E SILVA, J.: [*1*] Integração e derivação em espaços de Banach. Uni. Lisboa. Revista Fac. Ci. (2) A **1**, 117—166 (1950).

SEGRE, B.: [*1*] Forme differenziali e loro integrali. I. Calcolo algebrico esterno e proprietà differenziali locali. Istituto Nazionale di Alta Matematica. Rome: Docet (1951).

ŚLEBODZIŃSKI, W.: [*1*] Formes extérieures et leurs applications. II. Polska Akademia Nauk Monografie Matematyczne **40**. Warszawa: Państwowe Wydawnictwo Naukowe (1973).

SZÜCS, A.: [*1*] Sur la variation des intégrales triples et le théorème de Stokes. Acta Litt. Sci. Szeged. Sect. Sci. Math. **3**, 81—95 (1927).

TIENARI, M.: [*1*] Über die Lösung von partiellen Differentialgleichungen erster Ordnung nach der Methode der sukzessiven Approximationen. Ann. Acad. Sci. Fenn. A I **380** (1965).

WEYL, H.: [*1*] The method of orthogonal projection in potential theory. Duke Math. J. **7**, 411—444 (1940).

WHITNEY, H.: [*1*] Geometric integration theory. Princeton Math. Ser. **21**, Princeton (N.J.): Princeton Univ. Press (1957).

Index

721/36/72 V/12/6

Die Grundlehren der mathematischen Wissenschaften in Einzeldarstellungen mit besonderer Berücksichtigung der Anwendungsgebiete

Eine Auswahl

---

1. Blaschke: Elementare Differentialgeometrie,. 5. Aufl.
27. Hilbert/Ackermann: Grundzüge der theoretischen Logik. 6. Aufl.
67. Byrd/Friedman: Handbook of Elliptic Integrals for Engineers and Scientists. 2nd edition
68. Aumann: Reelle Funktionen. 2. Aufl.
76. Tricomi: Vorlesungen über Orthogonalreihen. 2. Aufl.
77. Behnke/Sommer: Theorie der analytischen Funktionen einer komplexen Veränderlichen. Nachdruck der 3. Aufl.
78. Lorenzen: Einführung in die operative Logik und Mathematik. 2. Aufl.
87. van der Waerden: Mathematische Statistik. 3. Aufl.
94. Funk: Variationsrechnung und ihre Anwendung in Physik und Technik. 2. Aufl.
99. Cassels: An introduction to the Geometry of Numbers. 2nd printing
104. Chung: Markov Chains with Stationary Transition Probabilities. 2nd edition
107. Köthe: Topologische lineare Räume. I. 2. Aufl.
114. MacLane: Homology. Reprint of 1st edition
116. Hörmander: Linear Partial Differential Operators. 3rd printing
117. O'Meara: Introduction to Quadratic Forms. 2nd printing
120. Collatz: Funktionalanalysis und numerische Mathematik. Nachdruck der 1. Aufl.
121./122. Dynkin: Markov Processes
123. Yosida: Functional Analysis. 3rd edition
124. Morgenstern: Einführung in die Wahrscheinlichkeitsrechnung und mathematische Statistik. 2. Aufl.
125. Itô/McKean jr.: Diffusion Processes and Their Sample Paths
126. Lehto/Virtanen: Quasiconformal Mappings in the Plane
127. Hermes: Enumerability, Decidability, Computability. 2nd edition
128. Braun/Koecher: Jordan-Algebren
129. Nikodým: The Mathematical Apparatus for Quantum Theories
130. Morrey jr.: Multiple Integrals in the Calculus of Variations
131. Hirzebruch: Topological Methods in Algebraic Geometry. 3rd edition
132. Kato: Perturbation Theory for Linear Operators
133. Haupt/Künneth: Geometrische Ordnungen
134. Huppert: Endliche Gruppen I
135. Handbook for Automatic Computation. Vol. 1/Part a: Rutishauser: Description of ALGOL 60
136. Greub: Multilinear Algebra
137. Handbook for Automatic Computation. Vol 1/Part b: Grau/Hill/Langmaack: Translation of ALGOL 60
138. Hahn: Stability of Motion
139. Doetsch/Schäfke/Tietz: Mathematische Hilfsmittel des Ingenieurs. 1. Teil
140. Collatz/Nicolovius/Törnig: Mathematische Hilfsmittel des Ingenieurs. 2. Teil
141. Mathematische Hilfsmittel des Ingenieurs. 3. Teil
142. Mathematische Hilfsmittel des Ingenieurs. 4. Teil
143. Schur: Vorlesungen über Invariantentheorie
144. Weil: Basic Number Theory